M. Sheridan
Boston 1983

Social Behavior of Female Vertebrates

Social Behavior of Female Vertebrates

EDITED BY

SAMUEL K. WASSER

Animal Behavior Program
Department of Psychology
University of Washington
Seattle, Washington

1983

AP

ACADEMIC PRESS

A Subsidiary of Harcourt Brace Jovanovich, Publishers

New York London
Paris San Diego San Francisco São Paulo Sydney Tokyo Toronto

ACADEMIC PRESS, INC.
111 Fifth Avenue, New York, New York 10003

United Kingdom Edition published by
ACADEMIC PRESS, INC. (LONDON) LTD.
24/28 Oval Road, London NW1 7DX

Library of Congress Cataloging in Publication Data

Wasser, Samuel K.
Social behavior of female vertebrates.

1. Vertebrates--Behavior. 2. Social behavior in
animals. I. Title. II. Title: Female vertebrates.
QL775.W37 1982 596'.056 82-11602
ISBN 0-12-735950-8

PRINTED IN THE UNITED STATES OF AMERICA

83 84 85 86 9 8 7 6 5 4 3 2 1

To my parents, Ethel and Frank

Contents

PART II. Interactions between the Sexes

Contributors

Numbers in parentheses indicate the pages on which the authors' contributions begin.

Sarah Blaffer Hrdy[1] (3), Department of Anthropology, Rice University, Houston, Texas 77251

Robert Boyd[2] (315), Division of Environmental Studies, University of California, Davis, California 95616

Luther Brown (39), Department of Biology, George Mason University, Fairfax, Virginia 22030

Jerry F. Downhower (39), Department of Zoology, Ohio State University, Columbus, Ohio 43210

Holly T. Dublin[3] (291), Wildlife Sciences Group, University of Washington, Seattle, Washington 98195

W. James Erckmann (113), Department of Zoology, University of Washington, Seattle, Washington 98195

William Irons (169), Department of Anthropology, Northwestern University, Evanston, Illinois 60201

Walter D. Koenig[4] (235), Museum of Vertebrate Zoology, University of California, Berkeley, California 94720

[1]Present address: Peabody Museum, Harvard University, Cambridge, Massachusetts 02138.

[2]Present address: School of Forestry and Environmental Studies, Duke University, Durham, North Carolina 27706.

[3]Present address: University of British Columbia, Department of Zoology, Vancouver, British Columbia V6T 1W5, Canada.

[4]Present address: Hastings Reservation, University of California, Carmel Valley, California 93924.

Susan Lumpkin (91), Department of Zoological Research, National Zoological Park, Smithsonian Institution, Washington, D.C. 20008

Ronald L. Mumme (235), Museum of Vertebrate Zoology, University of California, Berkeley, California 94720

Robert B. Payne (55), Museum of Zoology and Division of Biological Sciences, University of Michigan, Ann Arbor, Michigan 48109

Frank A. Pitelka (235), Museum of Vertebrate Zoology, University of California, Berkeley, California 94720

James K. Russell[5] (263), Department of Zoology, University of North Carolina, Chapel Hill, North Carolina 27514

Joan B. Silk[6] (315), Department of Anthropology, University of California, Davis, California 95616

Samuel K. Wasser (19, 349), Animal Behavior Program, Department of Psychology, University of Washington, Seattle, Washington 98195

Mary L. Waterhouse (19, 215), Department of Anthropology, University of Washington, Seattle, Washington 98105

George C. Williams[7] (3), Center for Advanced Study in the Behavioral Sciences, Stanford, California 94305

[5]Present address: Department of Zoological Research, National Zoological Park, Smithsonian Institution, Washington, D.C. 20008.

[6]Present address: Allee Laboratory of Animal Behavior, University of Chicago, Chicago, Illinois 60637.

[7]Present address: Department of Ecology and Evolution, State University of New York at Stony Brook, Stony Brook, New York 11794.

Preface

The seed for this book was first planted while I was conducting my dissertation research on yellow baboons in East Africa. Having just completed my general exams, I went to the field with several clear impressions. Theoretical and empirical research frequently focused on males and neglected females; and females rarely engaged in reproductive competition or affiliated with consexuals other than closest kin, but instead concentrated on choosing the best mates and taking care of their offspring or "allomothering" others'. Moreover, even mate choice was occasionally out of the females' control, because they were consistently dominated by more powerful males who could, at times, force reluctant females into copulating with them. Given these impressions, I was surprised to find female baboons thrashing each other, mishandling each other's infants, and resisting a considerable amount of intimidation by adult males. Males were not conforming to my expectations either. Baboons are highly sexually dimorphic and have a promiscuous, polygynous mating system. Yet, males were acting quite paternal and concerned over the welfare of newborns. Something was wrong!

Upon returning to the United States, I began to take a new look at the literature and was pleasantly surprised to find that a number of scientists had published views contrary to my earlier impressions. Suspecting that most behavioral biologists had impressions similar to my original ones, I decided to organize a symposium on female social strategies.

The Female Social Strategies Symposium was held at the annual Animal Behavioral Society meeting in Fort Collins, Colorado, and the response was tremendous. So many papers were submitted that the symposium quickly had to be changed from a half to a full day. Moreover, the lecture hall was packed for nearly every presentation. The seed had sprouted, and I hope that this book will be considered its fruit.

Although the majority of chapters in the book are elaborations of talks included in the symposium, not all symposium participants contributed their work. A few additional chapters were solicited from scientists who could not participate in the symposium, but whose research was relevant to the topics at hand. The book focuses primarily on the evolution of reproductive behaviors among females, including issues of mate choice and other influences of females on male reproductive behaviors, the evolution of mating systems, the coevolution of the sexes, sex-role reversal, female–female reproductive competition, maternal behavior, and ways in which females enhance the investment their offspring receive from others. However, other social behaviors that bear on the nature of affiliative associations between females are addressed as well. Although the symposium also included work on insects, the book has been confined to work on vertebrates, ranging from fish to human beings, because females have already received a fair share of attention in the social insect literature. In any case, the diverse group of theoretical and empirical issues, combined with the range of taxonomic groups covered, should make this book of value to scientists as well as lay people whose interests span a variety of fields.

I wish to thank the following individuals (some of whom are also contributors) for commenting on chapters included in this book: E. Ahern, D. Barash, M. Beecher, S. Blaffer Hrdy, R. Boyd, N. Burley, M. Daly, S. Emlen, M. Ficken, E. Fischer, J. Gittleman, G. Hausfater, J. Heerwagen, W. Hill, W. Irons, J. Kaufman, M. Kirkpatrick, J. Kurland, F. Pitelka, H. Power, J. Silk, R. Van Gelder, and J. Wittenberger. In addition, I would like to thank all of the symposium and book contributors for their participation, and the Animal Behavior Society for granting the symposium a full day's time. I gratefully acknowledge the encouragement and support provided throughout this project by D. Barash, M. Waterhouse, and the people at Academic Press. I am also grateful to Sarah Blaffer Hrdy for graciously agreeing to oversee the final phase of the book's production when I returned to Africa. Financial support for this work was provided by the Department of Psychology and the Graduate School of the University of Washington.

PART I

Introduction

1

Behavioral Biology and the Double Standard

SARAH BLAFFER HRDY

GEORGE C. WILLIAMS

*We don't ask what a woman does—we ask whom she belongs to.**

George Eliot, 1860

I. The Myth of the Passive Female

In much of the literature on behavioral biology, males are depicted as active, enterprising, and adventurous, as well as aggresive and promiscuous; females as passive, reactive, and coy. This persistent bias in describing the behavior of males and females has been particularly evident

*Conversation man-to-man, from *The Mill on the Floss* (Houghton Mifflin, reprinted 1961).

ISBN 0-12-735950-8

in reconstructions of primate evolution and popularizations concerning the human implications of research on the behavior of other animals. It promotes a prescriptive double standard that may be viewed as a continuation of nineteenth-century Victorian—and earlier—attitudes, which are echoed in the writings of twentieth-century researchers of both sexes. In its most extreme form, this dualism has been used as social Darwinist justification for inequitable social arrangements. There seem to be people who actually believe that

> the double moral standard which punishes an adulteress severely while often condoning the man can be defended on biological grounds. It increases a man's reproductive potential and it might be added that those who indulge in extramarital activities are those who are the "fittest" and most deserving to be biological fathers as they must possess a high degree of cunning and initiative, and often physical agility [Burton, 1976, p. 155; see also Morris, 1979].

The threat that biology will be (as it sometimes has been) used to justify social inequalities has fueled widespread condemnation of Darwinian analyses of social behavior (Chasin, 1977; Hubbard, 1979; Leibowitz, 1979; Lowe and Hubbard, 1979). In the resulting melee, distinctions between social Darwinism and Darwinism proper have been lost, as have distinctions between sociobiology and various older traditions in the behavioral sciences. Among the more vocal critics has been the Sociobiology Study Group from Science for the People, who charge that "Sociobiology cannot be divorced from its sexism." According to them, the very process of applying a Darwinian perspective is necessarily sex-biased:

> Not only are the postulated human universals sexist, but the asserted mode of their propagation in evolution is sexist as well . . . sociobiology carries with it the implication that human social behavioral traits evolved primarily through sexual selection on male traits. Much of "Man's" evolutionary history is really the history of "men". Female social behavior, which is not readily explainable on the basis of sexual selection, is—like the creation of Eve in Genesis—just an afterthought [Alper *et al.*, 1978, p. 485].

Yet there is a certain irony to this charge, evident to anyone who has seriously considered either the hymenopterist origins of sociobiology (Wilson, 1971, Chap. 23) or its recent record of elucidating female as well as male behaviors in a variety of animal groups. To many of us directly engaged in sociobiological research, what distinguishes it from earlier biological efforts to study behavior is its debt to *inclusive fitness theory* (Hamilton, 1964). This central concept of inclusive fitness (defined as the sum of an individual's own behavior plus the effect that his or her behavior has on the fitness of relatives) owes its origin to the overwhelmingly female-centered studies of the hymenopteran insects. The wasps, ants, and bees inhabit what is virtually a "female world" (Evans, 1963, p. 6). Typically, a single female (the queen) dominates colony life. For most of the year, only females—who do the work of the colony—are present. Males are short-lived, parasitical members of the society. Their primary—and sometimes only—role is as "winged sperm dispersers" that are located and then utilized by females for breeding purposes (Wilson, 1971). For animals generally, and not only for the Hymenoptera, the explanatory power of a Darwinian approach can be trained with equal potency upon the behaviors of either sex, as is amply demonstrated by the chapters in this volume.

Although it may be incorrect to claim that sociobiology is necessarily sexist, it would also be less than honest to pretend that no researcher using inclusive fitness theory has not also been influenced by older traditions that are biased by sexist preconceptions. It is worth keeping in mind the distinction between sociobiology and older intellectual traditions that have also involved comparisons between humans and the behavior of other animals. Lorenzian ethology in Europe and biological anthropology in America are examples of such traditions. In terms of primate evolution and human behavior, biological anthropology, particularly the work of Washburn and his students, has been extremely influential (Haraway, 1978) and has had an impact on sociobiological writings as well. Nevertheless, these represent separate traditions. Sociobiological interpretations of primate behavior, identifiable by at least implicit use of the inclusive fitness concept, have been published mainly in the last decade, and are numerous only after 1975.

Throughout the early (presociobiology) years of primatology, there was a strong tendency among field-workers to focus on male behavior. Males were identified as individuals, and disproportionate attention was directed to interactions involving adult males. The only exception was careful observations of mother–infant relations (Jay, 1963b; DeVore, 1963). Not surprisingly, such studies "showed" that it was males who were responsible for determining the group's social structure (see e.g., the critique by Leibowitz, 1979). The idea that "Order within most primate groups is maintained by a hierarchy, which depends ultimately primarily on the power of males" (Washburn and Hamburg, 1968, p. 464) was influential and long-lived, persisting long after the central role of female primates in determining social structure had been well documented for several cercopithecine species (Kawamura, 1958; Sade, 1967). In spite of findings on the relative permanence of female relationships—findings that made it clear that at least in cercopithecine and some colobine monkeys, females formed the core of the social group—the conviction persisted that males had to be the prevailing influence upon primate social life. In a recent recapitulation of these long-held views we are told,

> The important point is that it is *male* competition that must be brought under control to keep the social group functioning in an orderly way. Besides competing for food, resting places, and grooming or other forms of attention, males also compete for sexual access to females (females do not compete for similar exclusive rights to males). While virtually every female is assured of leaving offspring there is probably a great range in the number of offspring a given male might leave . . . Female groups, especially in rhesus and baboon societies, tend to be much noisier and more disorganized than male groups . . . . In a variety of nonhuman primate species, then, competition among males is more crucial (since leaving offspring is at stake). . . . The question then arises whether or not male primates (including humans) have some biological preadaptation for competition [Cronin, 1980, pp. 302–303].

Females were not thought to compete, or in fact to affect at all the basic structure of the group. A composite personality profile can be extracted from a 1963 essay entitled "The Female Primate" which appeared in *The Potential of Woman.* "Her primary focus is motherhood . . . her dominance status, associations with adult males, female companions, and daily activities to a large extent are a function of her status as a mother and her phase of the reproductive cycle" (Jay, 1963a, pp. 3–7).

In some cases, sweeping generalizations were drawn from very narrowly focused field studies. For example, one author describes a study of captive rhesus in which no males happened to be present: "The females were incapable of 'governing' the group and social tension and disorganization were constant. The introduction of but one adult male into the group corrected the situation immediately, and a more normal political and social pattern quickly returned" (Tiger, 1970/1977, p. 28).

What is relevant about these observations, the author concludes, "is that primate females seem biologically unprogrammed to dominate political systems, and the whole weight of the relevant primates' breeding history militates against female participation in what we can call 'primate public life' " (Tiger, 1970/1977, p. 28).

Stereotypes of the noncompetitive female—too preoccupied with mothering to have any impact on the social organization of her group—influenced other fields, including social psychology and economics (see Gilder, 1981, pp. 136–137). Even scholars deliberately careful to avoid sexual bias (e.g., Maccoby and Jacklin, 1974) were affected because until the 1970s, accurate information on female primates was meager, and the only alternative was to ignore primates altogether. Even when more balanced field studies were undertaken, beginning in the 1970s this information was slow to enter the mainstream of the literature in the social sciences. As late as 1978, in a volume specifically devoted to female dominance relations (*Female Hierarchy: An Evolutionary Approach*), old stereotypes persist, in particular the view that "females are not innately disposed to organize into hierarchies . . . primate males appear to be the archetypal 'political animal' . . . [so that] the male more often than the female is emotionally fulfilled by the hierarchy *per se*" (Abernethy, 1978, p. 132).

Coupled with the old contrast between political males and noncompetitive females are new explanations, explanations which do indeed enamate from the sociobiological tradition. Abernethy was drawing on this tradition when she added,

> Over evolutionary time there has been little functional advantage associated with the formation of female hierarchies. Insofar as the female usually has assumed responsibility for childrearing, and noting that there is little advantage to hierarchical organization when this is the task, a logical conclusion is that hierarchical behavior has never been a focus of selective pressure and that genetic predisposition toward the creation of hierarchy is not a major determinant of female behavior [p. 129].

Note that this argument does not make sense unless we accept one critical assumption. Only if we assume that there is no variation in female reproductive success, that one female is interchangeable with another, can one logically conclude that competitive behavior and the formation of hierarchies "has never been a focus of selection pressure" among females. This assumption can be laid at the doorstep of one widespread interpretation of sexual selection theory.

Sexual selection theory (Darwin, 1871; Bateman, 1948; Williams, 1966; Trivers, 1972) is one of the crown jewels of the Darwinian approach basic to sociobiology. Yet so scintillating were some of the revelations offered by the theory, that they tended to outshine the rest of the wreath, and to impede comprehension of the total design; in this instance, the intertwined, sometimes opposing, strategies and counterstrategies of both sexes which together compose the social and reproductive behavior of a species. Much of the focus on males within the sociobiological mainstream and in many earlier studies can be attributed to an infatuation with particular concepts inspired by sexual selection theory, and particularly, greater variation in male than in female zygote production (see discussions in chapters by Payne and Wasser, this volume).

A great deal of attention was directed to examples of great variance in male reproductive success; for instance, the difference between Moulay Ismail the Bloodthirsty—former emperor of Morocco who fathered some 888 newborns—and the poor commoner who fathered none at all. By contrast, the most fertile women would produce far fewer babies than Moulay could father, and the variation in baby-production between one woman and another would be far less. Hence, what we would regard as the best available introductory textbook in sociobiology has a chapter entitled "The Reluctant Female and the Ardent Male" with sections labeled "Nurturant Females" and "Competitive Males" (Daly and Wilson, 1978, pp. 58–59). These distinctions are based on the statement that "most adult females in most animal populations are likely to be breeding at or close to the theoretical limit of their capacity to *produce and rear young*. Among males, by contrast, there is always the possibility of doing better [italics added]." They go on to explain that males but not females can commonly *do better* by additional mating, and this may be largely correct. What is not correct is the clear implication that they have told the whole story and that there is no way for females to do better. They are not alone in this oversight: "If males are able to court one female after another, some will be big winners and others will be absolute losers, while virtually all healthy females will succeed in being fertilized" (Wilson, 1978, p. 125). Thus the ease that a female may have in achieving fertilization is equated with the ease in realizing her physiological potential for reproduction. The capacity to produce is substituted for the capacity to produce and rear.

Can there be any doubt that a substantial fraction of Mr. Bloodthirsty's children died young? We would wager that his favorite wives raised a larger proportion of their children than did his low-status concubines, and lived to see a larger number of thriving grandchildren. The Moroccan women may not have had to compete for sperm, but their successful reproduction could be achieved only with successful competition for other necessities. Any character that would have influenced their access to essential resources would have been subject to natural selection. The fallacy of measuring reproductive success in both sexes by mere zygote production is bad enough. Even worse is the assumption that it can be measured by mere copulatory performance, and that this will be the primary focus of selection on reproductive fitness.

This assumption is expressed with rare clarity in a recent summary of the primate evidence in a book entitled *Human Sociobiology: A Holistic Approach*:

> about 20% of the [rhesus macque] males (all high on the status hierarchy) performed about 80% of the copulations; whereas all estrous females tended to be impregnated. These data make it clear that only males are directly involved in differential selection among rhesus and most probably all the terrestrial and semiterrestrial primates. They are the major competitors, the big winners and the big losers [Freedman, 1979, p. 33].*

Laid bare here is a long-standing conviction: Explain male strategies for inseminating females and you have explained the reproductive strategy of the species. We believe that this sort of thinking has blinded many workers to important variables in the many steps a female must take to transform zygotes into independent offspring.

II. Sources of Variation in Female Reproductive Success

Genetic differences in behaviors of direct or indirect relevance to reproduction are known for a number of species of insects and a variety of vertebrates (Ehrman and Parsons, 1976). Unfortunately, the genetically best-understood examples are unlikely to represent the sort of raw material from which natural selection would be expected to produce adaptations. They are either gross abnormalities, like star-gazing in quail, or the product of disruptive selection for different optima, like the male–female difference in any dioecious population. Perhaps the most likely examples would be in variation in human personality traits, which adoption studies have shown to have moderate biological heritabilities and to conform to polygenic models.

Whatever their genetic basis, there is good reason to suppose that different life histories and different patterns of behavior have consequences for the number of surviving offspring that individual females and female lineages produce (Drickamer, 1974; Epple, 1975; Ralls, 1976; Sade *et al.*, 1976; Vehrencamp, 1977; Frame *et al.*, 1979; Dunbar, 1980; Wrangham, 1980; Blaffer Hrdy, 1981; Reiter *et al.*, 1981; J. Strassmann, 1981; Wittenberger, 1981). To the extent that these differences are reflected by differences in genotypes, selection will operate on females. With growing recognition of the importance of interfemale variation, researchers have begun to focus on behaviors that might expose females to natural selection. This volume is representative of that trend.

It has long been known for domestic cows and other animals that females may differ in inherent fertility. Less well known are the behavioral attributes that might affect fecundity by, for example, increasing access to resources that individuals can convert into offspring, or by enhancing ability to protect and rear the offspring produced. This volume pulls together representative examples of such female strategies* from a diverse array of birds, fish, and mammals. Obviously, complicated connections exist between female and male strategies. With three exceptions however, (the chapters by Irons, Waterhouse, and Erckmann) these complexities tend to be avoided by authors in this volume in order to focus on long-neglected female behaviors. It is important to keep in mind that this temporary compartmentalization of "male" and "female" strategies is a

***Strategy* is used here, and throughout this volume, as a general term to designate behaviors (which may be either rigidly stereotyped or flexible and opportunistic) for obtaining a particular result, usually the increase of the strategist's individual or inclusive fitness. No foresight or consciousness is necessarily implied.

heuristic device to be used pending a more consistently balanced approach in sociobiology and behavioral biology generally.

The book illuminates the critical spheres in which female behaviors contributing to variation in female reproductive success can be studied. These can be divided into five categories: female choice of mates, female elicitation of male protection and support, mothering styles and skills, competition with other females, and cooperation among females. As might be expected, there is considerable overlap among these categories.

A. Female Choice

The possibility that females actively choose mates with superior genotypes (Darwin, 1871; Trivers, 1978) has potentially enormous consequences for evolutionary theory. However, as Payne (this volume) points out, there is a shortage of evidence of female discrimination among male genotypes, and Taylor and Williams (in press) show the inadequacy of several proposed explanations for the evolution of female choice. That female choice may nevertheless be real and very important was recently indicated by experiments by Partridge (1980).

Whatever the basis of female discriminations (genotype, the female's history of association with the male, or whimsy), there is fairly good evidence (for primates, fish, and birds) that some females do indeed discriminate among potential consorts (Lindburg, 1975; Tutin, 1976; Burley, 1977; Seyfarth, 1978; chapters by Brown and Downhower, Payne, Silk and Boyd, Erckmann, this volume). Furthermore, females apparently use such conventional criteria as rank and body size, as well as less obvious criteria, such as breeding experience (Burley and Moran, 1979). In spite of the probable importance of female choice, little is known about the important differences in male genotypes or female abilities to distinguish among them. It may be that Symons' (1979, p. 203) conclusion that female choice is over-rated as an evolutionary force is supported by nothing stronger than an absence of information.

B. Female Elicitation of Male Support and Protection

The need for male assistance in rearing young, and hence for selection for female attributes that insure that males provide it, may be especially important in birds and mammals (Orians, 1969; Kleiman and Malcolm, 1981). To the extent that females must compete with other females for the resources that males control, or the services (such as carrying or feeding young) that males provide, a female's elicitation of male support will be an integral part of her competitiveness vis-à-vis other females. However, other strategies to elicit male tolerance and assistance for her young will be directed toward males with little reference to other females. In this realm, female sexual solicitations and mating histories may be the primary mechanisms (Benshoof and Thornhill, 1979; Blaffer Hrdy, 1979; Labov, 1980; Lumpkin, Irons, this volume). Attributes such as assertive female sexuality (defined as the readiness of a female to mate) or prolonged female sexual receptivity may have evolved as mechanisms to manipulate male behavior. In a somewhat less Machiavellian vein, Irons' chapter stresses reciprocal exchanges of benefits between mates in human societies. That is, women exchange sexual and other services for male support.

C. Mothering Styles and Skills

This category includes all mother–infant interactions initiated or permitted by the mother (such as holding, carrying, or suckling) that promote the survival and well-being of her infant. The concepts of parental manipulation and parent–offspring conflict have recently gained prominence in the thinking of sociobiologists (Alexander, 1974; Trivers, 1974). These ideas have given a special focus to some studies of maternal behavior in mammals, birds, and social insects. Related behaviors include maternal allocation of resources to offspring and such physiological attributes of the mother as milk production. In turn, pregnancy and lactation will be related to other spheres of activity, such as competition for resources (see chapters by Dublin, Silk and Boyd, Wasser, this volume).

There is a large literature on maternal behavior, but only very recently have researchers begun to utilize a holistic approach that takes into account the complex social and ecological contexts in which mothers actually rear their young (Altmann, 1980). In the case of long-lived animals where learning is especially important—as it is among primates—juvenile, subadult, and even adult behaviors that enhance maternal skills, in particular allomaternal caretaking practice, should also be subject to selection pressure. To the extent that allomaternal care contributes to infant survival, these behaviors may be explained in terms of kin selection and female cooperation (see chapter by Dublin, this volume). But for many animals, the situation is more complex than that, involving varying degrees of selfishness on the part of the caretakers (Blaffer Hrdy, 1977, Chap. 7; Brown and Brown, 1981; Russell, Wasser, this volume). In particular, the chapter by Wasser offers a balanced perspective on the underlying tension in many species of monkeys between cooperation and exploitation that characterizes relations between lactating mothers with young and the allomothers who attend them.

D. Competition with Other Females

Competitive interactions with conspecifics (including males, but especially other productive females) will be a critical component of female fitness whenever fecundity and the ability to rear young are limited by access to resources. In species in which maternal status influences the rank of offspring (either sons [Koford, 1963] or daughters [Sade, 1967; Hausfater, 1980]), the cumulative effects of maternal rank will be magnified over time. In social species in which female competition leads to female hierarchies, dominant females may compete on behalf of their offspring by eliminating competitors or even forestalling reproduction in the mothers of potential competitors (chapters by Wasser, Silk and Boyd, this volume).

It is important to note that a female's competitive status relative to other females may be the single most pervasive influence on her reproductive success, determining in some cases whether or not she breeds at all, and influencing such diverse aspects of her life as health and parasite load, the sex of the offspring she produces, treatment of these offspring by other group members, and so forth (Wasser, this volume). The chapters by Koenig *et al.* and by Erckmann (this volume) are of special interest, since the variance in female reproductive success among acorn woodpeckers and some shore birds is apparently greater for females than it is for males, and female–female competition is especially intense.

E. Cooperation among Females

Cooperation refers to behaviors in which all participating individuals derive some net benefit in inclusive fitness. In practice however, it is often difficult to weight costs and benefits to participants. Moreover it can be problematic to distinguish between apparent altruism and genetic selfishness. Nevertheless, some cases, such as communal suckling among coatis and elephants (chapters by Dublin, Russell, this volume) seem clear-cut cases of long-term cooperative relations among females, some or all of whom are close relatives. (But note superficially similar, yet exploitative, behaviors involved in enforced nursing among hierarchically ranked female wild dogs [Van Lawick, 1973].)

Once it is determined that all parties are in fact benefiting, the next question is whether apparently cooperative behavior arose through kin selection, manipulation, reciprocal altruism, group selection, or some combination of these. Russell's and Wasser's chapters (this volume) illustrate how field-workers go about answering such questions. It may also be useful to distinguish between individuals who cooperate with one another in the performance of specific tasks (e.g., one female babysits while another forages, and vice versa) and intragroup alliances, usually among related females, in defending resources or privileges against competing groups (see especially the recent model for female–female bonding proposed by Wrangham [1980] and the similar model proposed by Silk and Boyd, this volume).

In spite of the difficulty of documentation, the area of cooperative behaviors is probably crucial for those interested in human social evolution. The extensive female–female cooperation among humans described by Bernard (1981) far exceeds the more limited cooperative relationships which have been documented so far among other primates. With the exception of allogrooming and the less than perfectly cooperative behaviors of allomothers (see Wasser, this volume), most cooperation has been documented for females of one lineage competing with females of another. We do not often see among nonhuman primates the sort of task-oriented cooperation between individuals—often unrelated individuals—that is so richly developed in many human societies.

III. And Now That We Are Asking, What Is It That Females Do?

Modern sociobiological concepts may be partly responsible for the current transition, in the study of vertebrate social behavior, from the male-biased approaches that were so much the rule before 1970, to more balanced sorts of studies. With the realization that variance exists in female reproductive success, it has become clear that there is a great deal that a female can actively do to improve her fitness relative to the fitness of other females. From an evolutionary point of view, it was this variance—regardless of whether males or females have the greater general variance in reproductive success—that mattered.

Once the importance of variance in female reproductive success became apparent, researchers needed little extra impetus to spur them to study what it was that females were actually doing. Hence, behaviors of females in species in which they had been habitually neglected are beginning to be systematically recorded. The reproductive strategies of these animals are being examined from the perspective of both sexes, as is demonstrated by the chapters in this book. At this stage, it might not be amiss to do some soul-searching for other arbitrary biases that may impede our efforts to understand the evolution of social behavior. Perhaps we are currently handicapped by a shortage of thought and data on the behavioral development of juveniles, in both our own and other species. So many speculations on the evolution of human mental capabilities emphasize how useful intelligence might be for such traditionally adult male functions as warfare and big game hunting. Might there not be some more important considerations? One might be Hutchinson's (1965, p. 93) suggestions that "it would seem quite likely that human intellectual development is largely an adaptation to permit young individuals to learn how to behave in a population of individuals whose behavior is unusually dependent on nongenetic information" and that higher levels of adult intelligence are merely "a paedomorphic extension of a childish set of attributes."

Traditions of male-focused behavioral research have not merely meant a detour in our efforts to understand female behaviors, but also a temporary block for understanding the full complexity of animal mating systems. They have also led to a misuse of the biological evidence to bolster sexist preconceptions. Such unproductive attitudes, even after their exposure and abandonment by practicing scientists, may persist in the thinking of laypeople, and continue to have unwholesome effects on social and political thought. This justifiable fear, and the misapprehension that to describe behaviors as biological in origin is tantamount to saying that they are unalterable, have led to a widespread rejection of an evolutionary approach by those sympathetic to feminist goals. To a great extent the research of sociobiologists has been, and is being, prejudged. Some of the researchers who might be expected to be most perceptive and original in applying female perspectives to biology and to explaining female social behavior will be discouraged from even reading in this area, much less entertaining an evolutionary perspective in their own research.

Sociobiologists have not often addressed feminist concerns, but, as Seger (1981) has pointed out, the critics of sociobiology have in turn paid surprisingly little attention (considering all the published pages) to what sociobiologists are actually saying. The irony of this stalemate is that if a more accurate picture of female biology—as reflected in studies such as those compiled here—had been available before the pathways of communication collapsed, the reaction might have been a different one. One hopes that it is not too late and that books such as this one will be widely read, for among such studies lie the seeds for a much broader understanding of female nature than either Victorian biologists or twentieth-century feminists had previously recognized.

Acknowledgments

We thank J. Moore, J. Seger, B. Smuts, and S. Wasser for valuable criticisms.

References

Abernethy, V. Female hierarchy: An evolutionary perspective. In L. Tiger and H. Fowler (Eds.), *Female hierarchies.* Chicago: Beresford Book Service, 1978.

Alexander, R. D. The evolution of social behavior. *Annual Review of Ecology and Systematics,* 1974, 5, 325–383.

Alper, J., Beckwith, J., and Miller, L. G. Sociobiology is a political issue. In A. Caplan (Ed.), *The sociobiology debate.* New York: Harper & Row, 1978.

Altmann, J. *Baboon mothers and infants.* Cambridge, Mass.: Harvard Univ. Press, 1980.

Bateman, A. J. Intra-sexual selection in *Drosophila. Heredity,* 1948, **2**(3), 349–368.

Benshoof, L., and Thornhill, R. The evolution of monogamy and concealed ovulation in humans. *Journal of Social and Biological Structures,* 1979, **2**, 95–106.

Bernard, J. *The Female World.* New York: Free Press, 1981.

Blaffer Hrdy, S. *The langurs of Abu: Female and male strategies of reproduction.* Cambridge, Mass.: Harvard Univ. Press, 1977.

Blaffer Hrdy, S. The evolution of human sexuality: The latest word and the last. *Quarterly Review of Biology,* 1979, **54**, 309–314.

Blaffer Hrdy, S. *The woman that never evolved.* Cambridge, Mass.: Harvard Univ. Press, 1981.

Brown, J. L., and Brown, E. R. Kin selection and individual selection in babblers. In R. D. Alexander and D. W. Tinkle (Eds.), *Natural Selection and Social Behavior.* New York: Chiron Press, 1981.

Burley, N. Parental investment, mate choice, and mate quality. *Proceedings of the National Academy of Sciences,* 1977, **74**, 3476–3479.

Burley, N., and Moran, N. The significance of age and reproductive experience in the mate preferences of feral pigeons, *Columbia livia. Animal Behaviour,* 1979, **27**, 686–698.

Burton, R. *The mating game.* New York: Crown Publishers, 1976.

Chasin, B. Sociobiology: A sexist synthesis. *Science for the People,* 1977, May–June issue.

Cronin, C. Dominance relations and females. In D. R. Omark, F. F. Strayer, and D. G. Freedman (Eds.), *An ethological view of human conflict and social interaction.* New York: Garland Press, 1980.

Daly, M., and Wilson, M. *Sex, evolution and behavior.* North Scituate, Mass.: Duxbury Press, 1978.

Darwin, C. *The descent of man, and selection in relation to sex.* London: Murray, 1871.

DeVore, I. Mother–infant relations in free-ranging baboons. In H. Rheingold (Ed.), *Maternal behavior in mammals.* New York: Wiley, 1963.

Drickamer, L. A ten-year summary of reproductive data for free-ranging *Macaca mulatta. Folia Primatologica,* 1974, **21**, 61–80.

Dunbar, R. I. M. Determinants and evolutionary consequences of dominance among female gelada baboons. *Behavioral Ecology and Sociobiology,* 1980, **7**, 253–265.

Ehrman, L., and Parsons, P. A. *The genetics of behavior.* Sunderland, Mass.: Sinauer Associates, 1976.

Epple, G. The behavior of marmoset monkeys (*Calltihricidae*). In L. Rosenblum (Ed.), *Primate Behavior* (Vol. 4). New York: Academic Press, 1975.

Evans, H. *Wasp Farm.* New York: The Natural History Press, 1963.

Frame, L. H., Malcolm, J. R., Frame, G. W., and Van Lawick, G. W. Social organization of African wild dogs (*Lycaon pictus*) on the Serengeti Plains, Tanzania 1967–1978. *Zeitschrift für Tierpsychologie,* 1979, **50**, 225–249.

Freedman, D. *Human sociobiology: a holistic approach.* New York: Free Press, 1979.

Gilder, G. *Wealth and poverty.* New York: Basic Books, 1981.

Hamilton, W. D. The genetical evolution of social behavior. I, II. *Journal of Theoretical Biology,* 1964, **7**, 1–52..

Haraway, D. Animal sociology and a natural economy of the body politic: Parts I and II. *Signs,* 1978, **4**(1), 21–60.

Hausfater, G. *Long-term consistency of dominance relations in baboons* (*Papio cynocephalus*). Paper presented at the Eighth International Congress of Primatology, Florence, Italy, 1980.

Hubbard, R. Have only men evolved? In R. Hubbard, M. S. Henifin, and B. Fried (Eds.), *Women look at biology looking at women: A collection of feminist critiques.* Cambridge: Schenkman Press, 1979.

Hutchinson, G. E. *The ecological theatre and the evolutionary play.* New Haven: Yale University Press, 1965.

Jay, P. The female primate. In S. Farber and R. Wilson (Eds.), *The potential of woman. Man and Civilization Series.* New York: McGraw–Hill, 1963a.

Jay, P. Mother–infant relations in langurs. In H. Rheingold (Ed.), *Maternal behavior in mammals.* New York: John Wiley, 1963b.

Kawamura, S. The matriarchal social order in the Minoo-B troop: A study on the rank system of Japanese macaques. *Primates,* 1958, **1**, 149–156.

Kleiman, D. G., and Malcolm, J. R. The evolution of male parental investment in mammals. In D. Gubernick and P. Klopfer (Eds.), *Parental care in mammals.* New York: Plenum, 1981.

Koford, C. B. Ranks of mothers and sons in bands of rhesus monkeys. *Science,* 1963, *141,* 356–357.

Labov, J. B. Factors affecting infanticidal behavior in wild male house mice (*Mus musculus*). *Behavioral Ecology and Sociobiology,* 1980, *6,* 297–303.

Leibowitz, L. "Universals" and male dominance among primates: A critical examination. In R. Hubbard and M. Lowe (Eds.), *Genes and gender II.* New York: Gordian Press, 1979.

Lindburg, D. Mate selection in rhesus monkeys (*Macaca mulatta*) (Abstract). *American Journal of Physical Anthropology,* 1975, *42,* 315.

Lowe, M. and Hubbard, R. Sociobiology and biosociology: Can science prove the biological basis of sex differences and behavior? In R. Hubbard and M. Lowe (Eds.), *Genes and gender II.* New York: Gordian Press, 1979.

Maccoby, E. and Jacklin, C. N. *The psychology of sex differences,* 2 vols. Stanford, Calif.: Stanford Univ. Press, 1974.

Morris, S. Darwin and the double standard. *Playboy Magazine,* August 1979.

Orians, G. On the evolution of mating systems in birds and mammals. *American Naturalist,* 1969, *103,* 589–603.

Partridge, L. Mate choice increases a component of offspring fitness in fruit flies. *Nature,* 1980, *283,* 290–291.

Ralls, K. Mammals in which females are larger than males. *Quarterly Review of Biology,* 1976, *51,* 245–276.

Reiter, J., Panken, K. I., and LeBoeuf, B. J. Female competition and reproductive success in Northern elephant seals. *Animal Behavior,* 1981, *29,* 670–687.

Sade, D. Determinants of dominance in a group of free-ranging rhesus monkeys. In S. Altmann (Ed.), *Social communication among primates*. Chicago: Univ. of Chicago Press, 1967.

Sade, D., Cushing, K., Cushing, P., Dunaif, J., Figueroa, A., Kaplan, J., Lauer, C., Rhodes, D., and Schneider, J. Population dynamics in relation to social structure on Cayo Santiago. *Yearbook of Physical Anthropology*, 1976, *20*, 253–262.

Seger, J. Sociologists and sociobiologists get their lines crossed. *Nature*, 1981, *291*, 690.

Seyfarth, R. Social relationships among adult male and female baboons. *Behaviour*, 1978, *64*, 204–226.

Strassmann, J. Wasp reproduction and kin selection: Reproductive competition and dominance hierarchies among *Polistes annularis* foundresses. *The Florida Entomologist*, 1981, *64*(1), 74–88.

Symons, D. *The evolution of human sexuality*. London/New York: Oxford Univ. Press, 1979.

Taylor, P. D., and Williams, G. C. The lek paradox is not resolved. *Theoretical Population Biology* (in press).

Tiger, L. The possible biological origins of sexual discrimination. In D. Brothwell (Ed.), *Biosocial man*. London: The Eugenics Society, 1977. (Originally published, 1970.)

Trivers, R. L. Parental investment and sexual selection. In B. Campbell (Ed.), *Sexual Selection and the Descent of Man*. Chicago: Aldine, 1972.

Trivers, R. L. Parent–offspring conflict. *American Zoologist*, 1974, *14*, 249–264.

Trivers, R. L. *The logic of female choice*. Paper presented at the Annual Meeting of the Animal Behavior Society, Seattle, June 1978.

Tutin, C. *Sexual behavior and mating patterns in a community of wild chimpanzees* (*Pan troglodytes*). Unpublished doctoral dissertation, University of Edinburgh, 1976.

Van Lawick, H. *Solo: The story of an African wild dog puppy and her pack*. Glasgow: Collins, 1973.

Vehrencamp, S. L. Relative fecundity and parental effort in communally nesting ants, *Crotophaga sulcirostris*. *Science*, 1977, *197*, 403–405.

Washburn, S., and Hamburg, D. Aggressive behavior in Old World monkeys and apes. In P. Jay (Ed.), *Primates*. New York: Holt, Rinehart and Winston, 1968.

Williams, G. *Adaptation and natural selection: A critique of some current evolutionary thought*. Princeton: Princeton Univ. Press, 1966.

Wilson, E. O. *The insect societies*. Cambridge, Mass.: Belknap Press of Harvard, 1971.

Wilson, E. O. *On human nature*. Cambridge, Mass.: Harvard Univ. Press, 1978.

Wittenberger, J. A. *Animal social behavior*. Boston: Duxbury Press, 1981.

Wrangham, R. An ecological model of female-bonded groups. *Behaviour*, 1980, *75* (3–4) 262–300.

2

The Establishment and Maintenance of Sex Biases

SAMUEL K. WASSER

MARY L. WATERHOUSE

I. Why a Book on Female Social Behavior?

Females have commonly been viewed as the passive, subordinate sex, whose primary concern is bearing and nurturing offspring. This combined with the fact that most researchers have historically been males, or at least trained in the "male tradition," has resulted in a disproportionately large amount of behavioral research on vertebrates that is focused predominantly on males. Such a male focus has required many assumptions regarding the presence or absence of particular behaviors among females, a number of which have been questioned (see Lee, 1968, 1979; Leacock, 1972, 1980; Martin and Voorhies, 1975; Tanner and Zihlman, 1976; Liebowitz, 1978; Wittenberger, 1980; Adkins, 1980; Shields, 1980; Wrangham, 1980; Jacobs, 1981; Labov, 1981; Sanday, 1981; Blaffer Hrdy and Williams, this volume). For example, as the chapters in this book suggest, mate choice, kin-selected behavior, and parenting among females

SOCIAL BEHAVIOR OF
FEMALE VERTEBRATES

ISBN 0-12-735950-8

have at times been overemphasized, whereas female–female competition, reciprocity, and a variety of other female behaviors that contribute to individual and group functioning have been neglected or assumed unimportant. Without data on females, erroneous views regarding them have been particularly difficult to dispel, and the result has not been trivial. For example, our poor understanding of female behavior, and hence needs, has already detrimentally affected health care (Mendelson, 1981; Waterhouse, this volume), education (Giligan, 1979), and economic equality among women, as well as caused us to sidestep the importance of a number of female psychological (Maccoby and Jacklin, 1974; see also Hoffman 1977; Wittig and Petersen 1979) and physiological (Wingfield, 1980) processes.

We do feel that the basic evolutionary paradigm—that individuals will be adapted by natural selection to perform behaviors that maximize their own survival, health, and reproductive success, relative to that of other conspecifics—provides the most fundamental basis for analyzing sex differences in behavior (see especially Hrdy, 1981). This point has been all too often overlooked by critics of behavioral biology (e.g., Hubbard, 1979; Liebowitz, 1978), despite the broad and powerful application of evolutionary theory to the understanding of sex differences in species such as the social insects (e.g., Hamilton, 1964; Wilson, 1971; West-Eberhard, 1975; Trivers and Hare, 1976; Forsyth, 1980; Strassman, 1981a). Contrary to the critics' claims, the evolution of sex differences does not necessarily imply exploitation or superiority of one sex over the other. In fact, just the opposite may hold: since an individual's reproductive success per mating depends on that of its mate, individuals of both sexes should have evolved a variety of sex roles that also complement those of their mates. Such sex role complementarity would be particularly useful where parental care is shared and/or subsequent reproduction with the same mate is likely (e.g., hamlets in Fischer, 1981; skuas in Selander, 1966; Adelie penguins in Derksen, 1977; shorebirds in Erckmann, this volume; siamang in Chivers, 1974; black and white colobus in Dunbar and Dunbar, 1976; humans in Hearne, 1911, cited in Leacock, 1980; Aberle, 1961; Whyte, 1978). Indeed, the complementary nature of sex differences may be the basis for the origin of the sexes in the first place (see Parker *et al.*, 1972; Williams, 1975; Maynard Smith, 1978).

The importance of the natural selection paradigm to the study of sex differences, combined with the dangers of a highly probable male focus in the behavioral biology literature, speak to the need for work that puts such views into their proper perspective. Thus, the purpose of this book is to reevaluate prevalent views on the evolutionary establishment and maintenance of female social behavior and the coevolution of the sexes.

The book focuses on *females* in an attempt to fill some of the voids that have resulted from their previous neglect. This is a critical step if we are to integrate both male and female perspectives in a complementary manner. Although the involvement of males in behavior—such as mate choice among animals and parental care among mammals—has been similarly neglected, this is not detailed here because of the current need for a book that emphasizes female behavior (but see Conaway and Koford, 1965; Redican, 1976; Gouzoules, 1980; Kleiman and Malcolm, 1981).

In this chapter we outline some of the evolutionary and historical events believed to have led to a male focus in science. This includes a discussion of the processes leading to the evolution of greater size and/or aggressiveness of one sex (usually males) over the other, and is followed by a description of historical events believed to have led to a predominance of men in science and to a value system that makes the origins of sex differences vulnerable to misinterpretation. We then detail evolutionary arguments that provide alternatives to some of the major male-focused stereotypes found in the behavioral sciences literature. This is followed by a description of the remaining chapters that compose this book. We would like to make clear, however, that the intent of this book is not to criticize the scientists responsible for the development of the views we discuss. Indeed, their ideas often constituted vital transitions in the ways behavior has been viewed. What we do hope to accomplish is to make clear the need for all of us to step back and take a fresh approach to the study of the sexes.

II. Sex Differences Mediated by Sexual Selection

Some of the most apparent physical, behavioral, and morphological sex differences among a variety of species, including humans, have resulted from sexual selection. Sexual selection refers to selection for characteristics within a sex that confer on individuals who exhibit them some reproductive advantage relative to others who do not, or to those who exhibit them to a lesser degree. The process of sexual selection was first described by Darwin (1871) and later elaborated by a number of evolutionary biologists. The two major forms of sexual selection have been described as intrasexual competition for number of mates and mate choice (Darwin, 1871; Williams, 1966; Trivers, 1972; but see Wasser, this volume).

The relative importance of each form of sexual selection to one sex over the other depends on their mode of gamete production and parental investment patterns (Bateman, 1948; Williams, 1966; Trivers, 1972; but see Brown and Downhower, this volume). Mate choice has been said to be most important for the sex contributing the greatest amount of parental investment—usually females—because that sex has the most difficult time replacing such investment when it is wasted. Intrasexual competition for number of mates has been said to be usually most important for males because they tend to have a continuously available gamete supply, potentially enabling them to fertilize eggs from a number of mates over a comparatively short period of time. Males also tend to contribute less parental investment than do females, which allows males more time and energy to attempt to acquire mates. However, males' ability to acquire such fertilizations is limited by the smaller number of cyclically produced gametes and the generally greater parental investment contributed by females, hence the competition among males. Effects of these sex differences are particularly apparent among mammals because of gestation and lactation. The effects are also highly dependent on the species' ecology and mating system, and exceptions do occur (see Section V).

But among other things, these forces have resulted in sexual selection for large size and fighting ability among males of a variety of species, since these characteristics better enable them to compete for limited mates. Sexual selection also may have tended to make male competitive behavior more overt and eye-catching than comparable female behaviors (but see Sections V–VI). Thus, some of the most fundamental asymmetries between the sexes of a variety of species, including humans, appear to have arisen through sexual selection. Moreover, these asymmetries may have also influenced other sex differences in behavior and morphology. Researchers have pointed out that some physically and perhaps behaviorally dimorphic characteristics initially established by sexual selection may have been further exaggerated for nonsexually selected reasons. For example, male canine tooth size appears to have been further enlarged because it enhanced predator protection and troop defense in terrestrial old-world primates (Harvey *et al,* 1978; but see Wrangham, 1980). Sexual dimorphism in body size and bill shape and size among some bird species may have been further accentuated because it enhanced niche stratification and reduced competition for food between the sexes; where biparental care occurs, such niche stratification (and hence sexually dimorphic characteristics) may have also increased the range and total amount of resources that could be provided to the young by their parents (e.g., Selander, 1966).

The behavioral, physiological, and morphological sex differences that resulted from these forms of selection on humans have also influenced the nature of humans' sexual division of labor as well (Aberle, 1961; see also

Section III) and vice versa (Maccoby and Jacklin, 1974; Hoffman, 1977; Wittig and Petersen, 1979). However, our perceptions regarding the outcomes of these influences (e.g., male superiority versus male–female complementarity) have also been dependent on the values of researchers who have described such events. This was most clearly shown by the work of Lee (1968, 1979), Draper (cited in Lee, 1979), and Whyte (1978). Greater size and aggressiveness among men, combined with a reduced need to remain close to nursing young and therefore expose the young to risk while pursuing dangerous prey, have probably made hunting a more common role for men than women. However, earlier literature tended to weigh the relative contribution of men to subsistence more heavily than that of women because of the perceived value placed on hunting relative to gathering. Quantification of the relative workload contributions of *both sexes* showed the work input of men and women to be roughly equivalent, although women's work was often 2.5 times more productive than men's per person-hour (Lee, 1979). Similar value judgements regarding psychological sex differences among humans have led to an extremely unproductive polarization of the field with respect to the relative influences of heredity versus environment in the origins of these sex differences (cf. Money and Erhardt, 1972; Maccoby and Jacklin, 1974; Imperato-McGinley *et al*, 1979; and Rubin et al. 1981).

The tendency to consider females as the passive sex has even given an androcentric tone to theory regarding the origin of exclusively female phenomena. Consider for example the male orientation in the following current explanations for reproductive synchrony, concealed ovulation, continuous receptivity, and orgasm among women: Polygynous women synchronized their menstrual cycles to avoid inundating males with "contradictory information" from their independent cycles (Burley, 1979); estrus disappeared in women to facilitate male–male bonds (Etkin, 1963; Pfeiffer, 1969); loss of estrus evolved among women because it prolonged their period of sexual attractiveness to men, who provide them with meat in exchange for sex (Symons, 1979); concealed ovulation evolved to increase paternal certainty in humans and to force males into pair bonds (Alexander and Noonan, 1979); the female orgasm evolved to make women quiescent following copulation so as to prevent the male's sperm from leaking out of the vagina (Morris, 1967); the female orgasm evolved as a "by-product of selection for male orgasm" (Symons, 1979). For some alternatives to these male-focused views, see reviews by McClintock (1981) and Strassmann (1981b).

It seems appropriate here to ask, What historical events are believed to be responsible for such androcentric value judgments and a predominantly male focus in science? These events appear to be the same as those that strengthened the general economic and political roles of men relative to women—a progression that seems to have begun with the implementation of food production and storage techniques among hunter–gatherers approximately 10,000 years ago. (For those readers interested in events prior to this time, see Tanner and Zihlman [1976] and Lovejoy [1981].

III. Humans in Transition: The Establishment and Maintenance of Sex Biases in Science

Until 8000 B.C., population growth rates were extremely low, approximately .001% annually. However, during the early period of food production, the annual growth rate accelerated to about .1% on the average (Hassan, 1980). Humans then entered a period of nearly exponential growth that became particularly rapid during the period of industrialization (Hassan, 1980). This increase in family size is generally believed to have had economic as well as reproductive advantages for both sexes (Cohen, 1980; Hassan, 1980; Lee, 1980). For example, food production and storage techniques, along with a sedentary way of life, are believed to have increased the opportunity for relatively young children to help rear infants and contribute to subsistence. Both Omran (1971) and Caldwell (1977) suggest that this increase in fecundity was additionally important to offset mortality from an increase in the communicable diseases and food shortages due to drought that accompany sedentary life.

Recent work further indicates that the nature of the sexual division of labor changed once food production and storage techniques were implemented. In a relative sense, females are believed to have focused increasingly on child care and a variety of subsistence, medicinal, and other health-related activities that enabled them to remain close to home (see Brown, 1970; Nerlove, 1974; but see Draper, 1975), whereas males focused increasingly on agricultural modes of subsistence and on defense behavior. This division of labor presumably enabled both sexes to perform their combined roles with much greater efficiency, increasing the productivity of their economic unit (e.g., Whyte, 1978). However, individuals still had control over their own produce, and the nature of these sex roles eventually enabled men to accumulate more produce than could women (Engels, 1942; Leacock, 1972; Draper, 1975). Leacock (1972) asserts that the accumulation of goods that followed implementation of food production and storage techniques eventually undermined the communal relationships necessary for traditional hunter–gatherers, which resulted in the primacy of the family as the economic unit and the acquisition of private property. It was within this climate of individual or family gain that the relatively greater potential for accumulation of goods by the "father" of the family provided the economic basis for the decline of the political and social status of women. (An interesting contrast to these events, but one that supports the above economic arguments, is provided by a present-day diving village in Korea. Sex roles are reversed so that females, who are better adapted for diving, are the primary producers, while males are the primary caretakers of young. Yet, neither sex is dominant—"male-superiority ideology supported by the strict practices of ancestral memorial rituals, and female-dominated reality supported by economic and social power counterbalance the power of either sex" [Cho, 1979, p. 1].)

As population growth rates continued to increase, the need for social class specialization increased in order to maintain some source of social control (Adams, 1966). Throughout the Neolithic and Bronze Ages (8000–1200 B.C.), the sexual division of labor and relatively greater accumulation of wealth by males allowed male roles to change more rapidly than those of females (Engels, 1942; Leacock, 1972; see also Draper, 1975). Thus, as the social, political, and religious hierarchies grew, so did the positions of males within them. These trends were further exaggerated by the move towards commodity production, enabling new wealth in the form of slaves and herds to be accumulated by single individuals. Engels (1942, p. 119) states that this increase in wealth "made man's position in the family more important than the woman's, and . . . created an impulse to exploit this strengthened position in order to overthrow, in favor of his children, the traditional order of inheritance." Thus, he hypothesized that among such populations, inheritance moved from matrilineal to patrilineal descent, accompanied by strict social controls to ensure paternity certainty. Engels regards this as the "world historical defeat of the female sex." This tendency towards expansion, exploitation, and the accumulation of wealth had its first culmination in the Bronze Age (3000–1150 B.C.), in which we find the formation of kingdoms.

Basic economic change brought about the first conditions that strengthened the roles of males. The second factor developed during the later periods of antiquity, particularly during the Greek Classical Period. This factor can be defined as essentially political in nature.

In the eastern Mediterranean, the Bronze Age came to a sudden end in the twelfth century B.C. Thomas (1981) reviews the changes that transpired for Greece in the following "Dark Age" (1150–750 B.C.); similar conditions existed throughout the entire Near East (Griffeth and Thomas,

1981). The first half of the Dark Age was a time of devastation during which a dramatic decline in population size occurred; the latter half was a period of recovery. Although the actual cause of the devastation in the first half of the Dark Age is difficult to ascertain, it seems to have been associated with climatic change that led to severe drought, famine, and dislocation of trade. Kingdoms were broken up, populations dispersed, and "local village communities were thrown back on their own resources—both natural and human" (Thomas, 1981). Individuals in these small communities became highly dependent upon one another for survival. A successful individual was able to defend the property of others as well as his own and came to be regarded as the leader of his small community of kin and followers. Those communities were small—25–30 people—but from them developed the Classical form of societal organization: the city-state.

During the Classical Period (750–350 B.C.), Greece consisted of numerous small city-states, each striving to maintain its own independence. Since each city-state was simultaneously attempting to expand its own territory, defense was of paramount concern. As a result, those individuals defending the state were placed in highest esteem. The concept of citizenship was a product of Greek city-state organization and the possession of citizenship belonged exclusively to adult males. This is not to say that women were without rights and honor. Indeed, both parents had to be full members of a city-state by birth for a man to become a full citizen. Morever, women influenced social, political, and economic decisions indirectly (Kitto, 1951, Chap. 12).

Greeks of the Classical Period believed that citizens should be freed from constant attention to economic pursuits in order to better serve their state. Consequently, while citizens did farm the land and pursue various crafts, they were also allowed leisure. This leisure could be devoted to any number of concerns and it was during the Classical Period that abstract scientific and philosophic thought took root. Those individuals free to investigate the nature of the universe and the acquisition of knowledge were males. Although the city-states did not directly support the activities of individual scholars during the Classical Period, there was direct support for scholars of all interests during the following Hellenistic and Roman Periods. Science remained an avocation of, and/or supported by, wealthy males until the twentieth century, moving from a "rational" to an "empirical" science during the eighteenth century. Thus, men moved into the realm of science during a time when males were held in highest esteem, and this value system persisted through the growth of the empirical sciences (see Parsons, 1964; Jacobs, 1981). (In fact, one among several illustrations of the continued existence of such androcentric values can be found in a United Nations study released in 1980. It reveals that females presently make up half of the world's population and contribute two-thirds of the world's working hours; yet, females receive only 10% of the world's income and own only 1% of its property). We believe that these events, compounded by the effects of sexual selection (see Section III), were the primary factors that led to the male focus in science.

By the time people began to study the behavior of human beings, as well as that of other animals, males tended to be larger and stronger than females and often expressed their competitive behaviors more overtly than did females. Accompanied by a set of values that placed higher esteem on the roles of males than on those of females, scientists focused the majority of behavioral research on the male sex, thereby making androcentric views of females easy to create but difficult to dispel—a problem that persists today (see also Blaffer Hrdy and Williams, this volume).

IV. Some Evolutionary Alternatives to the Male Dominance Syndrome

One of the most unfortunate, misinterpreted stereotypes regarding the sexes that has resulted from the preceding combination of historical and evolutionary events is that males' large size and fighting ability often enable them to "dominate" females of their species. Dominance refers here to preferential access to, and control of, resources and sometimes even control over females themselves. Few stereotypes could make female behavior seem less interesting. Sanday (1981) has traced this stereotype as far back as the sixth century B.C. But even recent investigators continue to state or imply it (e.g., Divale and Harris, 1976; Symons, 1979). Such a stereotype is unfounded for a number of reasons. First, male size and dominance relative to females tend to be associated with the species mating system. Generally speaking, males tend to be larger than, and "dominant" over, females among polygynous species; they are of comparable size and inconsistently dominant over females among monogamous species (Hrdy 1981), and smaller than, and subordinate to, females in most polyandrous and some monogamous species (Ralls, 1976; Erckmann, this volume). Moreover, in the vast majority of cases, it is the female who plays the dominant role in determining which of these mating systems arises (Orians, 1969; Wittenberger, 1980; but see also articles by Erckmann, Irons, this volume). In still other species, females have eliminated their male conspecifics altogether, reproducing entirely through parthenogenesis (e.g., Oliver, 1971; Maslin, 1971), or through some other peculiar means. In the fish *Poeciliopsis monacha-lucida,* for example, the females' eggs are fertilized by males of another species (hybridogenesis). However, when their hybrid offspring begin to produce gametes of their own, the entire genetic contribution of the father is left scattered in the cytoplasm, eventually being resorbed or expelled so that only maternal chromosomes are passed to eggs (Schultz, 1977).

However, even if we ignore the above cases, we can still find a number of situations in which size and strength are not all-important in interactions between the sexes. For example, a variety of monogamous and polygynous species exist in which males are larger than females, but females either dominate males or the sex that is dominant is seasonally dependent, based on traditional ethological criteria (Smith, 1980; Hrdy, 1981). Large size and strength are not always effective in competition between males either (reviewed by Rubinstein, 1980). In fact, males of some species use small size and "sneakiness" to give them a competitive advantage over larger males who have difficulty catching them or keeping them away from breeding territories (e.g., Gross and Charnov, 1980, for sunfish; Howard, 1981, for bullfrogs).

The male dominance assumption also ignores the potentially powerful influence that female choice of mates has had on the evolution of a variety of male characteristics (e.g., Cox and LeBoeuf, 1977; Halliday, 1978; but see articles by Brown and Downhower, Payne, this volume), as well as the importance to male fitness of a variety of roles that females perform (e.g., Landes, 1937; Dunbar and Dunbar, 1976; Whyte, 1978; Leacock, 1980; Fischer, 1981); even in cases in which males are said to dominate females, males still tend to expend the greater amount of energy and risk in attempting to attract their mates. Thus, females often appear able to choose the males for whom they will perform their valued roles, and males appear to be selected to meet the females' criteria in order to qualify as their recipients. This also suggests that although brute force by males may occasionally be effective in short-term associations with females, it is unlikely to be so in long-term ones in which males can benefit from the females' presence (see Wasser, 1982). In fact, a nice example of the ineffectiveness of brute force in the latter case is suggested by the evolution of emotions such as "love" among humans. It seems unlikely that such strong emotion could have evolved if one sex could consistently take advantage of the other. This is particularly the case considering how vulnerable this emotion can make *each* mate to the other.

Females have coevolved with males throughout the evolutionary history of sexual species, and have also been selected to maximize their own personal fitnesses. It is therefore most probable that they have evolved effective means of coping with, as well as complementing, the opposite sex. The problem is that we have consistently failed to look for such means. The same can be argued for a variety of social interactions among females, especially with respect to female–female competition and cooperation (see Section VI). In essence, we have simply failed to study females adequately, and from the proper perspective, to be able to support some of the things we have said about them.

V. What Lies Ahead?

Alternative views regarding the sexes are now surfacing, in part due to an increase in both the numbers of female researchers and to the influence of the women's movement (Jacobs, 1981). Scientists are being forced to look more carefully, and often from different perspectives, at the roles females play in both female–male, and female–female interactions. New data are being gathered on the involvement of both sexes in roles not previously examined with respect to that sex, and long-standing theories are now being critically reevaluated, extended, or changed altogether. Essentially, some of the unfortunate consequences of failing to recognize alternative hypotheses (Bateson and Hinde, 1976) are now being rectified, creating the potential for research that combines both male and female perspectives so as to better understand the whole. Although we still have a long way to go, it appears that the stage is now set for what Kuhn (1970) has called a "scientific revolution," some effects of which can be seen in the collection of chapters that compose this book. Thus, the chapters that follow are purposely diverse, nearly all reflecting fresh approaches to the study of female behavior.

Part II of this book focuses on the effects of female behavior on that of males, and on the coevolution of the sexes. Chapters by Brown and Downhower and by Payne both reveal the complexities and limitations of female mate choice, as well as treat the difficulties of ascribing the evolution of a given male characteristic to any of the known mediators of sexual selection. Brown and Downhower's chapter models the trade-offs females must make when choosing mates; they then test their model with empirical data on the mottled sculpin. Payne's chapter addresses whether female choice or male–male competition shapes bird song by males, comparing two species of finches, the North American indigo bunting and the African indigobird. Lumpkin's chapter reveals a less obvious means by which females may take advantage of and influence male behavior in ring doves. She examines how females influence male investment in courtship behavior.

Chapters by Erckmann, Waterhouse, and Irons each take a more coevolutionary approach to the study of the sexes. Erckmann offers a fresh perspective on the evolution of mating systems in shorebirds, focusing to a large degree on species in which polyandry and sex-role reversal have evolved. He provides a variety of tests of his own coevolutionary view, as well as of other views of mating systems now prevalent in the literature. Waterhouse and Irons each address the coevolution of sex roles and behavioral strategies among humans, including ways in which female–male, and female–female roles conflict with and/or complement one another. Waterhouse reinterprets past work on the patrilocal system of the Taiwanese, where women have previously been interpreted as highly subordinate to men. She also points out some of the consequences such value judgements can have on the direction of present-day transitions among the Taiwanese, particularly those revolving around their medical system. Irons, on the other hand, deals primarily with the large variety of strategies generally employed by women in order to maximize the amount of nurturance their offspring obtain from their mates, kin, and other group members. The strategies females use are also shown to have a dramatic influence on the type of mating system and the residence patterns their culture adopts.

Part III focuses on reproductive competition and cooperation among female vertebrates and the ways these two forces may be in conflict. It reveals a much more elaborate nexus of social behavior among females than has been implied by past literature. For example, reproductive competition has been reputed to be rare among females, especially among those who invest heavily in their offspring (Williams, 1966; Trivers, 1972), and even recent work by careful biologists has assumed little or no variance in female reproductive success (e.g., Daly and Wilson, 1978; Freedman, 1979; Wade and Arnold, 1980). All chapters in Part III of this book dispel this idea, revealing elaborate female–female competition among "cooperatively breeding" acorn woodpeckers (Koenig, Mumme, and Pitelka), coatis (Russell), elephants (Dublin), baboons (Wasser), and macaques (Silk and Boyd). Wasser's chapter also points out that past sexual selection theory regarding intrasexual reproductive competition has been a quantity-versus-quality argument, focusing on competition that increases the number of conceptions an individual can obtain. We have largely ignored reproductive competition that increases the survival and the subsequent reproductive capabilities of offspring. The former type of competition is particularly important for males (Bateman, 1948; Williams,

1966; Trivers, 1972), whereas the latter type is particularly important for females (Wasser, this volume). Thus, emphasis on the former type of competition has resulted in a conspicuous absence of theoretical and empirical work on female–female competition.

The past literature also implies that reciprocal associations among distant or nonrelated female mammals are unimportant compared to those among close kin, especially for species having predominantly male dispersal (see reviews by Greenwood, 1980 and Wrangham, 1980). However, Russell's and Wasser's chapters deal with the complex interplay between kin selection and reciprocity among female coatis and baboons, respectively. Together, their chapters show the association between distant kin or "nonkin" may be common, under a broad range of conditions. Dublin's chapter also addresses reciprocity. However, her chapter, as well as those of Koenig, Mumme, and Pitelka, and of Silk and Boyd, gives greater attention to the complexities of cooperative interactions between close kin. All of these chapters identify a considerable degree of flexibility in female association patterns. Of particular interest are the ramifications such behaviors are shown to have on patterns of mate choice, alloparental care, cooperative breeding, and the interplay between female–female cooperation and competition.

In summary, a large body of past work on females and on the evolution of sex roles in general, has been carried out from a male perspective, with a primary focus on males. The bases for this male perspective and focus appear to be largely a consequence of sexual selection, coupled with the sex-role, economic, and political changes humans have undergone beginning with the onset of food production and storage techniques, approximately 10,000 years ago. These sex-role, economic, and political changes culminated during the Classical Period of Greek history—the era in which scientific thought was born. It was primarily men who were free to move into the realm of science at that time, bringing with them a male-focused set of values. These values persisted throughout the development of the empirical sciences as well. Males were more frequently studied than were females, generating androcentric biases that were easy to create and difficult to dispel. Although this male focus in science seriously affected the applicability and sometimes the validity of theory regarding both sexes, accumulating data on females are currently forcing scientists to step back and take a fresh approach to the study of the sexes. This reassessment is revealing a rich array of female behavior as well as new ways to approach the coevolution of the sexes. Given the relevance of such new views to the fields of anthropology, sociology, psychiatry, psychology, political science, medicine, zoology, education, and others, it seems likely that a new scientific revolution has already begun and that a complementary integration of male and female perspectives will soon be possible. It is our hope that the chapters in this book can provide a step in this direction and that we can all look forward to the exciting directions that lie ahead.

Acknowledgments

We wish to thank the following people for commenting on this manuscript: D. Barash, L. Carter, I. DeVore, J. Heerwagen, W. Hill, S. Hrdy, L. Iglitzin, S. Jacobs, P. Lunneborg, E. Smith, G. Williams, and E. Winans. Special thanks go to Carol Thomas for tutoring us on the history of ancient Greece and the origin of the city-state.

References

Aberle, D. F. Matrilineal descent in cross cultural perspective. In D. M. Schneider and K. Gough (Eds.), *Matrilineal kinship*. Berkeley: Univ. of California Press, 1961.

Adams, R. *The evolution of urban societies*. Chicago: Aldine, 1966.

Adkins, E. K. Genes, hormones, sex and gender. In G. W. Barlow and J. Silverberg (Eds.), *Sociobiology: Beyond nature/nurture?* Boulder, Colo.: Westview Press, 1980.

Alexander, R. D., and Noonan, K. M. Concealed ovulation, parental care, and human social evolution. In N. H. Chagnon and W. Irons (Eds.), *Evolutionary biology and human social behavior: An anthropological perspective*. N. Scituate, Mass.: Duxbury, 1979.

Bateman, A. J. Intra-sexual selection in drosophila. *Heredity*, 1948, *2*, 349–368.

Bateson, P. P. G., and Hinde, R. A. Conclusion—on asking the right questions. In P. P. G. Bateson and R. A. Hinde (Eds.), *Growing points in ethology*. Cambridge, Mass.: Cambridge Univ. Press, 1976.

Brown, J. K. A note on the division of labor. *American Anthropologist*, 1970, *72*, 1073–1078.

Burley, N. The evolution of concealed ovulation. *American Naturalist*, 1979, *114*, 835–858.

Caldwell, J. C. The economic rationality of high fertility: an investigation illustrated with Nigerian data. *Population Studies*, 1977, *31*, 5–27.

Chivers, D. J. The siamang in Malaya: A field study of a primate in tropical rain forest. *Contributions to Primatology*, 1974, *4*. (Monograph)

Cho, H. *An ethnographic study of a female diver's village in Korea: focused on the sexual division of labor*. Unpublished doctoral dissertation, University of California, Los Angeles, 1979.

Cohen, M. N. Speculations on the evolution of density measurement and population regulation in *Homo sapiens*. In M. N. Cohen, R. S. Malpass, and H. G. Klein (Eds.), *Biosocial mechanisms of population regulation*. New Haven, Conn.: Yale Univ. Press, 1980.

Conaway, C. H., and Koford, C. B. Estrous cycles and mating behavior in a free-ranging band of rhesus monkeys. *Journal of Mammology*, 1965, *45*, 577–588.

Cox, C. R., and LeBoeuf, B. J. Female incitation of male competition: a mechanism in sexual selection. *American Naturalist*, 1977, *111*, 317–335.

Daly, M., and Wilson, M. *Sex, evolution, and behavior*. N. Scituate, Mass.: Duxbury, 1978.

Darwin, C. *The descent of man, and selection in relation to sex* (Vols. 1 and 2). New York: Appleton, 1871.

Derksen, D. V. A quantitative analysis of the incubation behavior of the Adelie penguin. *Auk,* 1977, *94,* 552–556.

Divale, W. T., and Harris, M. Population, warfare, and the male supremacist complex. *American Anthropologist,* 1976, *78,* 521–538.

Draper, P. !Kung women: contrasts in sexual egalitarianism in foraging and sedentary contexts. In R. R. Reiter (Ed.), *Toward an anthropology of women.* New York: Monthly Review Press, 1975.

Dunbar, R. I. M., and Dunbar, E. P. Contrasts in social structure among black-and-white colobus monkey groups. *Animal Behaviour,* 1976, *24,* 84–92.

Engels, F. *The origin of the family, private property, and the state.* New York: Intern. Publ. Co., 1942.

Etkin, W. Social behavioral factors in the emergence of man. *Human Biology,* 1963, *35,* 299–310.

Fischer, E. A. Sexual allocation in a simultaneously hermaphroditic coral reef fish. *American Naturalist,* 1981, *117,* 64–82.

Forsyth, A. Worker control of queen density in hymenopteran societies. *American Naturalist,* 1980, *116,* 895–898.

Freedman, D. G. *Human sociobiology: a holistic approach.* New York: Free Press, 1979.

Giligan, C. Woman's place in man's life cycle. *Harvard Educational Review,* 1979, *49,* 431–450.

Gouzoules, H. The alpha female: observations on captive pigtail monkeys. *Folia Primatologica,* 1980, *33,* 46–56.

Greenwood, P. J. Mating systems, philopatry, and dispersal in birds and mammals. *Animal Behaviour,* 1980, *28,* 1140–1162.

Griffeth, R., and Thomas, C. (Eds.), *The city state in five cultures.* Santa Barbara: ABC Clio, 1981.

Gross, M. R., and Charnov, E.. L. Alternative male life histories in bluegill sunfish. *Proceeding of the National Academy of Science, U.S.A.,* 1980, *77,* 6937–6940.

Halliday, T. R. Sexual selection and mate choice. In J. R. Krebs and N. B. Davies (Eds.), *Behavioural ecology: an evolutionary approach,* Sunderland, Mass.: Sinauer, 1978.

Hamilton, W. D. The genetical evolution of social behavior, I, II. *Journal of Theoretical Biology,* 1964, *7,* 1–52.

Harvey, P. H., Kavanaugh, M., and Clutton-Brock, T. H. Sexual dimorphism in primate teeth. *Journal of the Zoological Society, London,* 1978, *186,* 475–485.

Hassan, F. A. The growth and regulation of human population in prehistoric times. In M. N. Cohen, R. S. Malpass, and H. G. Klein (Eds.), *Biosocial mechanisms of population regulation.* New Haven, Conn.: Yale Univ. Press, 1980.

Hoffman, M. L. Sex-differences in empathy and related behaviors. *Psychological Bulletin,* 1977, *84,* 712–722.

Howard, R. D. Male age–size distribution and male mating success in bullfrogs. In R. D. Alexander and D. W. Tinkle (Eds.),, *Natural selection and social behavior: recent research and new theory.* Concord, Mass.: Chiron Press, 1981.

Hrdy, S. B. *The woman that never evolved.* Cambridge, Mass.: Harvard Univ. Press, 1981.

Hubbard, R. Have only men evolved? In R. Hubbard, M. S. Henefin, and B. Fried (Eds.), *Women look at biology looking at women. A collection of feminist critiques.* Boston: G. K. Hall and Co., 1979.

Imperato-McGinley, J., Peterson, R. E., Gautier, T., and Sturla, E. Androgens and the evolution of male-gender identity among male pseudohermaphrodites with 5a-reductase deficiency. *New England Journal of Medicine,* 1979, *300,* 1233–1237.

Jacobs, S. E. Memorandum on gender roles in U.S. society. New York: Ward Cannel, Crown and Bridge Publ., 1981.

Kitto, H. D. F. *The greeks.* Baltimore: Penguin, 1951.

Kleiman, D. G., and Malcolm, J. R. The evolution of male parental investment in mammals. In D. J. Gubernick and P. H. Klopfer (Eds.) *Parental care in mammals.* New York: Plenum, 1981.

Kuhn, T. S. *The structure of scientific revolutions* (2nd ed.). Chicago: Univ. of Chicago Press, 1970.

Labov, J. B. Pregnancy blocking in rodents: adaptive advantages for females. *American Naturalist,* 1981, *118,* 361–371.

Landes, R. The Ojibwa. In M. Mead (Ed.), *Cooperation and competition among primitive peoples.* New York: McGraw–Hill, 1937.

Leacock, E. B. Introduction. In F. Engels, *The origin of the family, private property, and the state.* New York: Intern. Univ. Press, 1972.

Leacock, E. B. Social behavior, biology and the double standard. In G. W. Barlow and J. Silverberg. *Sociobiology: Beyond nature/nurture?* Boulder, Colo.: Westview Press, 1980.

Lee, R. B. What hunters do for a living, or, how to make out on scarce resources. In R. B. Lee and I. DeVore (Eds.), *Man the hunter.* Chicago: Aldine, 1968.

Lee, R. B. The !Kung San: men, women, and work in a foraging society. London/New York: Cambridge Univ. Press, 1979.

Lee, R. B. Lactation, ovulation, infanticide, and women's work: a study of hunter–gatherer population regulation. In M. N. Cohen, R. S. Malpass, and H. G. Klein (Eds.), *Biosocial mechanisms of population regulation.* New Haven, Conn.: Yale University Press, 1980.

Liebowitz, L. *Females, males, families: a biosocial approach.* N. Scituate, Mass.: Duxbury, 1978.

Lovejoy, O. The origin of man. *Science,* 1981, *211,* 341–350.

Maccoby, E. M., and Jacklin, C. N. *The psychology of sex differences.* Stanford, Calif.: Stanford University Press, 1974.

Martin, M. K., and Voorhies, B. *Female of the species.* New York: Columbia Univ. Press, 1975.

Maslin, T. P. Parthenogenesis in reptiles. *American Zoologist,* 1971, *11,* 361–380.

Maynard Smith, J. *The evolution of sex.* London/New York: Cambridge Univ. Press, 1978.

McClintock, M. K. Social control of the ovarian cycle and the function of estrous synchrony. *American Zoologist,* 1981, *21,* 243–256.

Mendelson, R. S. *Mal(e) practice: how doctors manipulate women.* Chicago: Contemporary Books, 1981.

Money, J. and Erhardt, A. A. *Man and woman, boy and girl.* Baltimore: Johns Hopkins Press, 1972.

Morris, D. *The naked ape.* London: Jonathan Cape, 1967.

Nerlove, S. B. Women's workload and infant feeding practices: a relationship with demographic implications. *Ethnology,* 1974, *2,* 207–214.

Oliver, J. H. Introduction to the symposium on parthenogenesis. *American Zoologist,* 1971, *11,* 241–243.

Omran, A. The epidemiologic transition: a theory of the epidemiology of population change, Part I. *Milbank Memorial Fund Quarterly,* 1971, *49,* 521–531.

Orians, G. H. On the evolution of mating systems in birds and mammals. *American Naturalist,* 1969, *103,* 589–603.

Parker, G. A., Baker, R. R., and Smith, V.G. F. The origin and evolution of gamete dimorphism and the male–female phenomenon. *Journal of Theoretical Biology,* 1972, *36,* 529–553.

Parsons, T. *Essays in sociological theory* (Rev. ed.) New York: Free Press, 1964.

Pfeiffer, J. E. *The emergence of man.* New York: Harper & Row, 1969.

Ralls, K. Mammals in which females are larger than males. *Quarterly Review of Biology,* 1976, *51,* 245–276.

Redican, W. K. Adult male–infant interactions in nonhuman primates. In M. E. Lamb (Ed.), *The role of the father in child development.* New York: Wiley, 1976.

Rubenstein, D. I. On the evolution of alternative mating strategies. In J. E. R. Staddon (Ed.), *Limits to action: the allocation of individual behavior.* New York: Academic Press, 1980.

Rubin, R. T., Reinisch, J. M., and Haskett, R. F. Postnatal gonadal steroid effects on human behavior. *Science,* 1981, *211,* 1318–1324.

Sanday, P. R. *Female power and male dominance: on the origins of sexual inequality.* London/New York: Cambridge Univ. Press, 1981.

Schultz, R. J. Evolution and ecology of unisexual fishes. In M. K. Hecht, W. C. Steere, and B. Wallace (Eds.), *Evolutionary Biology* (Vol. 10). New York: Plenum, 1977.

Selander, R. K. Sexual dimorphism and differential niche utilization in birds. *Condor,* 1966, *68,* 113–151.

Shields, S. A. Nineteenth-century evolutionary theory and male scientific bias. In G. W. Barlow and J. Silverberg (Eds.), *Sociobiology: Beyond nature/nurture?* Boulder, Colo.. Westview, 1980.

Smith, S. M. Henpecked males: the general pattern in monogamy? *Journal of Field Ornithology,* 1980, *51,* 55–64.

Strassmann, B. I. Wasp reproduction and kin selection: Reproductive competition and dominance hierarchies among *Polistes annularis* foundresses. *The Florida Entomologist,* 1981a, *64,* 74–88.

Strassmann, B. I. Sexual selection, parental care, and concealed ovulation in humans. *Ethology and Sociobiology,* 1981b, *2,* 31–40.

Symons, D. *The evolution of human sexuality.* London/New York Oxford Univ. Press, 1979.

Tanner, N., and Zihlman, A. Women in evolution, Part I.: Innovation and selection in human origins. *Signs,* 1976, *1,* 585–608.

Thomas, C. The greek polis. In R Griffeth and C. Thomas (Eds.), *The city-state in five cultures.* Santa Barbara: ABC Clio, 1981.

Trivers, R. L. Parental investment and sexual selection. In B. Campbell (Ed.), *Sexual selection and the descent of man (1871–1971).* Chicago: Aldine, 1972.

Trivers, R. L., and Hare, H. Haplodiploidy and the evolution of the social insects. *Science,* 1976, *191,* 249–263

Wade, M. J., and Arnold, S. J. The intensity of sexual selection in relation to male sexual behavior, female choice, and sperm precedence. *Animal Behaviour,* 1980, *28,* 446–461.

Wasser, S. K. Reciprocity and the trade-off between associate quality and relatedness, *American Naturalist,* 1982, *119,* 720–731.

West-Eberhard, M. J. The evolution of social behavior by kin selection. *Quarterly Review of Biology,*. 1975, *50,* 1–33.

Whyte, M. K. *The status of women in preindustrial societies.* Princeton, N.J.: Princeton Univ. Press, 1978.

Williams, G. C. *Adaptation and natural selection: a critique of current evolutionary thought.* Princeton, N.J.: Princeton Univ. Press, 1966.

Williams, G. C. *Sex and evolution.* Princeton, N.J.: Princeton Univ. Press, 1975.

Wilson, E. O. *The insect societies.* Cambridge, Mass.: Harvard Univ. Press, 1971.

Wingfield, J. C. Fine temporal adjustment of reproductive functions. In A. Epple and M. H. Stetson (Eds.), *Avian Endocrinology.* New York: Academic Press, 1980.

Wittenberger, J. F. Group size and polygamy in social mammals, *American Naturalist,* 1980, *115,* 197–222.

Wittig, M. A., and Petersen, A. C. *Sex-related differences in cognitive function.* New York: Academic Press, 1979.

Wrangham, R. C. An ecological model of female-bonded primate groups. *Behaviour,* 1980, *75,* 262–300.

PART II

Interactions between the Sexes

3

Constraints on Female Choice in the Mottled Sculpin*

LUTHER BROWN

JERRY F. DOWNHOWER

I. Introduction

Mate selection has been proposed as one of the primary organizers of the social systems of vertebrates (e.g., Orians, 1969; Emlen and Oring, 1977; Altmann *et al.*, 1977). Since various arguments suggest that females are typically the choosier sex (e.g.,. Trivers, 1972; Halliday, 1978; Parker, 1979), female choice of males as mates has frequently been considered a primary determinant of social structure. In one way or another, most analyses of female choice assume that natural selection has resulted in the evolution of choice patterns that allow females to choose the best male as a mate. However, the concept of the "best" male may have multiple

*This work was supported by grants from the Ohio State University Instruction and Research Computer Center, the Ohio State University Graduate School, and the National Science Foundation (DEB 76-02657, J.F. Downhower, principal investigator).

SOCIAL BEHAVIOR OF
FEMALE VERTEBRATES

ISBN 0-12-735950-8

definitions. A female might choose a male who is best in an absolute sense, that is, the male may be the best of all possible available males (e.g., Davies, 1978, p. 332). Alternatively, the best male may be defined in a more relative or political fashion. Thus, a given male may be best with respect to his neighboring competitors (e.g., Brown, 1980) or he may be the best male available at a given time in the breeding season (e.g., Pleszczynska, 1978).

Regardless of whether the concept of the best male is used in an absolute sense, a relative sense, or an undefined sense, use of the concept implies assumptions about the information available to and perceived by females when choosing mates, as well as information regarding the availability and desirability of males as determined by investigators. In most cases, the distributions of males available to females remain unknown or unstudied. More importantly, female sampling abilities remain obscure for almost all species.

The mottled sculpin (*Cottus bairdi*) is a small freshwater fish with a breeding biology that allows examination of some of the consequences of female sampling during mate selection. We examined selected aspects of sculpin mate selection and propose that female sculpins choose males that are almost never absolutely best, and only relatively best in an extremely restricted sense. Mate selection by female sculpins appears to represent a compromise in which females make the best of an apparently poor-choice situation.

II. The Mottled Sculpin

The mottled sculpin is common in high-gradient streams throughout much of North America (Scott and Crossman, 1973). The general biology of this fish has recently been reviewed by Ludwig and Norden (1969) and Downhower and Brown (1979, 1980). At the beginning of the spring spawning season, adult male sculpins occupy cavities or burrows beneath rubble on the streambed. Males defend their individual burrows, but do not defend the surrounding area. When females approach a male's burrow they are courted and may either spawn with the male or leave the burrow without spawning. When females do spawn, they lay all of their eggs in a discrete, hemispherical egg mass. After spawning, the female leaves the nest while the male remains with the eggs and courts additional females. Females thus breed with only one male per year, whereas males may breed with as many as twelve females per year (Downhower and

Brown, 1980), and the courtship success of a male can be determined by simply counting the number of egg masses in his nest.

Several aspects of sculpin reproduction are important to the following discussion. Male sculpins locate nests beneath existing rubble on the streambed and then remain at their nest sites during the breeding season. Gravid female sculpins are thus confronted with an array of potential mates and nesting sites that is both static and determined by streambed topography. Male sculpins are responsible for the care of the eggs and fry immediately after hatching. Finally, certain males mate with large numbers of females, whereas others remain unmated, suggesting that particular males are chosen more frequently as mates (see Brown, 1980; for the related topic of male choice of females, see Downhower and Brown, 1981).

III. Methods

Breeding patterns were examined by seeding a stretch of Anderson Creek in Champaign County, Ohio, with artificial spawning sites. Sites were composed of rectangular slate tiles supported on steel pins driven into the streambed. Six tile sizes were used during the 1976 breeding season (all sites were 14.5 cm long, widths were 4, 7, 10.5, 14.5, 18, and 21.5 cm). Other tile size combinations have been used in other years (see Brown, 1978). Tiles were sampled at 3–4-day intervals during the entire breeding season. Sampling involved individually isolating each tile from the surrounding stream with a capture box (Downhower and Brown, 1977) and capturing all fish resident beneath the tile. Sizes, sexes, and numbers of fish at each tile were recorded. Photographs were taken of each nest on each sample data. These methods have been used for several years (Brown, 1978; Downhower and Brown, 1977, 1980). The present analysis is restricted to data collected in 1976.

The geometry of egg masses and the sizes of individual eggs prohibited analysis of the survival of single eggs. Photographs of the nest, however, did allow determination of the fates of egg masses as a whole. Sequences of nest photos for each nest were analyzed, and entire egg masses were scored as successes if at least one egg actually hatched, or as failures if all of the eggs either died or disappeared prior to hatching. Since each egg mass represented the entire annual reproductive output of a single female, we were able to make a gross estimate of female reproductive success at hatching.

IV. Results

The hatching success of an egg mass was directly proportional to the size of the male occupying the nest (Fig. 1). Large males succeeded in hatching a greater proportion of their egg masses than did smaller males. Females thus affected their reproductive success by choosing a large mate (Downhower and Brown, 1980). This observation alone suggests that females might prefer large mates, and this suggestion is supported by the facts that larger males bred earlier in the season and mated with greater numbers of females than did smaller males (Brown, 1978, 1980; Downhower and Brown, 1979, 1980, 1981). Females could thus increase their breeding success by choosing large males and actually appeared to select large males as mates.

If male size were the only factor influencing hatching success, then the linear relationship between male size and hatching (Fig. 1) suggests that females should choose the largest possible males as mates. But choice of

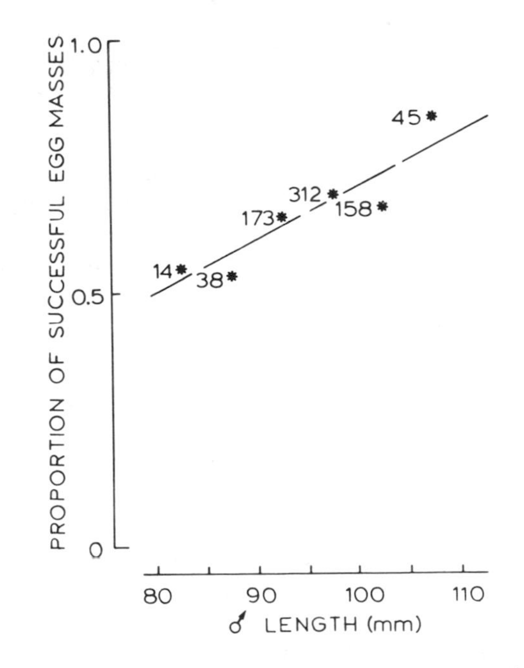

Fig. 1. The proportion of successfully hatching egg masses as a function of guardian male size for the 1976 breeding season. The number of egg masses is shown next to the asterisks. (Source: Downhower and Brown, 1980.)

the largest male involves two essential problems: before the largest male can be chosen, he must be located and identified. These are sampling problems for the females that are made difficult by two aspects of sculpin biology. Gravid females are sluggish swimmers, presumably because of their greatly egg-distended abdomens. Secondly, the spatial distribution and the size distributions of males residing in the study site change dramatically from year to year. Spatial distributions of males change because the distributions of nest sites vary as a consequence of spring and late winter flooding. Size distributions also change. For example, in 1975, approximately 10% of the breeding males in the study population exceeded 100 mm in total length (Fig. 2). One year later almost 30% were larger than 100 mm, and 2 years later less than 10% exceeded 100 mm. Gravid females thus may not know the location or identity of large males.

Thus, female sculpins are confronted by a sampling problem, and sampling requires time. Time represents a cost in terms of hatching success: hatching success was inversely proportional to oviposition date. Almost 80% of the egg masses deposited during the first week of the 1976 spawning season were successes, while almost all ovipositions 2–3 weeks later were failures (Fig. 3).

Two selective factors can thus be identified as affecting female choice in sculpins. Although females can increase their reproductive success by locating and identifying a large mate, they may decrease their reproductive success if they delay reproduction while searching for a large male. These factors allow estimation of the consequences of various types of female choice mechanisms. The benefits of any given choice mechanism are those deriving from the size of the male chosen by that mechanism. The costs of a given choice are those deriving from the amount of time, or the number of males evaluated, before a suitable male is located.

There are a number of ways that a female might go about locating a large mate (see especially Otte, 1974; Janetos, 1980). Only a few possible mechanisms will be considered here. At one extreme, females might look for males who are best in an absolute sense. Such females might evaluate all available males before picking the biggest possible mate. This type of choice is extremely costly, however, since the number of males that would have to be evaluated is directly proportional to the number resident in the breeding habitat (Fig. 4). For example, if females possess total recall of male qualities and sample each male only once, then the number of spawning sites that must be visited before the best male is located and returned to is equal to the number of sites occupied by males plus one. On April 10, 1976 (the date of peak spawning in that year), 244 males occupied the study site; thus, a female making this type of choice would have had to visit an average of 245 nesting sites. If females remember male

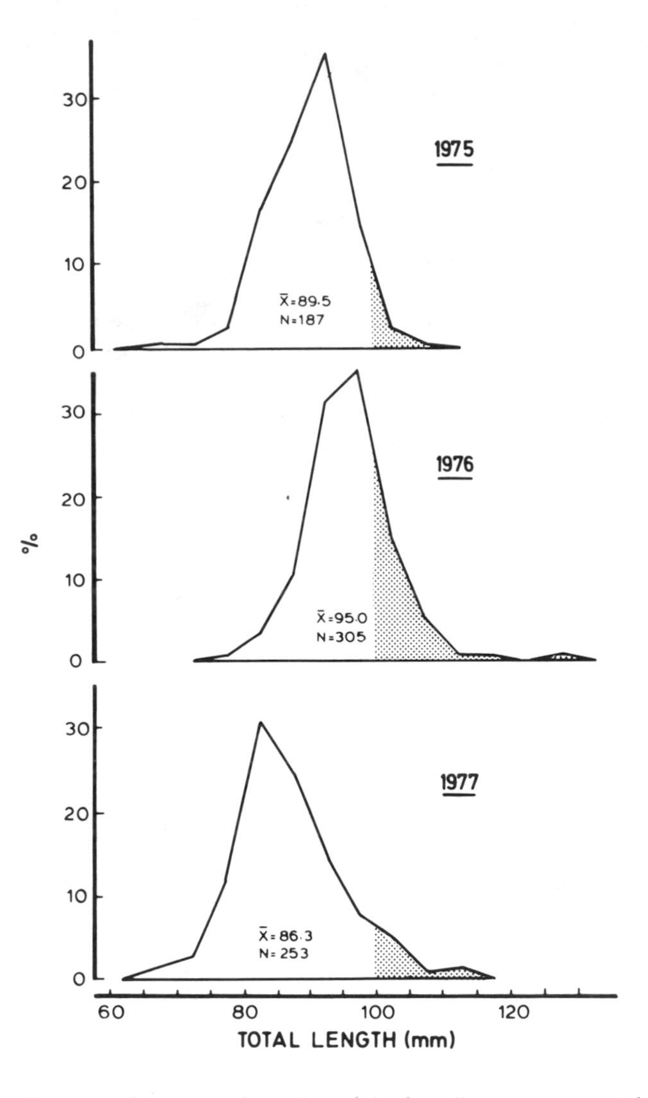

Fig. 2. Frequency histograms for male sculpins breeding at experimental nesting sites in Anderson Creek during the 1975, 1976, and 1977 spawning seasons. Shaded areas indicate those males larger than 100 mm total length.

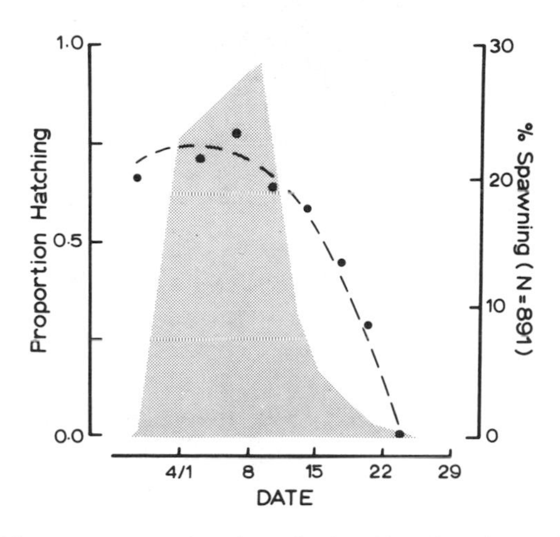

Fig. 3. Hatching success as a function of oviposition date during the 1976 spawning season ($r_s = -.90$). The shaded curve indicates the intensity of spawning on each sample date.

qualities but sample randomly with replacement, then the number of males that must be sampled is even higher. Before she was 95% confident that she had found the best male, a female breeding on April 10 would have had to visit 734 males (Fig. 4). Absolute choice mechanisms thus entail high costs in terms of time and demand that females have complex memory abilities and be capable of complicated sampling movements.

An alternative to locating the best possible male is location of a male who is large relative to his neighbors. The simplest such relative choice involves a sample of two males. For example, a female might spawn with a male if he were larger than the last male she encountered. This type of choice model is analogous to that developed by Mitchell (1975) in his analyses of beetle oviposition tactics. It is a minimum assumption model in that it assumes only that females move minimum distances (between nearest neighbors) when comparing males and that females can remember the size of the last male encountered during their search. This model can be conveniently developed through computer simulations based on the known distributions of nesting sites in the study population and the sizes of males occupying each site on each date (Brown, 1980). In its simplest form, the simulated female picked a tile randomly and compared the size of a male occupying the tile to his nearest neighbor. If the nearest neighbor was larger than or equal to the first male, then a simulated spawn was recorded for the neighbor. If the nearest neighbor was smaller than the first male, the comparison routine continued comparing nearest neighbors

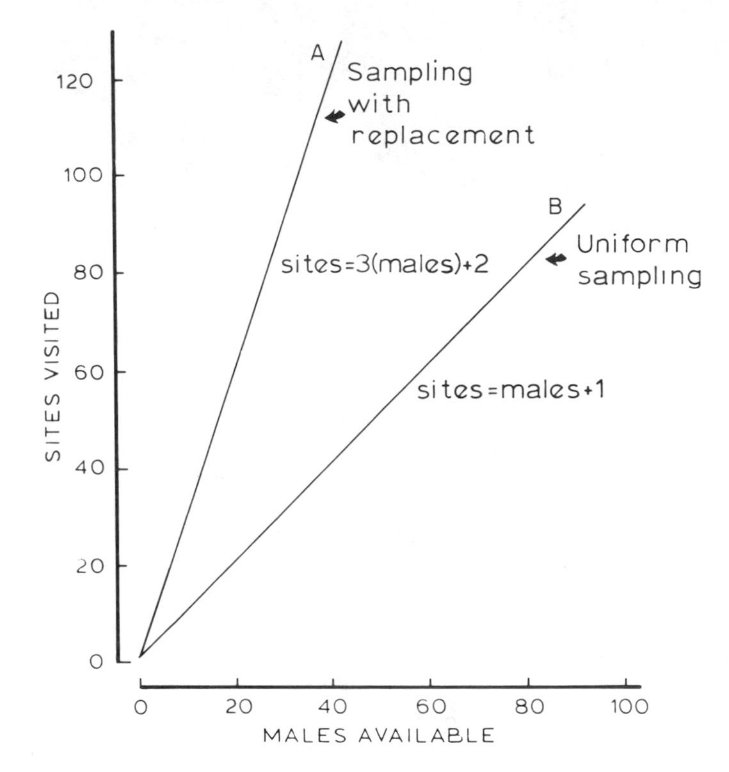

Fig. 4. The relationships between the number of suitors in a population and the number of sites a female must visit before locating and returning to the best of all possible mates. Line A was generated under the assumption that females sample males randomly with replacement and are 95% confident that they have visited all available males before picking their mate. Thus, $(1 - .95) = 1 - 1/\text{males available}^{sv}$, where sv stands for sites visited. Line B was generated under the assumption that females sample each male only once. Both lines assume that females possess total recall of male qualities and that all visited breeding sites are occupied by mature males.

until a larger or equal-sized male was located. An abbreviated flow chart of the simulation routine is given in Fig. 5.

The foregoing relative choice model can be expanded in several different ways, two of which are shown in Fig. 6. One set of expansions involved varying the number of males sequentially compared by a female before choosing a mate. For example, a female might evaluate only one male before spawning. She would thus spawn randomly. Alternatively, she might compare two males and mate with the second if he were larger than the first. Continuing the expansion, she might evaluate three males and mate with the third if he were larger than the second who was larger than

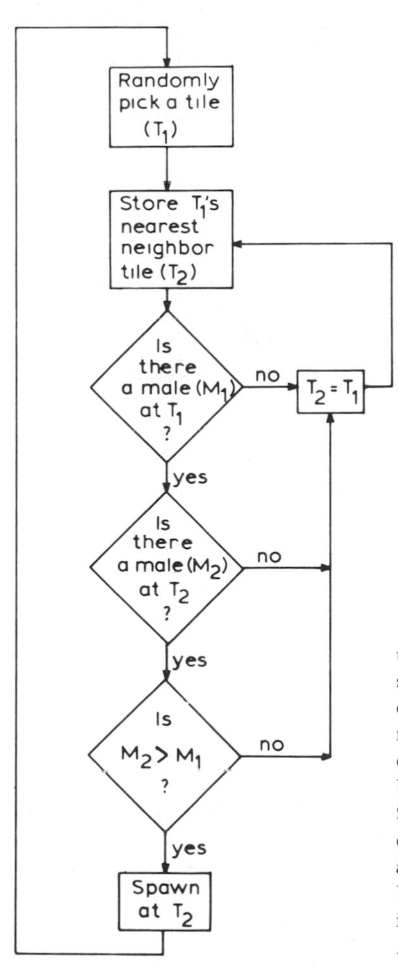

Fig. 5. Flow chart of the computer program used to simulate spawnings. The program used to simulate spawnings was written in Fortran and was composed on 407 statement cards. Random numbers were generated using the UNI subroutine of the pseudorandom number package "Super-Duper" (McGill University School of Computer Science). The program was run on an IBM 370/168 computer, with funds provided by the Instruction and Research Computer Center, The Ohio State University. The distributions of tiles and males used in all simulations were those observed in the Anderson Creek study site on April 10, 1976.

the first. Because each of these expansions is based on sequential comparisons, all are minimal assumption models in that they require female memory only of the last male encountered before each comparison. A second set of expansions involved simple comparisons between two neighboring males but varied the amount by which an acceptable mate exceeded the size of his neighbor. For example, a female might mate with a male who was larger by 5 mm than the last male she encountered, or some other fixed amount.

The consequences of these various degrees of female selectivity and two types of choice model can be evaluated through computer simulations.

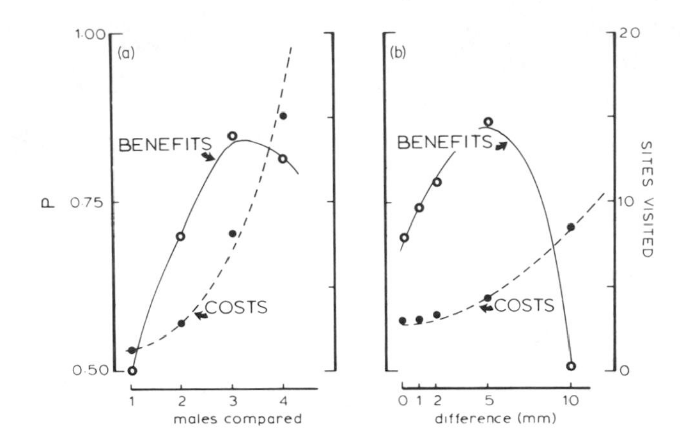

Fig. 6. The benefits (solid lines) and costs (dashed lines) of female selectivity for two types of relative choice mechanisms. Benefits are given in terms of the probability that the male selected as a mate will be larger than or equal to the average-size male in the population. Costs are given in terms of the number of breeding sites that must be visited by a female before choosing a mate. Costs and benefits for each degree of female selectivity were determined empirically by computer simulation using the known distributions of spawning sites and associated males in the Anderson Creek study site on April 10, 1976, and based on the assumption that females move among nearest neighboring nest sites when searching for a mate (see text). In the first set of models, (a), the degree of female selectivity was determined by the number of sequentially larger males compared by a female. Thus, a female might have mated with the first male encountered (males compared = 1). By extension, she might have mated with a male only if he were larger than or equal to the last male that she had encountered, that is, his nearest neighbor (males compared = 2), or if he were larger than or equal to the last male who was in turn larger than or equal to his nearest neighbor (males compared = 3). In the second set of models, (b), females mated with a male only if he were larger than this predecessor by some fixed amount (0, 1, 2, 5, or 10 mm).

These simulations are based on the known distributions of males and nesting sites in the study population on April 10, 1976, and are analogous to the simulation routine presented in Fig. 5. The costs of each choice process are proportional to the number of breeding sites visited by a simulated female before a spawn was recorded. The benefits of each choice process are proportional to the probability that the mate chosen by that process is larger than or equal to the mean size of available males (Fig. 6).

Several aspects of these analyses deserve emphasis. If a female bred with the first male she encountered, she would, on the average, pick a male of average size. This reflects the fact that the distribution of available males was approximately normal. A female making a very simple relative comparison (i.e., mating with a male who was larger than the last male encountered) would greatly increase the chances that her mate was larger than average and would do so with a very small increase in the associate cost, sampling only one or two additional sites before mating. Moderate degrees of selectivity (e.g., requiring that a suitable mate be larger than his predecessor by 5 mm) virtually assure that a chosen male will exceed the population mean. Extreme degrees of selectivity in either model are counterproductive: variances in chosen male sizes increase so dramatically that the probability of a chosen male being larger than average actually declines.

The applicability of these simulated choice processes can be evaluated by comparing the frequency distributions of matings per male generated by the models with matings actually observed in the field. Such comparisons reveal that a model in which a female mates with a male larger than or equal to the last male encountered produces patterns of breeding like those actually observed. Comparisons for each of the six tile sizes in the study stream on April 10, 1976 are presented in Fig. 7. Further comparisons of observed breeding patterns with those generated by this model appear in Brown's (1980) paper.

Two conclusions can be drawn from these analyses. Choice of a male who is absolutely best (biggest) may take a long time and involve complicated sampling. The consequent delay in oviposition date, and subsequent decrease in hatching success may offset any advantage gained by mating with the best male. While the time required for a female to evaluate a male as a mate remains unknown, several lines of evidence suggest that females spent approximately one week searching for males in 1976 (Brown, 1978). Because females move among nests primarily at night and courtship activities may continue for several hours before oviposition or rejection of a male (Smyly, 1957; Savage, 1963; Brown, 1978), females appear to sample only a few males prior to spawning. Choice of a relatively large male is possible even when sampling is very restricted if females evaluate males with respect to their neighboring competitors.

Secondly, comparisons of the spawning patterns generated by several simple models of female choice with those patterns observed in the field suggest that female sculpins make an extremely limited type of relative choice. The observed patterns are consistent with a model in which females compare adjacent males and mate with a male who is larger than the last male encountered.

Selection of a mate poses a problem to female sculpins. While choice of a large male increases reproductive success, females that delay reproduction while searching for a large male may actually decrease their reproductive success by prolonging the search. This is a problem that may be faced by many female vertebrates. The timing of reproduction is known to be an important determinant of reproductive success in many birds,

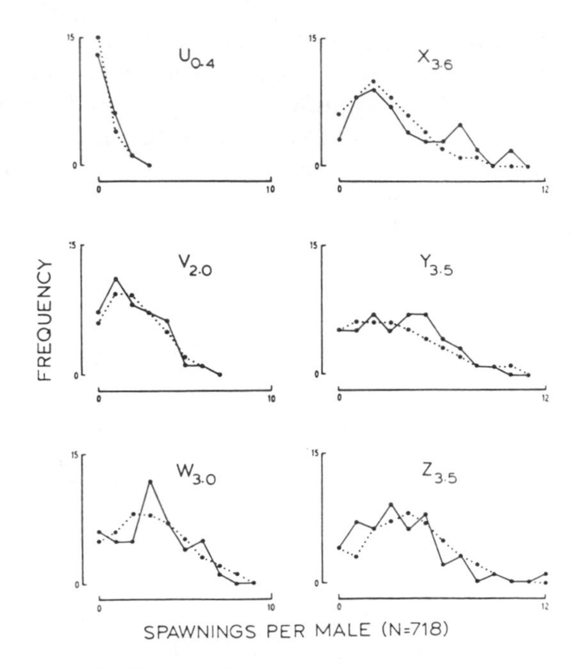

Fig. 7. Frequency distributions of spawnings per male observed in the study population (solid lines) and generated by the female choice model (dotted lines) in which a female mated with a male who was larger than or equal to the last male that she had encountered. The distributions were observed April 10, 1976. Breeding site sizes are arranged alphabetically from smallest to largest. The mean number of observed spawnings per male for each tile size is given as a subscript. In no case are observed and modeled distributions different (χ^2 test, $p = .05$). On the same date, simulated random spawning differed from the observed distributions for three of the tile sizes. (Source: Brown, 1980.)

mammals and fishes (e.g., Sadlier, 1965; Lack, 1968; Lowe-McConnell, 1975). The breeding biology of the individual organism will determine the nature of the relationship between timing and breeding (Fig. 8). Timing may be very critical to particular organisms (e.g., Blaxter, 1974) or it may be relatively unimportant. At the same time, the movement abilities of females and the dispersion patterns of males will determine the rate at which females can move between males. If males are widely dispersed or females have limited mobility, female sampling may be time-consuming. If males are clumped or females are highly mobile, females may be able to visit large numbers of males rapidly (Fig. 9). Reproductive timing and sampling ability can be thought of as interacting to determine the optimal number of males that a female evaluates before choosing a male and the time spent evaluating each male (Fig. 10). If timing is not critical to reproduction, or females are highly mobile, or males are clumped, then a

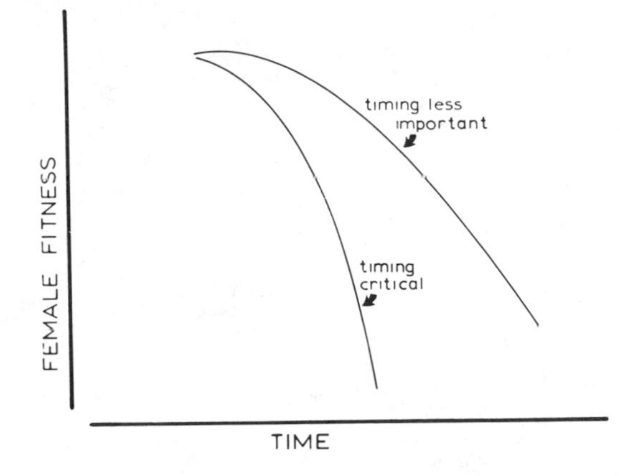

Fig. 8. Possible relationships between reproductive timing and success. Note that relationships considered here deal with the end of breeding seasons. Extremely early breeding may also be detrimental for some organisms.

female may be able to visit a large number of males before breeding, evaluate complex male characteristics, and possibly choose the male who is locally best. This is evidenced in lek-breeding organisms in which males come together and are evaluated more or less simultaneously by females. In lek situations, individual males account for the vast majority of matings, suggesting that these males are judged locally best by many females (e.g., Wiley, 1973; Shepard, 1975; Bradbury, 1977; Davies, 1978). Similarly, flight may allow many birds to rapidly sample available males and choose breeding situations that are locally best (e.g., Orians, 1969; Altmann *et al.*, 1977; but see Payne, this volume).

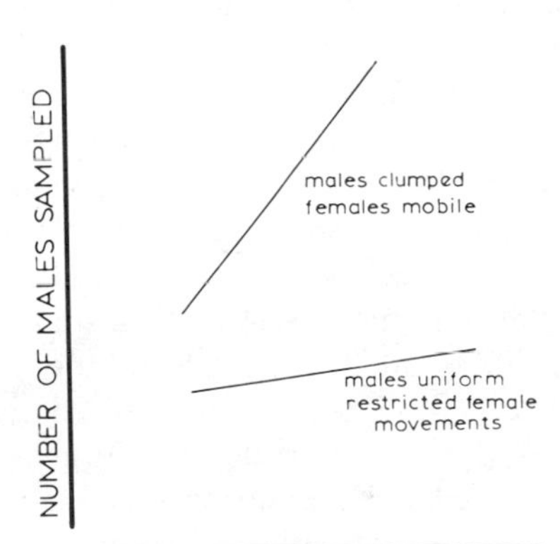

Fig. 9. Possible relationships between female movement abilities, male dispersion patterns, and the rate at which females can evaluate potential mates.

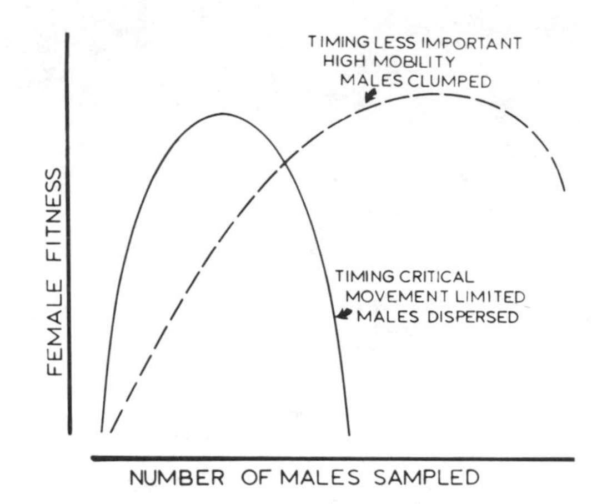

Fig. 10. Summary of the interactions between reproductive timing, female movement abilities, male dispersion patterns, and the number of males that a female may evaluate before picking a mate.

If reproductive timing is more critical to success or if females are less mobile or males more widely spaced, then females may make more restricted choices. For example, female yellow-bellied marmots (*Marmota flaviventris*) breed in grassy male territories separated by inhospitable areas. Males are widely dispersed. The fact that female fitness declines with increasing harem size in this species (Downhower and Armitage, 1971) may indicate that the costs of locating and identifying a better male outweigh the benefits of remaining with an existing mate.

Differences in the importance of reproductive timing, movement abilities, and dispersion patterns characteristic of populations thus affect female choice mechanisms and patterns. While limits to selectivity have been recognized for many years (e.g., Fisher, 1930), general models for the mating systems of broad taxonomic groups (e.g., Orians, 1969; Altmann *et al.*, 1977; Emlen and Oring, 1977) may have minimized the importance of interspecific sampling differences while attempting to provide general evolutionary schemes. Thus, movement and male dispersion patterns may interact with reproductive timing to allow females of some species the relative omniscience necessary to chose the absolutely best mating situations (e.g., Orians, 1969; Altmann *et al.*, 1977). At the same time, these factors may prevent females of other species from knowing or evaluating all available options. Analysis of the factors affecting reproductive success in any species is critical to an understanding of the patterns of female choice apparent in the breeding population. For example, differences in movement ability may be largely responsible for the failure of avian models to describe adequately mating patterns among mammals (Ralls, 1977).

In summary, two selective forces seem to be important for female choice patterns in mottled sculpins. Although females can increase their reproductive success by locating and identifying a large mate, they may decrease their success by delaying oviposition while extending the search. Locating and identifying a male who is best in an absolute sense requires extreme sampling effort and takes a great deal of time. Females can, however, locate a large male in a very short amount of time by choosing a relatively high-quality mate. The breeding patterns of female sculpins suggest that females make a restricted type of choice and thus mate with a male who is best in only a limited sense. Similar factors may affect selection in many other vertebrates. When males are locally abundant and females can easily and rapidly sample large numbers of males, female choice may allow selection of an absolutely best male as a mate. When males are dispersed and when movement among males is difficult or time-consuming, female choice may be more limited.

Acknowledgments

Many people assisted us during the field portions of this research. We are especially grateful to K. Heiber, S. Miller, C. Schlichting, J. Moreo, J. Gaudion, M. Hoeltzer, R. Yost, G. Staples, R. Pedersen, D. Rouvet, J. Blackburn, V. Scuilli, L. Mylroie, and M. Brown.

References

Altmann, S. A., Wagner, A. A., and Lenington, S. Two models for the evolution of polygyny. *Behavioral Ecology and Sociobiology*, 1977, *2*, 397–410.

Blaxter, J. H. S. (Ed.) *The early life history of fishes.* New York: Springer Verlag, 1974.

Bradbury, J. W. Lek mating behavior in the hammer-headed bat. *Zeitschrift für Tierpsychologie*, 1977, *45*, 225–255.

Brown, L. *Polygamy, female choice, and the mottled sculpin, Cottus bairdi.* Unpublished doctoral dissertation, The Ohio State University, 1978.

Brown, L. Patterns of female choice in mottled sculpins (Cottidae, Teleostei). *Animal Behaviour*, 1980, *29*, 375–382.

Davies, N. B. Ecological questions about territorial behavior. In J. R. Krebs and N. B. Davies (Eds.), *Behavioral Ecology*. London: Blackwell, 1978.

Downhower, J. F., and Armitage, K. B. The yellow-bellied marmot and the evolution of polygamy. *American Naturalist*, 1971, *105*, 355–370.

Downhower, J. F., and Brown, L. A sampling technique for benthic fish population. *Copeia*, 1977, *1977*, 403–406.

Downhower, J. F., and Brown, L. Seasonal changes in the social organization of a population of mottled sculpins, *Cottus bairdi*. *Animal Behaviour*, 1979, *27*, 452–458.

Downhower, J. F., and Brown, L. Mate preference of female mottled sculpins, *Cottus bairdi. Animal Behaviour,* 1980, *28,* 728–734.

Downhower, J. F., and Brown, L. The timing of reproduction and its behavioral consequences for mottled sculpins, *Cottus bairdi.* In R. D. Alexander and D. W. Tinkle (Eds.), *Natural Selection and Social Behavior.* New York: Chiron Press, 1981.

Emlen, S. T., and Oring, L. W. Ecology, sexual selection and the evolution of mating systems. *Science,* 1977, *197,* 215–223.

Fisher, R. A. *The genetical theory of natural selection.* Oxford: Clarendon Press, 1930.

Halliday, T. R. Sexual selection and mate choice. In J. R. Krebs and N. B. Davies (Eds.), *Behavioural Ecology.* London: Blackwell, 1978.

Janetos, A. C. Strategies of female mate choice: A theoretical analysis. *Behavioral Ecology and Sociobiology,* 1980, *7,* 107–112.

Lack, D. Ecological Adaptations for Breeding in Birds. London: Chapman and Hall, 1968.

Lowe-McConnell, R. H. *Fish communities in tropical freshwaters.* New York: Longman, 1975.

Ludwig, G. M., and C. R. Norden. Age, growth and reproduction of the northern mottled sculpin (*Cottus bairdi bairdi*) in Mt. Vernon Creek, *Milwaukee Publ. Mus., Occasional Paper No. 2,* 1–67.

Mitchell, R. The evolution of oviposition tactics in the bean weevil *Callosobruchus maculatus* (F.). *Ecology,* 1975, *56,* 696–702.

Orians, G. H. On the evolution of mating systems in birds and mammals. *American Naturalist,* 1969, *103,* 598–603.

Otte, D. Effects and functions in the evolution of signalling systems. *Annual Review of Ecology and Systematics,* 1974, *5,* 385–417.

Parker, G. A. Sexual selection and sexual conflict. In M. S. Blum and N. A. Blum (Eds.), *sexual selection and reproductive competition in insects.* New York: Academic Press, 1979.

Pleszczynska, W. K. Microgeographic prediction of polygyny in the lark bunting. *Science,* 1978, *201,* 935–937.

Ralls, K. Sexual dimorphism in mammals: Avian models and unanswered questions. *American Naturalist,* 1977, *111,* 917–938.

Sadlier, R. M. F. S. The relationship between agnostic behavior and population changes in the deermouse, *Peromycsus maniculatus* (Wagner). *Journal of Animal Ecology,* 1965, *34*(2), 331–352.

Savage, T. Reproductive behavior in the mottled sculpin, *Cattus bairdi* (Girard). *Copeia,* 1963, *1963,* 317–325.

Scott, W. B., and Crossman, E. J. 1973. *Freshwater fishes of canada.* Fish Research Board Canada, Bulletin 184, 826–830.

Shepard, J. M. Factors influencing female choice in the lek mating system of the ruff. *Living Bird,* 1975, *14,* 87–111.

Smyly, W. J. P. The life-history of the bullhead or miller's thumb (*Cottus gobio* L.). *Proceedings of the Zoological Society of London,* 1957, *128,* 431–453.

Trivers, R. L. Parental investment and sexual selection. In B. Campbell (Ed.), *Sexual selection and the descent of man.* Chicago: Aldine, 1972.

Wiley, R. H. Territoriality and non-random mating in the sage grouse, *Centrocercus urophasianus. Animal Behavious Monograph,* 1973, *6*(2), 87–169.

4

Bird Songs, Sexual Selection, and Female Mating Strategies*

ROBERT B. PAYNE

I. Introduction

Bird songs, like other animal signals, are behaviors that affect the singer's social relations with other individuals of the species. Song as a social signal presumably has evolved by selection insofar as it has increased the fitness of the singers relative to others within their own

*My research has been supported by National Science Foundation grants GB29017X, BMS75-03913, BNS78-03178, and BNS81-02404.

SOCIAL BEHAVIOR OF
FEMALE VERTEBRATES

ISBN 0-12-735950-8

species. The responsiveness of other individuals likewise has been selected to increase their own fitness. Song has evolved by a process of competitive genetic selection rather than by a non-Darwinian process, such as optimizing information exchange and communication among members of a "cooperating" species (Dawkins and Krebs, 1978). Although bird song evolution can be explained in general by selection, the selective pressures on the competing behavior strategies of the sexes and the compromises to these strategies remain largely unknown. Most catalog lists of the "functions" of bird songs point out territorial defense against other males and attraction of a female (Hinde, 1956; Brown, 1975). Thus, song may have been shaped in evolution primarily either by the female, in choosing among different males, or by males, in intrasexual competition. Both mate choice by the females and intrasexual competition among the males may have shaped song together, and the epigamic and competitive aspects of sexual selection may explain the major features of bird song evolution completely, without reference to other species.

Bird songs are commonly thought to be important to females in mate selection at the species level. Most birds are recognizable to species members by their song, and this has naturally led us to think that species differences in song are the result of selection for birds to recognize their own species and to distinguish them from others. It has sometimes been assumed that the main evolved function of bird song was to assure mating barriers between species and that selective pressure towards the elaboration of song involved mainly the lack of genetic success of individuals that mated outside their own species limit. In fact, reviews of bird song and mate selection have equated mate selection with species recognition (Marler, 1960; Mayr, 1963; Lanyon, 1969). This view can be questioned. In our own human species, which we know better than any other, it is unlikely that sexual and ecological behavior is shaped by selection in the evolutionary sense mainly to avoid interbreeding with other primate species. Why should it be so with the birds? Probably it is not, and certainly in a developmental sense, song is normally shaped within the species. In the species that have been studied, all songbirds learn the details of their songs from other birds of their species, either within their kinship group, after dispersal from the family, or both before and after dispersal (Marler, 1975; Thielcke, 1977; Kroodsma, 1979a; Payne, 1981a, 1981b). But birds can learn the songs of other species, and many cases are known of birds singing the song of another species in the wild (e.g., Lemaire, 1977). Even interspecies hybrids generally have the song of one parental species, not the mixed or intermediate kind of song to be expected through recombination of species-specific genetic determinants (Payne, 1980a). If bird songs had been selected mainly at the level of species distinctiveness, then one would expect a developmental pathway that was more dependent on species-specific genetic determinants than what we observe in nature.

The structure of bird song is also subject to certain nonsocial selective factors, such as acoustic transmission properties in different habitats (Chappuis, 1971; Jilka and Leisler, 1974; Morton, 1975; Marten and Marler, 1977; Marten *et al.*, 1977; Wiley and Richards, 1978; Bowman, 1979; Hunter and Krebs, 1979). These engineering problems may apply to simple features of the physical environment in the propagation of different sound frequencies, but they do not explain the great diversity of the temporal and frequency characteristics that we use ourselves to distinguish bird songs, nor do they explain the fine structure of bird songs.

Alternative explanations of the evolution of song in birds have been focused recently upon song dialects. Naturalists have long known that not all birds in the same species sound alike. Different songs in neighboring local populations of certain species have been recognized for more than 200 years (Barrington, 1773; Darwin, 1871; Armstrong, 1963). In the best-documented case, local songs of white-crowned sparrows (*Zonotrichia leucophrys nuttalli*) have been recorded in coastal central California for over 20 years, and some local song populations have retained their local song tradition for as long as 40 years (Blanchard, 1941; Baptista, 1975). The local song differences are learned by the young bird, and birds from one population can match the dialect songs from another population if they hear the song when they are young (Marler and Tamura, 1962; Marler, 1970). But the evolutionary significance of these local song differences is a matter of dispute. First, they may simply result from historical founding incidents when the original population had a song slightly different from that of the source population, and copying errors may have accumulated over the generations in the cultural transmission of song. If historical accidents explain the origin of these song dialects, then they may have no particular evolutionary significance and need not be explained in terms of evolutionary selection. Second, they may reflect acoustic adaptations to local habitats. However, in the bird species studied, the different local song populations do not consistently correspond with local habitat patches (Baker, 1975; King, 1972; Payne, 1978b; 1981a), and when they do, the song differences are not obviously adapted to sound transmission in each different habitat (Bowman, 1978; Nottebohm, 1975). Finally, the tendency of neighbors to match songs may result from sexual selection, and the social interactions among birds may determine the relative success of the different songs (Payne, 1981b; 1982).

In his discussion of the evolution of traits that increase an animal's success in mating, Darwin (1871) recognized two distinct processes, both of which he termed "sexual selection." Sexual characters may evolve by sexual selection either through females choosing certain males as mates over other males by their elaborate characters, such as songs or plumage, or through competition among males for mates. Song as a sexual character seems likely to have evolved by sexual selection. The question asked here is whether competition among males accounts for the form of song, whether mate choice by females has shaped songs, or whether both forms of sexual selection are involved.

If a female can assess a male's genetic quality by hearing his song, then she might choose among males by comparing their songs and mating with the male that matches her own genetic background. In Darwin's terms, a female might choose a male and mate with him by choosing his song as the most "beautiful" or otherwise attractive song, independently of how other males respond to the song. Assuming again that song is correlated with male genetic quality, males with this song trait would then leave their genes, including any for song behavior, in disproportionate numbers in the next generation. In birds with song dialects, it has been argued that a female would be at an advantage if she mated with a male with a song like the one her own father sang, rather than a different style of song of the same species, for she would then be choosing a male with similar genes that may be coadapted for the same local habitat where she herself had been raised successfully (Nottebohm, 1972; Payne, 1981a).

In its other form, sexual selection results from the competitive edge of certain males over others in gaining and holding a social position, a territory, or other resources. If song differences among local males are associated with any genetic differences, then the gene frequencies may change with time as the competitively successful males gain more resources or access to mates. The intrasexual competitive model fits Darwin's example of stags fighting among themselves—a female would mate with a certain male not because he was inherently more "beautiful" but because he had successfully won out in competition with other males. She might even choose him on the basis of his antlers (or, in the case of birds, on the basis of song), but only because such traits would provide information about his competitive success among his peer males. In the songbirds, a male may gain a competitive advantage if he matches the songs of his older neighbors, because other males differentiate among individuals by their songs (Falls, 1969; Emlen, 1971; Goldman, 1973; Richards, 1979). Imitation may be a strategy of mimicry that deceives other males into mistaking the imitator for the older territorial neighbor. A male may gain an advantage in maintaining a territory if he sounds like another local male by adapting his song to the realities of the social structure.

I have previously contrasted these two hypotheses in the form of different models of selection for song dialects (Payne, 1981a). In this chapter, I recognize that the processes apply more broadly to distinguishing between the two forms of Darwin's sexual selection, and I test some predictions of each form to find whether bird song can be explained as a result of either female choice of mates, male competition, or both. Females are of course involved with the males at some point in any sexually reproducing species; the question is whether they take an active role in shaping male social signals or take a passive role in following the outcome of male competition. Female mating strategies may involve (*a*) direct comparison of the males themselves, or (*b*) indirect comparison of the males by responding to the position of the males in the social organization resulting from male–male aggressive behavior or to their resources after they have successfully established a territory and have maintained it against their competitors.

We face two main tasks in testing whether either form of sexual selection can explain the evolution of bird song. First, we must determine whether there is any nonrandom difference in the mating success of different males on which seleciton may act. Second, we must attempt to distinguish between the two forms of sexual selection, the one involving an active choice of a mate by a female, independent of any influence of other males, and the other involving the outcome of male–male competition. The following sets of predictions may prove useful in both of these problems.

II. Sexual Selection

Sexual selection is generally presumed to explain the elaborate courtship structures and behavior of the sex, usually the male, that has the smaller risk in its investment in mating. The behavioral difference in the sexes often seems to reflect a different gamesmanship of the sexes, with males taking risks and females being coy and playing hard-to-get (Williams, 1975; Maynard-Smith, 1978). Necessary conditions for evolutionary selection are genetic variation and differential breeding success. We generally lack information about the first, but information and logic are becoming available on the second condition. Wade and Arnold (1980) have adapted an earlier theoretical population-genetics model of Crow (1958) that provides an index of the intensity of selection based on the variation in breeding success among individuals in a population. The degree to which the observed variance exceeds the distribution of mating success if all individuals were creating equally is an index of potential sexual selection (Wade and Arnold, 1980). Note that this index gives only an upper limit to any possible evolutionary selection—if no genetic differences were associated with the differences in mating success, then no evolutionary sexual selection would occur.

We may test the following predictions of the idea that sexual selection has shaped the song behavior of birds by asking whether there is any evidence of uneven mating success among males and whether song differences reflect genetic differences. Sexual selection theory leads to the following predictions:

1. Mating success of individual males is nonuniform.
2. Mating success of individual males varies nonrandomly, with a frequency distribution that departs from a random distribution, such as a Poisson or binomial distribution.
3. Among species or populations, the variance in individual mating success is directly related to the mating system.
4. Song is restricted to males, the sex with the higher variance in mating success.
5. Song differences among individuals are associated with genetic differences.

A. Male Competition in Sexual Selection of Bird Song

Once we find that sexual selection has shaped bird songs, we can ask how this has occurred. If individually competitive interactions among males are responsible for song, then we should expect to find evidence in support of the following predictions:

1. *Context* Song is used mainly in competitive or aggressive situations, in advertising exclusive territories, in chasing other males, or in fights during the establishment of a socially ordered relationship within a population.
2. *Development* Song is acquired by matching part or all of the songs of older males or neighbors, or by improvisation, in such a way that it benefits the singer through intimidation, or through deceiving other males of his identity.
3. *Repertoire size* Songs should be most diverse in birds that are most successful in their aggressive interactions with other males.
4. *Stereotypy and predictability* Songs used in aggressive behavior are more stereotyped within an individual than are those used in nonaggres-

sive behavior, if recognition of individuals or of their rank and status is advantageous to the singer.

5. *Conformity* Birds that sing songs that do not conform to usual local standards are unsuccessful in competition with other males.

6. *Dialects* In species with song differences among local populations, males respond differently to local songs and to foreign songs, and males with local songs have a better chance of establishing or defending a territory against other local males.

B. Mate Choice by Females in Sexual Selection of Bird Song

What can we expect to observe if mate choice has been important in a direct way in shaping the evolution of song? Predictions can be made, but few are mutually exclusive of the male competition model. Nevertheless, a few predictions for comparison of the mate choice model with the male competition model may be useful. If song is shaped in measure by females choosing a mate among local males by differences in their songs, then we might expect to observe the following:

1. *Context* Song is used mainly in sexual situations, with males singing to females and courting them with song, arousing the females' behavior, endocrinology, or ovarian development.

2. *Development* Song develops with a pathway from the genome in such a way that birds with different genomes develop different songs as a result, even when the birds have equivalent experiences in hearing songs and in associating songs with certain social behaviors. Only if this is true will females gain any information about the genetic quality of a male in relation to other males by hearing his song. This prediction seems to have been overlooked by naive panselectionists who assume female omniscience, and that females can gain information of genetic makeup of males and can thereby compare "genetic quality" among males from their songs.

3. *Repertoire size* Repertoire size varies with the intensity of sexual selection, and is greater in species with nonmonogamous mating systems than in species with monogamous mating, as a rule. Also, repertoire size within a species varies with the mating success of the male. These predictions assume that females respond directly to the number of song types they hear from the different males, and choose the male with the most song types.

4. *Stereotypy and predictability* Songs used in sexual behavior are more constant than songs used in other contexts, because females may identify certain males by their song and may mate repeatedly with them.

5. *Conformity* Males with odd songs are avoided by females.
6. *Dialects* In species with local song dialects, females respond differently to the different dialect songs and choose a mate with some constancy within their own song dialect.

Certain predictions can be tested most effectively in species with local song dialects; for example, the social position and the mating success of males within a song-matching clique can be compared with their songs. Color-marked populations have been observed for a few species. We can compare the song behavior with both male social interactions and female choice of mates, both within and among populations in these dialectal birds. These birds also give us a test of the predictions on species with different mating systems. I have tested most of these predictions in some detail in two species and shall compare the field observations of each with the behavior predicted from the models of sexual selection by mate choice and male competition.

III. A Comparison of Two Species

Indigobirds are small finches found locally in nonforested parts of Africa. Each indigobird species is a species-specific brood parasite. It lays its eggs in the nest of a different species, the host species, which rears the young parasite. The widespread village indigobird *Vidua chalybeata* lays its eggs in the nest of the red-billed firefinch *Lagonosticta senegala*. The host species then rears the young indigobirds along with its own (Payne, 1977a; 1977b). Other species of indigobirds parasitize other species of firefinches, and most other kinds of parasitic finches (there are about 14 in all) also are species-specific (Payne, in press). The host-specific parasite mimics the songs of the host species. The village indigobird and some other *Vidua* species also have local song dialects in which both their mimetic and nonmimetic songs (the songs that do not match those of the host species) differ among local populations. Each male has a repertoire of about 20 song types. All song types are shared in fine detail among neighbors but differ among populations. Some songs are given in certain circumstances, and all populations studied have distinct songs given when a female approaches the male, when the male courts a female, when he chases another male, and when he initiates a bout of singing (Payne, 1979b; Payne, unpublished data, 1972–1979). In addition, the context-associated songs and the other song types are all given by males in advertising song during which the male may sing continuously for as long as 60 min.

The social organization is a dispersed lek, with neighboring males usually spaced several hundred m apart on their song trees or "call-sites" (Payne and Payne, 1977). Each female in the local song population repeatedly visits the call-sites and males through the breeding season. Each male sings and courts the female at his call-site. After visiting several males, the female returns to one and mates with him. Females then lay in the nest of a host firefinch while the males remain on the call-sites to attract more females. Female parasitize several host nests in a season, leaving 20 or more eggs strategically laid in a series of host nests (Payne, 1977a; Payne and Payne, 1977). One male usually gains more than half of all the matings in a local song population; a few other males get the other matings, and most males do not mate at all within a breeding season. The males that share the same songs compete for the same females, and females visit and appear to sample comparatively most of the singing males in the song neighborhood. We have made field observations on the village indigobirds for more than 2000 hr between 1972 and 1979 at Lochinvar National Park, Zambia.

Indigo buntings *(Passerina cyanea)* are small finches common in eastern North America. Like most small songbirds of the northern temperate areas, they usually form pairs, with one female associating with one male for a breeding season. A few males have more than one mate (Carey and Nolan, 1975; 1979). Females build the nest, incubate the eggs, and usually provide all the insect food for their young. A few males help feed the young, especially after the young leave the nest. Many first-year males do not attract a mate locally, though they may disperse repeatedly during their first breeding season and may be successful in attracting a female for the first time as late as July, eight weeks after they have returned to their breeding area (Payne, 1981b, 1982). Many males have a distinctive song unlike that of other males (Thompson, 1970; Emlen, 1971), but in the hundreds of color-marked individuals that I have observed in Michigan, most match the song of a neighbor (Payne, 1981b, 1982). First-year males learn their song in their first breeding season after they settle on a territory next to another bunting and match his song. Birds with two or more neighbors with different songs usually match the song of only one, most often the bird with whom they most intensively interact with aggressive chasing across their common territory boundary. In experimental conditions, a first-year male consistently copies the song of the bird with which it interacts rather than the song of a bird that it can merely hear (Payne, 1981b). This song matching, together with the tendency of birds to return to the territory where they learned their song, leads to neighbors sharing the same song. Song dialects are very local and may involve only two birds; the largest song neighborhood in my populations had 18 birds.

About 600 color-marked birds were observed from 1977 through 1982 for a total of about 6000 hr. Observations included tape-recording songs, finding nearly all nests, and sampling behavior of the individual buntings.

Table I summarizes the major characteristics of the social organization and song of these two species of indigos. The two species differ in social organization, parental care, and mating system, and also in their systematic affinities—the old-world Ploceidae (Viduinae) versus the new-world Emberizidae (Cardinalinae). However, they are similar in demography and sexual dimorphism, they overlap in the range of the neighborhood size of their song populations, and so they are comparable for a test of the relationship of mating strategies and song behavior.

IV. Evidence for Sexual Selection

The number of mates or matings were determined in two populations each for village indigobirds and indigo buntings. The success of males is summarized in Tables II–IV. In each species, the success of males is unequal. In both populations observed in the indigobirds, one male gained

TABLE I
Comparison of Social Organization, Mating System, and Song in Two Species of Indigos with Song Dialects

Characteristic	Village indigobird	Indigo bunting
Social organization	Dispersed lek	Territorial
Mating system	Promiscuous	Mainly monogamous
Proportion of adult males mating/year	<50%	>90%
Parental care	None (brood parasite)	Mainly by female
Habitat	Open bush, grass	Old fields, shrubs, swamps
Annual survival (adult males)	55%	59%
First-year male plumage	Same as adult male	Variable, female-like to adult-male-like
Weight (males, females)	13 gm, 12 gm	14 gm, 12.5 gm
Song mimicry	Inter- and intraspecific	Intraspecific only
Number of song neighbors (males)	1–20 (mean about 14)	1–18 (mean about 4)
Song repertoire	20	1
Singing time (min/hr)	18.4	27.5
Singing rate (songs/hr)	160	113

more than half of all matings, and most males did not mate. In the buntings, most males had one female, but a few males were bigamous, and a few had three or four females. The variance relative to the mean is higher for the promiscuous indigobird than for the usually monogamous indigo bunting. As a test for the potential sexual selection, we can compare the observed evenness of matings with that expected if all males were equally successful. The departure from evenness, and the potential for selection, is higher in the promiscuous indigobirds.

This difference in fitness among males might result in large measure from stochastic events and not from song or genetic differences. The frequencies expected from a binomial distribution were compared with the observed distributions to test whether this random distribution might explain the observed mating success of males. The variance of a binomial distribution is less than the mean (Grieg-Smith, 1957; Sokal and Rohlf, 1969). The observed variance in mating success of the indigobirds was twice as great as the mean (Table II). However, the variance in the buntings was less than the mean (Table III). The observed variance in mating success was significantly greater (*F*-test) than expected from a binomial distribution in the indigobirds but not in the buntings. In the indigobirds, there were more observed cases of males with no mates, and of males with many matings, than expected with a binomial distribution. In the buntings, there was no strong evidence in the data that a female tended to select a male on the basis of whether he was already mated. Analysis of the time of pair formation shows that the females that mated with an already-mated male did so after the first few weeks of the breeding season. These females may have mated earlier with another male. Comparison of

TABLE II
Variation in Male Breeding Success in Two Populations of Village Indigobirds

Population	Males (*N*)	Matings per season (*N*)[a]						Mean $\bar{x}$	Variance s^2	I_m[b]
		0	1–10	11–20	21–50	51–100	>100			
Junction, 1973	16	10	1	1	1	1	1	23.0	2261.0	4.27
Cowpie, 1976	20	16	0	2	1	0	1	8.8	668.0	8.63

[a]Estimated from number of matings observed per day multiplied by number of days male was active, from Tables 5 and 6 of Payne and Payne (1977), but including also the 30% of males within the local population that were not observed to sing on a call-site. This estimate of nonsinging males comes from the number of males banded near Junction call-site in 1972 and 1973 and never seen to sing or mate (Payne and Payne, 1977).

[b]$I_m = s^2/\bar{x}^2$ indicates the total index of selection on males (Wade and Arnold, 1980).

TABLE III
Variation in Male Mating Success in Two Populations of Indigo Buntings

Population	Males (N)	Females per male (N) 0	1	2	3	4	Mean $\bar{x}$	Variance s^2	I_m
George Reserve, 1980	92	12	67	13	0	0	1.011	0.2716	.27
Niles, 1980	51	4	33	10	4	1	1.533	0.6205	.26

the observed matings with expected matings in both of these species thus shows that matings were nonrandom, that is, that females were actively selecting certain males. In the indigobirds, females probably mated with the most successful male independently of whether or not another female mated with him, whereas in the buntings, mate choice by the females was probably constrained by the previous matings of a male. Even with the nonrandom mating in the usually monogamous buntings, however, a few males had two or more mates and two polygynists were each seen mating synchronously with two females—ORXW and RXRY each flew back and forth from one end of the territory to the other during early July 1980, mating repeatedly with two different females that both nested in the territory.

An index of the potential for sexual selection can be estimated as a function I_m equal to the variance in success divided by the square of the mean success (Crow, 1958; Wade and Arnold, 1980). The index estimates the potential change in the mean fitness of individuals in a population in one generation if fitness differences were entirely heritable, that is, due to genetic differences. Tables II–IV indicate I_m, the index of potential for sexual selection in males in the two bird species. The potential for sexual selection in males is much greater in the promiscuous indigobirds than in the usually monogamous buntings. The value for the indigobirds is similar to that of another polygynous bird, the red-winged blackbird (*Agelaius*

TABLE IV

Population	Males (N)	Young fledged 0	1	2	3	4
George Reserve, 1980	82	41	1	11	19	6
Niles, 1980	48	15	4	6	12	5

phoeniceus) (Payne, 1979a; Wade and Arnold, 1980). For the indigo buntings, the number of young fledged from the nest indicates a higher index of selection than do the number of mates. The difference in I_m is due mainly to mortality differences and not fecundity differences. In more than 90% of nests examined with eggs, the clutch size was either 3 or 4; nests with 1 or 2 eggs may have had an egg lost by removal by a brood parasitic cowbird (*Molothrus ater*), and no nests were found with more than 4 bunting eggs. Most eggs and young that were unsuccessful were lost either to predators or to cold, rainy weather. It is unlikely that a large proportion of the differences in success after mating were related to heritable differences among the males or among the females. Although the results suggest that most potential for selection occurs after mate choice by the females, there is still considerable potential for sexual modes of selection.

The distribution of mating success and breeding success shows the potential for sexual selection in male indigobirds and buntings. Females have a lower variance in breeding success than males. This general trend is evident in the indigobirds (Payne and Payne, 1977), indigo buntings (Payne, unpublished data, 1978–1982), and other polygynous birds (Payne, 1979a). The difference in the variance of success between the sexes is related to the differences in their behavior. Males spend much time singing and fighting among themselves over a territory, and the cost of this aggressive behavior is paid because the most successful males are so much more successful than the least successful males. Successful male indigobirds and indigo buntings may gain an advantage in mating by their singing, which may take up half of their daytime behavior (Table I). The potential payoff differential for the females is not correspondingly greater, and with the lower payoff stakes the females show fewer costs and risks in their social behavior. Aggressive behavior between consexuals is common in males but not in females. In more than 2000 hours of field observation of the indigobirds, I have seen only four instances of aggressive behavior among females. Females occasionally happen to arrive at the call-site of a singing male at the same time, when the male is off the site, and when they do one may crouch towards the other female and supplant it from its

Variation in Breeding Success in Two Populations of Indigo Buntings

per male (*N*)					Mean	Variance	
5	6	7	8	9	$\bar{x}$	s^2	I_m
0	3	1	0	0	1.573	3.245	1.31
1	3	1	0	1	2.31	4.632	.87

perch. No direct contact has been seen. A female occasionally chases another from the nest of the host, and females may compete for host nests in which to lay (Morel, 1973). But female indigobirds are usually unobtrusive and sneak into the nests of their hosts alone and without overt aggression (Morel, 1973; Payne, 1977a). Female indigobirds commonly feed together, and perhaps a female benefits by seeing another fly away when a predator approaches. Alarm calls are rarely given by any particular finch, perhaps because associating birds are not close kin and lack a social family history from which they can learn which birds are related (Payne, 1977b). In the buntings, I have seen only one case of a female chasing another from her nesting area; the female chased later nested on another territory. Females occasionally respond to each other when one is attracted to the chipping calls that another female gives when her nestlings or fledglings are disturbed, but usually the chipping does not bring in another female. The responding female does not attack the predator (e.g., the fieldworker), and she may be behaving in a selfish manner by gaining information about the identity and location of the disturbance. I have seen no aggressive behavior between females nesting on the same territory even when the male is mating with both of them on the same day, and I have never heard a female sing. Nolan (1958) reported a female singing but did not examine the bird in the hand and the bird may have been a brown-plumaged first-year male.

The secretive, nonsocial, nonaggressive behavior of females in the indigobirds and indigo buntings contrasts sharply with the behavior of the males. This can be explained by proposing that the sex with the higher variance in mating success has the most to gain by taking risks in exposing itself to singing, and so it does, whereas the sex with less variance does not have much to gain, so does not sing. How much "risk" the advertising males are subjected to is unknown. Male birds generally have a higher chance of surviving from year to year than do females (Payne and Payne, 1977; Payne, unpublished data, 1977–1982), and I have never seen a singing bird being attacked or killed by a predator. Field observations indicate no social strategy of females toward each other, and the females appear to ignore or to avoid each other in the breeding season. In some species where the female sings, she cooperates with the male in excluding other pairs from their common territory (Payne, 1971).

The sex-limited singing by male indigobirds and buntings and the differences in their mating and breeding success show a potential for sexual selection, but there is no evidence that the individual differences in song or in breeding success are associated with genetic differences. Although we cannot verify all the assumptions of the model of sexual selection, the observations are consistent with the view that song has evolved by sexual selection.

V. Mate Choice by Females or Male Competition?

Field observations on the village indigobirds and the indigo buntings allow a comparative test of some predictions developed about the nature of past sexual selection as a force shaping the songs of birds.

A. Village Indigobirds

Village indigobirds have a large song repertoire of about 15 song types that are population-specific and contain few or no notes that mimic the host species, and another 5–10 songs that match and mimic the vocalizations of their local firefinch host species (Payne, 1973a; Payne, 1979b). Both classes of songs are used by males advertising on their call-sites. Two nonmimetic song types are associated with sex: one is given when the female flies to the site to visit or mate with the male, and the other is given as the male displays over the perched female in a mating attempt. Mimicry is interspersed with the nonmimetic song types during long bouts of advertising and is apparently the cue used by females to choose among local species of indigobirds. At Lochinvar Park, there are two species of indigobirds, each mimicking and parasitizing a different species of host. By mimicking the host species, a male advertises the host species that reared him, and a female is attracted to the song of males that sound like her own foster father. The mimicry songs are evidently learned in part by the young birds while under care of their foster species (Payne, 1973a; Payne, 1980a). The host-mimetic songs are not used to deceive the host species (Payne, 1973a; 1977a; 1977b; 1979b; 1980a; Sullivan, 1976). I have seen the hosts visiting the indigobird call-site only infrequently and have never seen any apparent deception at the nest. The mimetic songs are used by the territorial males in advertising their own early experience in being reared by the mimicked foster species.

Indigobirds sing both in sexual and male–male aggressive contexts, but in most instances when song was associated with a social interaction, the interaction was with another male (Payne, 1979b). All song types are given during territorial advertisement. These appear to repel other males from the call-site. When the singing male leaves his tree to feed or to visit other call-sites, other males visit his own unprotected site. Certain song types are given particularly during chases of intruding males or after the male returns to the call-site from a chase. The song types associated with aggressive behavior were given more frequently than were the song types associated with sexual behavior (Payne, 1979b). These observations suggest that song is used more as an aggressive signal to other males than as a

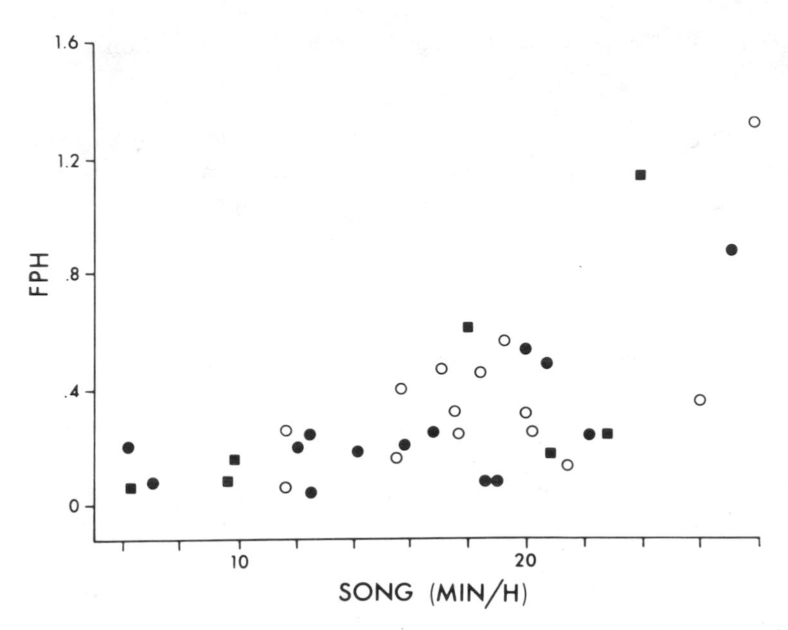

Fig. 1 Comparison of mean song activity of a male village indigobird (mean minutes/hour during which the bird sang at least once) with the mean visiting rate of female indigobirds (females per hour, or FPH) to the male's call-site. Open circles represent individual males in the Junction dialect in 1973, closed circles the cowpie dialect in 1976, and closed squares the other dialects in 1976. For an illustration of the relationship between female visiting rate and male mating success, see Fig. 16, Payne and Payne (1977).

sexual signal to females and that song behavior in general has been shaped by male–male competition as well as by female choice of mates.

Song is directly related to differences in breeding success among male indigobirds. Males that are actively present on their call-sites for more minutes a day attract proportionally more female visitors and have more matings than other males. Males that sing more minutes per day also gain more female visits and more matings than males that sing less (Fig. 1). Song itself may be directly responsible for this difference in success, insofar as no habitat feature was as consistently associated with male success as was his song (Payne and Payne, 1977).

The association of song and sexual success may involve male–male competition rather than a simple cause-and-effect interaction between males and females. Males that were less successful in attracting females and in mating spent more time away from their call-site. These males were seen visiting the call-sites that were more successful than their own to a greater degree than were the most successful males. Thus, the less successful males may have been splitting their time between advertising on their call-sites and attempting to get a better site elsewhere. But on an hour-to-hour basis there was no evidence that mating reinforced signing behavior (Payne and Payne, 1977). On the other hand, both the time spent singing and song rate were closely associated with male breeding success and with the frequency of female visits, whereas the total time that the male was perched on his call-site was not (Payne and Payne, 1977). The correlation of mating success with the quantity of song provides evidence for song as a sexually selected behavior in the indigobirds.

Several observations point to the evolution of some aspects of the male indigobird song through female choice rather than through the male–male aggressive function of song. In particular, the mimicry of the song of the host is suggested to be a result of female mate choice.

Males are unselective in responding to indigobird song, even of other species. In areas where two or more species live together in the same habitat, males of different species are dispersed as a single species, with the nearest-neighbor distances being independent of the species of the indigobird neighbor (Payne, 1973a). The call-sites are defended against both conspecifics and males of other indigobird species (Payne, 1973a; Payne, 1980a). Male *Vidua* attempt to mate with all female *Vidua* regardless of their species, and any male is potentially a sexual spoiler (Payne, 1980a). In experiments and in playbacks used to capture males for color-marking, I have found that male indigobirds will be as readily attracted to the tape-recorded songs of their own species as to songs of remote conspecific populations, or songs of other indigobird species.

In contrast, females are highly selective in their response to certain songs. Female indigobirds in breeding condition are equally likely to approach the tape-recorded nonmimetic songs of their own local population, another population, or another indigobird species. But they are selective in approaching the song of the mimicry of the host of their own species and not the mimicry of the host of other indigobird species (Payne, 1973a). The female indigobirds, in approaching the songs that mimic their own host species, may be using the mimicry song as a guide to a male with an upbringing similar to their own. Different indigobird species have different colors and patterns in the mouths of the begging young, with each species mimicking as a nestling the mouth pattern of the host species. It would benefit a female to mate with an individual male that had been reared by the same host species she herself had, insofar as this choice of a mate would ensure that her young would benefit from mimicking the begging releasers of the same host species.

Experiments were carried out to test whether the sexual condition of female indigobirds is affected directly by hearing the mimetic song. Captive females from Zimbabwe were paired with one *Vidua chalybeata* and one *V. purpurascens* in a cage maintained in soundproof acoustical chambers. In two trials, the females were exposed to tape-recorded song of the

TABLE V
Ovarian Development in Female Indigobirds (*Vidua* spp.) Hearing Their Own Species' Song Mimicry and Another Species' Song Mimicry

Trial	Song	Female	Ovarian follicles, diameter (mm)
1	*V. chalybeata*	*V. chalybeata*	1.2, 1.0, .9
		V. purpurascens	.7, .7, .7
2	*V. chalybeata*	*V. chalybeata*	1.0, 1.0, .9
		V. purpurascens	.9, .9, .8
3	*V. purpurascens*	*V. chalybeata*	.8, .8, .7
		V. purpurascens	.8, .8, .7
4	*V. purpurascens*	*V. chalybeata*	.8, .8, .7
		V. purpurascens	.9, .9, .8

red-billed firefinch (*Lagonosticta senegala*) as mimicked by a male *V. chalybeata* from the same population as the experimental females. In the other two trials, the females were exposed to song of the pink firefinch (*L. rhodopareia*) as mimicked by a male *V. purpurascens*. Birds were kept on 12-hr photoperiods and were exposed to two 15-min recorded sessions a day on alternate days for 12 days of playback, then they were laparotomized and their ovarian follicles were examined and measured. In one trial, the female *V. chalybeata* that heard her own species mimicry developed enlarged ovarian follicles that were up to five times the volume of the other female. In three of four cases, the female that heard her own species mimicry developed larger follicles than her control (Table V); in the fourth case there was no difference. The results suggest that females respond physiologically as well as behaviorally to mimicked song.

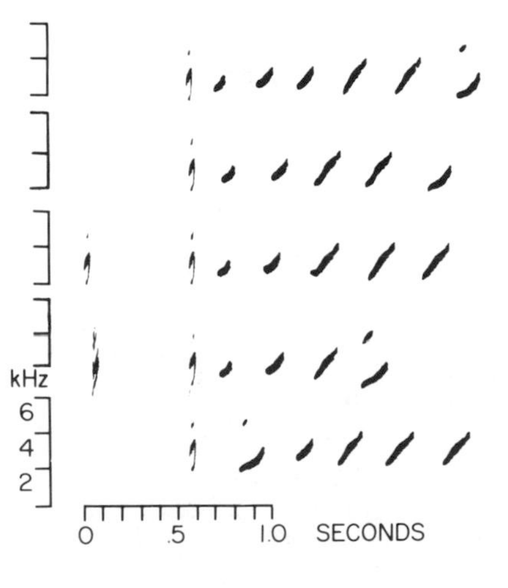

Fig. 2 Songs of an individual male red-billed firefinch (*Lagonosticta senegala*) from Lochinvar National Park, Zambia.

The diversity of the song mimicry of an individual male indigobird indicates that female behavior has influenced male song in an evolutionary sense. Although both male and female host firefinches rear the young brood parasite indigobird together with their own young, only the male firefinch sings. Each *L. senegala* that I have recorded repeatedly in captivity from the population at Lochinvar National Park has only a single song, unvarying except for dropping or adding a note or two at the end (Fig. 2). If a male indigobird learned only the song of his father and not other firefinch songs, he would have only a single mimetic song type. In fact, each wild male indigobird sings several mimetic songs (Fig. 3). Individual differences in mimicry song repertoire among the male indigos are not related to differences in their mating success, because all males share the same songs, and sing them in the same sequence. The number of mimicry songs given by one bird indicates that males may learn the songs of several individual firefinches and not only the song of the individual foster father.

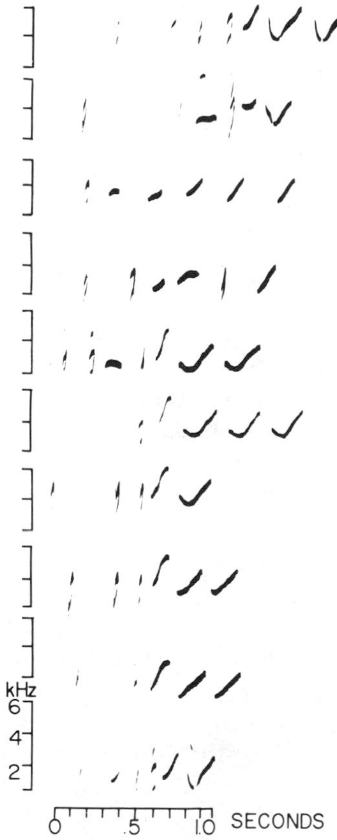

Fig. 3 Host-species mimetic songs of an individual male village indigobird (*Vidua chalybeata*), recorded in Lochinvar National Park, Zambia. Note the greater variation in the songs of the indigobird than in an individual male of the host species, the red-billed firefinch (see Fig. 2).

In experiments with captive birds, I have determined that indigobirds mimic both the songs of other individual firefinches of their host species and also of other male indigobirds that mimic the same host species. The diversity of mimetic songs in a male's repertoire does not seem to be explained by male aggressive behavior, insofar as other songs are given in aggressive circumstances (Payne, 1979b), and as indigobird species that mimic different species of hosts are nevertheless interspecifically territorial (Payne, 1973a, 1980a). On the other hand, the diversity can be explained by sexual selection acting through female mate choice. Males with several mimetic songs would be likely to have at least one song that sounds like the song of a local female's foster father, and these males may attract more females than would males that mimic only one individual of the host species. Female behavior in response to song may have been selected for a certain perceptual generalization, within the limits imposed by the indigobird's species-specific brood parasitism. Females respond to the song by approaching, and are then courted when the singer is the mimic indigobird, and follow the singer and search for a nest when the singer is the song-model firefinch. Even if females do generalize on the song pattern of their own foster father enough to recognize a host species' song given by another firefinch, it would seem advantageous for a male indigobird to advertise with a diversity of firefinch songs in order to enhance his attracting females. In this case, the songs selected by the female in mate choice do not follow some arbitrary stylistic "beauty" as Darwin (1871) suggested, but rather advertise a trait useful to the female.

Sexual selection by female choice also seems to have shaped the form of the male's sexual songs in a way that illustrates the importance of male–male competition to the female. Most indigobird song types begin with 1–4 notes that mimic the alarm call of the host species. The other notes in these songs are harsh, chattery, have multiple inflections, and do not match any notes of the host species (Payne, 1979b). The nonmimetic songs that lack this introductory mimetic note are of two functional kinds: the aggressive songs given in male–male conflict situations, and the two sexual song types given when the female approaches the male on his call-site and when the male displays over the female. These two classes of song begin with chatter notes. The general features of the songs used in the most intensive sexual circumstances most closely resemble the songs given in male combat (Payne, 1979b). The structural similarity of these two song classes suggests that females are attracted by male aggressive behavior. This outcome is what one would expect if males competed among themselves and females simply read out the social order established among the fighting males to set their own priorities for mating, cueing by the same signals that the males use among themselves. Several other promiscuous species of birds in which the female does not live socially with the male also have ritualized male aggressive signals in their courtship behavior (Hjorth, 1970; Payne and Payne, 1977), and with this nonfamily life style, the female mating strategy appears to be: choose a male that has won in competition with other males.

Changes in the fine details of each song type from year to year allow a test of the prediction that the sexual songs or the aggressive songs should be more constant across years than the other songs. All song types of a local song population usually change slightly from year to year, even within the same bird (Payne, 1979b; Payne, 1980b). I determined indices of similarity for each song type in one song population (Junction) at Lochinvar National Park across a period of successive years, from 1972 through 1975. For each nonmimetic song type in each year, I measured the duration and carrier frequency for the first notes and the intervals between notes. The phenetic difference between corresponding songs in successive years was determined by calculating a product-moment correlation coefficient r of the standardized variables, and the index of similarity was then taken as $1 - r$. For the 16 song types, there was no tendency for the sexual songs to be less variable across years than for the aggressive songs or for the other songs. Nor were the aggressive songs either more or less similar across years than were the other songs. The comparison provides no evidence in support of either the mate choice model or the male competition model of sexual selection of song.

The interaction between experience and genetic determinants in the development of song in the indigobirds may set certain constraints on the usefulness of comparing the songs of different males in choosing a mate. Indigobirds are not unusual in having such a constraint; since most songbirds learn their songs, song provides no genetic information to the females. Nevertheless, the details of song development in these brood parasites indicate that song may not only provide no clues that would be useful in detecting special genotypic information about any male, it also may even mislead a female about a male's species identity. Song details in the indigobird are learned from the foster parents and are then generalized in such a way that the male continues to recognize and match the songs of other individuals that fit the general song features of the foster species. In several species of parasitic finches, occasional birds are reared in the "wrong" host species' nest and develop the "wrong" set of songs which attract not females of their own species, but females that are associated with that set of songs (Payne, 1973a). Wild interspecies hybrids likewise develop the songs of one host species and the nonmimetic songs of that same species' usual parasite, and they lack any song of the other parental species (Payne, 1980a). Males that disperse from one area to another may switch their songs to match those of the new population, even as adults (Payne, 1981a). A female indigobird would gain no infallible information about the relative genetic quality of her mate from song alone. Not only would song give no clue to his natal population, it would not even give unfailingly reliable information about his species.

Female indigobirds generally visit males within their own dialect area, but this may be a matter of effective use of their time as much as attraction to a certain set of songs. Males that share their songs occupy an area of 5–20 km^2, and some individual females visit males across this large area (Payne and Payne, 1977). In our study of color-marked females in 1973, we found no instances of males in this area that did not match the songs of their neighbors. The observation that experimental captive females are attracted with equal frequency to the songs of their own dialect and to other dialects (Payne, 1973a) suggests that the females do not discriminate very much among the males on the basis of their local song dialects.

Although differences in the singing behavior of local male indigobirds are closely associated with differences in their mating success, it is by no means the case that songs have evolved mainly by sexual selection outside the arena of male competition. Indigobirds sing both in aggressive encounters among males and in sexual interaction with females. Song repertoire, song structure, and the sequence of song types are all involved in advertising a territorial male to competing males (Payne, 1979a), and the songs that are used in the most intensive sexual interactions with females are similar in structure to the songs used in the aggressive chases and attacks among the males. Female choice of the most aggressive males, in these birds with a nonfamily life-style, apparently is of the male that is most successful in competing with other males. The songs that mimic the host species are not used in aggressive contexts and are most likely an evolutionary result of female mate choice. Hence, sexual selection has shaped the evolution of the songs of the indigobirds both through intrasexual competition and the female strategy of choosing a mate with an early upbringing that is similar to her own.

B. Indigo Buntings

Indigo buntings sing most intensively early in the season while they are establishing territories. When a male has attracted a female, especially during the days of courtship and mating, the male dramatically ceases his daylong singing and sings only occasionally (Thompson, 1972; Payne, unpublished data, 1977–1982). In courtship, the male buzzes softly and the female twitters; the male does not sing during the mating bouts (Thompson and Rice, 1970; Payne, unpublished data, 1977–1982). After the female has laid her clutch, the male continues singing and sometimes attracts a

TABLE VI
Mating Success of Male Indigo Buntings During Two Years in Relation to Whether They Matched a Neighbor's Song

Age	Song matching	Mated females (N)		Nests (N)	
		0	1+	0	1
Adult	Yes	5	71	12	63
	No	6	14	6	14
	G		7.04		1.85
	df		1		1
	p		<.01		ns
First-year	Yes	37	48	41	43
	No	26	8	6	14
	G		11.05		17.64
	df		1		1
	p		<.01		<.0001

second female to his territory. An adult bunting usually has only one song. First-year males arrive on the breeding ground with either a variable song or a fixed song unlike the local adults', and in a few days to a few weeks they usually change their song to match a territorial neighbor's. The context of singing suggests males may initially attract females by their song, but they do not court with song, and a male sings largely in territorial interactions with other male buntings.

Bird song is generally thought to be used in mating, but there is little information on the comparative mating and breeding success of males as a function of any differences in their song. In the buntings there is a strong association of song and mating success. First-year males that match a neighbor's song have a significantly higher chance of obtaining a mate and a nest than do first-year males that do not match a local song type (Table VI). The advantage occurs in males that are present for at least 10 days in the study area and is not limited to territorial, as opposed to nonterritorial, males, though the males that match an older neighbor are more likely to establish a territory and to hold the territory for a long time (Payne, 1982).

Observations suggest that song matching is important both in establishing and maintaining a territory and in the success of the male in attracting and mating with a female after he has a territory. By following individually marked males through a season, I have found that some first-year males move from one area to another and have a series of territories, and each time they change neighborhoods they change their song to match that of their new neighbors. Thus the matching of a neighboring adult is not only a means to acquire an adult-like song, it is a strategy to mimic a certain individual. As a test of the mode of sexual selection for song behavior, we may ask whether this intraspecific song mimicry is directed towards other males or towards the females.

If song mimicry is sexual mimicry and is directed toward the females, then we would expect to see cases of a female leaving her old mate for a new male with the same song or allowing another male that sounds like her own mate to mate with her. During 1978 and 1979, we followed 103 banded females but found only 2 that changed mates within a season. Both mated with a neighbor with a song that did not match their first mate. Between seasons, most females returned to the same territory where they bred the year before. A few females changed areas and mates even when the mate with whom they had nested successfully the year before returned, and in these cases, the female mated with a male with a song unlike her former mate's. In only one case did the female (RYGY) mate with a male having a song like her former mate. Of 13 females that returned from 1978 to 1979, 5 mated with their old mates in the old territory; 3 mated with different males in their old territory (including 2 with a song unlike their old mates'); 1 mated with another male with her old mate's song in a neighboring territory (her old mate returned also); and 3 settled in a remote territory and mated with males with a song unlike their former mates'. In all cases, the old mates or other males with that same song returned. Females, then, generally return to their old territories regardless of the males occupying them, and do not tend to remate with males that mimic their former mates.

In 1980, a first-year male (WOGX) mated with a female who had nested in the previous year with the adult (XOOR) whose song he mimicked, but WOGX changed his song to match this older male only after the female had paired and laid her eggs. The timing of the song change indicates that the male did not gain by his song imitation in influencing his mate's behavior, though he may have gained an advantage among the local males.

We have seen five females mating outside the "pair" bond in a nesting, including cases of a female welcoming and mating with a visitor in her home territory when her mate was not present and cases of active solicitation and mating with other males on their territories. In all instances, the extrapair matings were with established territorial neighboring males. In no instance was the neighbor a song sharer as well as a mate sharer. The observations do not support the concept that songmatching in buntings is a form of sexual mimicry. It may, however, be involved in intrasexual aggressive mimicry (Payne, 1982).

Another observation suggests that song may be involved in mate choice by the female. One male drastically altered his adult song, and the change presented a natural experiment in testing whether a male with an

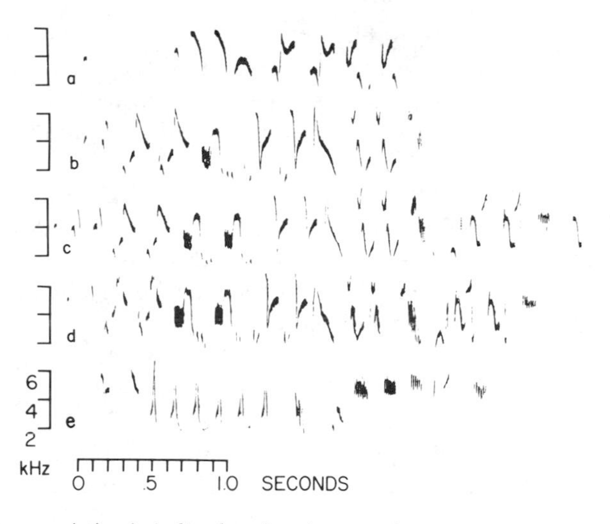

Fig. 4 Song variation in indigo buntings in a population at Niles, Michigan. (a,b) Songs of two neighboring adult buntings illustrating the difference between birds not in the same song group, (c) song of GOGX, a first-year male that copied the song of bird (b), (d) song of GOGX as a second-year male before his song changed; (e) song of GOGX as a second-year male after his song changed to one that did not closely resemble adult bunting songs.

odd song would attract a mate. Male GOGX returned in 1980 to his territory of 1979. In the earlier year he was a first-year male that matched the song of a local dialect. During June 1980, his song changed from a normal bunting song with several kings of paired notes like those of buntings in other populations (Shiovitz and Thompson, 1970; Thompson, 1970) to one that I did not at first recognize as a bunting (Fig. 4). He had two nesting females before his song changed. One nest was deserted when two eggs of a brown-headed cowbird (*Molothrus ater*) were laid in it; the other was successful, and two young fledged a week after the male's song changed. He actively maintained his territory with this unusual song for the following six weeks but did not attract another mate, though two of his territorial neighbors each attracted another female and had a nest. However, several other males had normal bunting songs and did not attract females or have late nests, and GOGX's lack of late nesting might not be due to his song. His success in keeping a territory during a time when new territories were still being established locally was likely due to his active chasing of other males as well as to his earlier priority on the territory.

Would the songs of indigo buntings provide useful information to females about the genetic quality of males? Probably they would not, because the males learn their songs, and differences in song can be explained entirely as differences in individual experience. Most first-year males that I have banded in the field sooner or later match the song of a neighboring adult. All young birds that I have raised together with live song tutors have learned to match the song of the social tutor (Payne, 1981b). Some males change songs repeatedly in a season, matching a series of males. In addition, in more than 200 young banded as babies, only 5 (2 males and 3 females) have returned to the study area in a later breeding season through 1980. None of them settled in the territories of their fathers' song group, and none of the males had a song like their fathers' song. The one male observed that did have an abnormal song (GOGX) did not have "bad genes" insofar as he mated and produced fledglings in 2 years. The locally distinctive song types that comprise the local dialects have a short period of survival. Finally, the samples of songs recorded over 5 years at the George Reserve in Michigan show that about half of the songs disappear over a period of 4–5 years, a time corresponding to a half-life of about 3 generations of indigo buntings (Payne *et al.*, 1981). Thus the dispersal, the learned development of the songs, and the relatively short persistence of the local song traditions all indicate that the differences in the songs in male indigo buntings would provide no information useful in assessing the relative genetic quality of the males as mates to the females.

The role of song in mate selection by female indigo buntings is probably related to the role of song in male competition. Males sing in aggressive interactions with other males, and males establish and hold territories with their song. So a female, upon hearing a male singing long and loud, might initially be attracted to him on the chance that he had gained control over a territory suitable for her nesting. She might be cued by the same song traits that would cue another male that the territory was under control. The observation that males do not sing much on the days when they are mating suggests that, if song attracts females, there is counterselection by the potential interference of any other male attracted to a male that advertises his receptive mating female. Neighboring males in fact often do approach a copulating pair on the resident male's territory and are sometimes successful in mating with the female themselves. When a mating male does sing, it is when he sees and chases one of these intruding male neighbors. Thus, the reproductive competition among males appears to set limits to the degree to which song is used in sexual behavior and to the possible role of females in choosing a male by his song.

VI. Discussion

A. How Has Sexual Selection Shaped Bird Song?

A direct comparison of these two species of indigos suggests that birds with a higher measured potential for sexual selection (the promiscuous–

polygynous indigobirds) have more elaborate songs and have also undergone more obvious shaping of their songs by female choice than have the more monogamous species (the indigo buntings). Nevertheless, both the context and the individual development of song point to the importance of male–male interactions.

The mode of sexual selection in shaping the song behavior of these two species of indigos appears to reflect the influence of male competition for mating sites more than the female choice of males by their song. The importance of male–male competitive interactions in shaping song behavior of indigobirds is suggested by the elaboration of the repertoire, the greater association of songs with territorial behavior among males than with courtship of females, the shifting of songs from year to year, and the accumulating changes in song traditions from year to year even in the sexually winning birds in the dialect populations. However, the elaboration of songs that mimic the host species and the form of the songs that are used to attract and court the females both can be explained as a result of female mate choice in these promiscuous birds. In the monogamous indigo buntings, song is apparently shaped through male–male interactions and no aspects of song are obviously related only to female choice. In the buntings, there is no clear evidence of sexual selection by the process of female choice of mates on the form of the song. In both species, the selective advantage to a male in having a song that matches another male appears to be gained through his competitive relationships with other males and not directly through increased attractiveness to a female.

The learning of song imposes a major constraint on the mating strategy of the female by concealing the genetic identity of the individual males. Especially in species where males match precisely the songs of their older neighbors, as in the indigobirds and the indigo buntings, a female probably could not distinguish a male from his model or his song imitator. Song mimicry involves the possibility of "cheating" on individual recognition. This may benefit the male that mimics either by allowing him ready access to a traditional mating site when the older stud disappears (we observed a replacement male village indigobird mating during his first day on the site [Payne and Payne, 1977]) or by deterring other territorial competitors from the area where they had already been chased by the older song model (this may be the basis of the advantage to the song mimics in the indigo buntings [Payne, 1982]). In either case, aggressive song mimicry would be effective if other birds associated the local song with the territorial dominance of the established song model and avoided nearby areas where they heard the same song.

In principle, a female might still be able to compare the genetic quality of different males by listening to their songs if certain males were better song learners than others, and if the song differences among males were not due simply to differences in experience. However, there is no evidence for this. All songbirds are known to learn their songs regardless of their mating system and thus regardless of the intensity of sexual selection. In the indigobirds, all males sampled locally had identical song repertoires, and nearly all males sang the corresponding song so nearly identically that I could not distinguish among the males even with close examination of many audiospectrograms (Payne, 1973a; 1979b; 1980b). In the buntings, all first-year males that I tested experimentally performed well in copying the songs of older males with whom they interacted (Payne, 1981b). Also, in buntings observed in the field, those first-year males that did not copy a neighbor's song showed no morphological differences from those that were good imitators (Payne, 1982). Thus, there is no morphological evidence of any genetic difference in the song mimics and the song individualists. Although the idea that females might be a driving force for the evolution by sexual selection of song learning ability in males is appealing, there is no evidence for it.

The suggestion that song dialect differences might cue a female to the genetic identity of individual males and lead to genetic assortative mating (Nottebohm, 1972) is being disproven for most dialectal birds (Payne, 1981a). Song differences are not closely associated with genetic differences either within or between populations in white-crowned sparrows (*Zonotrichia leucophrys nuttalli*) (Baker, 1975); individuals born in one dialect area may breed in another (Baker and Mewaldt, 1978), and males with songs that differ from those of their neighbors usually end up mating successfully with local females regardless of their "mismatched" song (Baptista, 1975).

Mate choice by the female has probably shaped the form of song in birds that sing different kinds of songs in close-contact sexual situations and in territorial advertisement. The acoustic properties of sexual songs seem to be adapted for transmission at short range (Morton, 1975; Morton, 1977). Sexual songs have been described in several families of birds, mainly in monogamous species (Hall, 1962; Baptista, 1978).

However, not all songbird species have special songs used only in sexual circumstances, and not all species with more than one song type use the songs in different contexts (Howard, 1974; Verner, 1976; Payne, 1979b; Reid, 1979; Catchpole, 1980; Garson, 1980). The diversity of a male's songs are sometimes related to his mating success. Howard (1974) found that mockingbirds (*Mimus polyglottos*) with a more diverse repertoire mated earlier in the season. He suggested that the difference was probably caused by male–male interactions insofar as these males had richer territories and females were apparently attracted to the territories rather than directly to the song repertoires themselves. These males were apparently better able to hold and defend the better territories against other males. Although breeding success was not measured, early mating is thought to be advantageous for several reasons (Fisher, 1958), including the increased time for renesting later in the season. In these birds, sexual selection for song elaboration apparently works by way of male–male competition and not directly by an independent pathway of female choice.

In the red-winged blackbird (*Agelaius phoeniceus*), the evidence available suggests that females select a good nesting habitat rather than a particular male (Lenington, 1980; Orians, 1980). Males differ in the number of songs in their repertoire, but the difference is not closely tied to their mating success (Smith and Reid, 1979). Male blackbirds court females with a high twitter rather than with a song; some males mate with no vocal preliminaries at all (Payne, 1979a). A female may mate with more than one male (Bray *et al.*, 1975). The independence of the sperm or genetic quality of the real-estate owner where they nest from that of the biologically effective mate further complicates the analysis of sexual selection of song in these promiscuous females.

Sexual selection may help explain the more elaborate songs of polygynous species compared with their monogamous species relatives in some other birds. In North American wrens, the polygynous species have larger song repertoires than the monogamous species (Kroodsma, 1977), but in European warblers of the genus *Acrocephalus*, the monogamous species have larger song repertoires than the polygynous birds (Catchpole, 1980). In the wrens, males of all species are aggressive and spend much time advertising their territories against other males as well as announcing their availability to females. In their observations of marsh wrens (*Cistothorus* spp.), Verner (1976) and Kroodsma (1979b) both emphasize the competitive context of male song matching and do not describe any apparent female responses to the complex song repertoires. In the sometimes polygynous reed warbler (*A. scirpaceus*), song rate decreases after pairing but song duels persist through the breeding season among territorially interacting males; in contrast, in the monogamous sedge warbler (*A. schoenobaenus*), the male stops singing after he is paired and uses visual threats rather than song in his later territorial interactions with his neighbors (Catchpole, 1973). In neither species were the responses of females to song observed or experimentally tested. Epigamic sexual selection may be involved in the evolution of the large song repertoire in the sedge warblers insofar as those males with a large song repertoire mated earlier, even though they did not necessarily gain a territory earlier in the season (Catchpole, 1980). The disparate results among these groups of birds point out the necessity of observing both intrasexual and intersexual behavioral interactions in marked individuals, and caution against accepting general statements before more kinds of birds have been studied.

We have no generally satisfactory hypothesis to account for the elaboration of song into a large repertoire in certain species, and for the occurrence of a single song type in others. The large repertoire of some species (such as wrens) may be explained in terms of Maynard Smith's (1974) "war of attrition" model, where only the males with better territories can afford the time to advertise elaborate songs. By advertising with a large number of songs, they may make it too costly in time for their competitors to spend hours listening to them so as to learn their songs and match them (E. S. Morton, personal communication, 1981). The model explains song diversity in terms of male–male competition, but why this mode of sexual selection leads to such dramatic differences in song repertoires in related species is unknown.

Females respond to male song in the few species in which they have been tested. The indigobird species *Vidua chalybeata* and *V. purpurascens* both approach specifically the mimetic calls of their own species and not those of the other species (Payne, 1973a). Also, in the parasitic paradise whydahs (*V. paradisaea*), the females approach selectively the songs that their males mimic and not those mimicked by closely related species (Payne, 1973b). In this last species, the females sometimes develop their ovaries in response to song over several days up to the stage of ovulation and egg laying (Payne, 1973b). Kroodsma (1976) found that in Belgian Wasserschlager canaries (*Serinus canarius*), a domesticated strain that has been genetically selected by man for its elaborate song, the females exposed to tape recordings of large repertoires built their nests more quickly and laid more eggs than did females exposed to smaller song repertoires. The differential response of males to these song differences is not known either in the captive strain or in wild canaries. Female brown-headed cowbirds (*Molothrus ater*) respond sexually to male song by assuming a copulation-soliciting posture, and they appear to discriminate between the songs of naive males and the songs of males that were reared with adult song (King and West, 1977). These observations suggest that females have evolved the ability to respond to different songs differently; if so, then the perceptive females could drive sexual selection by their choice of males that had more elaborate songs.

B. Sexual Selection or Species Isolation?

In all of the cases discussed so far, the evolution of male song behavior can be explained by sexual selection involving female mate choice, and it seems unnecessary to invoke species isolation as a teleonomic function of the elaborate songs. In contrast, the evolution of bird song has earlier been described mainly in terms of the benefit gained from having species-specific mating signals (Marler, 1960; Mayr, 1963). The proposition that species isolation was the main selective basis for elaborate species-specific signals led to many studies of the differences in the territorial response of males to the songs of their own species and other species (e.g., Lanyon, 1969). Nevertheless, it seems just as likely that the individual histories of the species have shaped the evolution of song by sexual selection as by the advantage of discriminating against other birds of other species. The territorial approach of a male to a song of another of his species within his territory is a far different context than the sexual approach of a female to a potential mate, and the territorial approach paradigm has been over-interpreted to show evidence for sexual reproductive isolation.

An earlier set of predictions of the species-isolating mechanism function of bird songs included the following: (*a*) different species have different songs, (*b*) song is used in sexual context, (*c*) males that sing the "wrong" song should attract the corresponding "wrong" females, and (*d*) females should respond in a sexual context only to the songs of their own species (Payne, 1973b). The same predictions apply to the hypothesis that songs have evolved by sexual selection, and the predictions fail to resolve the issue of whether either hypothesis explains the evolution of song or of any aspect of song. I suggest the following questions as tests of the two major hypotheses to explain the evolution of song:

1. Is there a correlation between species-distinctiveness of song and the mating system? The reproductive isolation (RI) model predicts none; the sexual selection (SS) model predicts a greater elaboration, hence, an accompanying greater difference in song in the nonmonogamous species.
2. Are songs more elaborate where more closely related species coexist? Predictions: RI, yes; SS, no.
3. Are songs of different species more distinct in sympatry than where only one species occurs? Predictions: RI, yes; SS, no.

In certain groups, such as the wrens, there appears to be a correlation between song diversity and song sequencing with the mating system in the direction of SS (prediction 1), but this is not true of others (Catchpole, 1976; Catchpole, 1980). Comparisons of related species in and out of local sympatry generally indicate no evidence of reproductive character displacement (Thielcke, 1969; Grant, 1972; Payne, 1973a) in song or in other characters and do not support the RI model predictions 2 or 3. We should seriously consider sexual selection as the main directive force within each species that leads to the elaboration of song and to the independent course of song development in each species that leads to species differences. This explanation extends only slightly Mayr's (1963) suggestion that species behavior differences arise in isolation either by chance or as by-products or coevolved responses of other environmentally selected responses. It is also the logical conclusion of Darwin's theory of sexual selection as it applies to closely related species (Thornhill, 1980).

VII. Summary

Song behavior in birds is probably shaped by two forms of sexual selection: (*a*) competition among males for resources and mating sites and (*b*) attraction of females and choice of males as mates. A set of predictions was developed to test for each of these Darwinian modes of sexual selection. Field observations on two species, the promiscuous brood-parasitic African village indigobird (*Vidua chalybeata*) and the usually monogamous North American indigo bunting (*Passerina cyanea*), were made and predictions of the two models of sexual selection were tested. The male–male competition model explained more aspects of song than did the female mate choice model. The host mimicry of the songs of the indigobirds is best explained by the female choice model, as is the form of the sexual songs in this species. Song in the monogamous bunting seems explained mainly by the advantage a male has in mimicking the song of his established neighbors, and the species has local song neighborhoods of song-mimicking males. The greater elaboration of song in the indigobirds is predicted from their much higher index of sexual selection, as measured in their mating success variance.

Sexual selection may explain the species differences among songbirds in song without recourse to a hypothesis of selection for behavioral isolating mechanisms between species. Predictions are suggested for distinguishing between these evolutionary processes in shaping bird song.

The importance of male song learning to the strategy of mate choice by female birds involves the potential for a male "cheating" or lying about his individual identity. Especially in the dialectal birds where one male copies the song of his neighbor, interindividual mimicry as an adaptive male song strategy makes it impossible for a female to compare the "genetic fitness" of males. This may explain the common observation that females choose habitats, not genetic mates.

Acknowledgments

I thank Eugene S. Morton for critically reading the manuscript.

References

Armstrong, E. A. *A study of bird song.* London/New York: Oxford Univ. Press, 1963.
Baker, M. C. Song dialects and genetic differences in white-crowned sparrows (*Zonotrichia leucophrys*). *Evolution*, 1975, *29*, 226–241.
Baker, M. C., and Mewaldt, L. R. Song dialects as barriers to dispersal in white-crowned sparrows, *Zonotrichia leucophrys nuttalli*. *Evolution*, 1978, *32*, 712–722.
Baptista, L. F. Song dialects and demes in sedentary populations of the white-crowned sparrow (*Zonotrichia leucophrys nuttalli*). *University of California Publications in Zoölogy*, 1975, *105*.
Baptista, L. F. Territorial, courtship and duet songs of the Cuban grassquit (*Tiaris canora*). *Journal für Ornithologie*, 1978, *119*, 91–101.
Barrington, D. Experiments and observations on the singing of birds. *Philosophical Transactions of the Royal Society of London*, 1773, 442–451.
Blanchard, B. D. The white-crowned sparrows (*Zonotrichia leucophrys*) of the Pacific seaboard: Environment and annual cycle. *University of California Publications in Zoölogy*, 1941, *46*, 1–178.
Bowman, R. I. Adaptive morphology of song dialects in Darwin's finches. *Journal für Ornithologie*, 1979, *120*, 353–389.
Bray, O. E., Kennelly, J. J., and Guarino, J. L. Fertility of eggs produced on territories of vasectomized red-winged blackbirds. *Wilson Bulletin*, 1975, *87*, 187–195.
Brown, J. L. *The evolution of behavior.* New York: Norton, 1975.
Carey, M., and Nolan, V. Polygyny in indigo buntings: A hypothesis tested. *Science*, 1975, *190*, 1296–1297.
Carey, M., and Nolan, V. Population dynamics of indigo buntings and the evolution of avian polygyny. *Evolution*, 1979, *33*, 1180–1192.
Catchpole, C. K. The functions of advertising song in the sedge warbler (*Acrocephalus schoenobaenus*) and the reed warbler (*A. scirpaceus*). *Behaviour*, 1973, *46*, 300–320.
Catchpole, C. K. Temporal and sequential organization of song in the sedge warbler (*Acrocephalus schoenobaenus*). *Behaviour*, 1976, *59*, 226–246.
Catchpole, C. K. Sexual selection and the evolution of complex songs among European warblers of the genus *Acrocephalus*. *Behaviour*, 1980, *74*, 149–166.
Chappuis, C. Un example de l'influence du milieu sur les émissions vocales des oiseaux: l'evolution des chants en forêt equatorial. *Terre Vie*, 1971, *25*, 183–202.
Crow, J. F. Some possibilities for measuring selection intensities in man. *Human Biology*, 1958, *30*, 1–13.
Darwin, C. *The descent of man and selection in relation to sex.* New York: Appleton, 1871.
Dawkins, R., and Krebs, J. R. Animal signals: Information or manipulation? In J. R. Krebs and N. B. Davies (Eds.), *Behavioural ecology, an evolutionary approach.* Sunderland, Mass.: Sinauer Associates, 1978.
Emlen, S. T. The role of song in individual recognition in the indigo bunting. *Zeitschrift für Tierpsychologie*, 1971, *28*, 241–246.
Falls, J. B. Functions of territorial song in the white-throated sparrow. In R. A. Hinde (Ed.), *Bird vocalizations.* London/New York: Cambridge Univ. Press, 1969.
Fisher, R. A. *The genetical theory of natural selection.* New York: Dover, 1958.
Garson, P. J. Male behaviour and female choice: Mate selection in the wren? *Animal Behaviour*, 1980, *28*, 491–502.
Goldman, P. Song recognition by field sparrows. *Auk*, 1973, *90*, 106–113.
Grant, P. R. Convergent and divergent character displacement. *Biological Journal of the Linnean Society*, 1972, *4*, 39–68.
Greig-Smith, P. *Quantitative plant ecology.* London: Butterworths, 1957.

Hall, M. F. Evolutionary aspects of estrildid song. *Symposia of the Zoological Society of London*, 1962, *8*, 37–55.
Hinde, R. A. The biological significance of the territories of birds. *Ibis*, 1956, *98*, 340–369.
Hjorth, I. Reproductive behavior in tetraonidae, with special reference to males. *Viltrevy, Swedish Wildlife*, 1970, *7*, 184–596.
Howard, R. D. The influence of sexual selection and interspecific competition on mockingbird song (*Mimus polyglottus*). *Evolution*, 1974, *28*, 428–439.
Hunter, M. L., and Krebs, J. R. Geographical variation in the song of the great tit (*Parus major*) in relation to ecological factors. *J. Anim. Ecol.*, 1979, *48*, 759–785.
Jilka, A., and Leisler, B. Die Einpassung dreier Rohrsängerarten (*Acrocephalus schoenobaenus, A. scirpaceus, A. arundinaceus*) in ihre Lebensräume in Bezug auf das Frequenzspektrum ihrer Reviergesänge. *Journal für Ornithologie*, 1974, *115*, 192–212.
King, A. P., and West, M. J. Species identification in the North American cowbird: Appropriate responses to abnormal song. *Science*, 1977, *195*, 1002–1004.
King, J. A. Variation in the song of the rufous-collared sparrow, *Zonotrichia capensis*, in northwestern Argentina. *Zeitschrift für Tierpsychologie*, 1972, *30*, 344–373.
Kroodsma, D. E. Reproductive development in a female songbird: Differential stimulation by quality of male song. *Science*, 1976, *192*, 574–575.
Kroodsma, D. E. Correlates of song organization among North American wrens. *American Naturalist*, 1977, *111*, 995–1008.
Kroodsma, D. E. Aspects of learning in the ontogeny of bird song: Where, from whom, when, how many, which, and how accurately? In G. Burghardt and M. Bekoff (Eds.), *Ontogeny of behavior*. New York: Garland, 1979a.
Kroodsma, D. E. Vocal dueling among male marsh wrens: Evidence for ritualized expressions of dominance/subordinance. *Auk*, 1979b, *96*, 506–515.
Lanyon, W. E. Vocal characters and avian systematics. In R. A. Hinde (Ed.), *Bird vocalizations*. London/New York: Cambridge Univ. Press, 1969.
Lemaire, F. Mixed song, interspecific competition and hybridisation in the reed and marsh warblers (*Acrocephalus scirpaceus* and *palustris*). *Behaviour*, 1977, *63*, 215–240.
Lenington, S. Female choice and polygyny in redwinged blackbirds. *Animal Behaviour*, 1980, *28*, 347–361.
Marler, P. Bird songs and mate selection. In W. E. Lanyon and W. N. Tavolga (Eds.), *Animal sounds and communication* (Publ. No. 7). Washington, D.C.: American Institute of Biological Sciences, 1960.
Marler, P. A comparative approach to vocal learning: Song development in white-crowned sparrows. *Journal of Comparative and Physiological Psychology*, 1970, *71*(2), 1–25.
Marler, P. On strategies of behavioural development. In G. Baerends, C. Beer, and A. Manning (Eds.), *Function and evolution in behaviour*. London/New York: Oxford Univ. Press (Clarendon), 1975.
Marler, P., and Tamura, M. Song "dialects" in three populations of white-crowned sparrows. *Condor*, 1962, *64*, 368–377.
Marten, K., and Marler, P. Sound transmission and its significance for animal vocalization. I. Temperate habitats. *Behavioral Ecology and Sociobiology*, 1977, *2*, 271–290.
Marten, K., Quine, D., and Marler, P. Sound transmission and its significance for animal vocalization. II. Tropical forest habitats. *Behavioral Ecology and Sociobiology*, 1977, *2*, 291–302.
Maynard Smith, J. The theory of games and the evolution of animal conflicts. *Journal of Theoretical Biology*, 1974, *47*, 209–221.
Maynard Smith, J. *The evolution of sex*. London/New York: Cambridge Univ. Press, 1978.
Mayr, E. *Animal species and evolution*. Cambridge: Belknap Press, 1963.

Morel, M.-Y. Contribution à l'etude dynamique de la population de *Lagonosticta senegala* L. (estrildides) à Richard-Toll (Sénégal). Interrelations avec le parasite *Hypochera chalybeata* (Müller)(viduines). *Mémoires du Muséum National d'Histoire Naturelle Série A: Zoologie* (Paris), 1973, *78*, 1–156.

Morton, E. S. Ecological sources of selection on avian sounds. *American Naturalist*, 1975, *109*, 17–34.

Morton, E. S. On the occurrence and significance of motivation-structural rules in some bird and mammal sounds. *American Naturalist*, 1977, *111*, 855–869.

Nolan, V. Singing by female indigo bunting and rufous-sided towhee. *Wilson Bulletin*, 1958, *70*, 287–288.

Nottebohm, F. The origins of vocal learning. *American Naturalist*, 1972, *106*, 116–140.

Nottebohm, F. Continental patterns of song variability in *Zonotrichia capensis*: Some possible ecological correlates. *American Naturalist*, 1975, *109*, 605–624.

Orians, G. H. Some adaptations of marsh-nesting blackbirds. *Princeton Monographs in Population Biology*, 1980, *14*.

Payne, R. B. Duetting and chorus singing in African birds. *Ostrich Supplement*, 1971, *9*, 125–145.

Payne, R. B. Behavior, mimetic songs and song dialects, and relationships of the parasitic indigobirds (*Vidua*) of Africa. *Ornithological Monographs*, 1973a, *11*.

Payne, R. B. Vocal mimicry of the paradise whydahs (*Vidua*) and response of female whydahs to songs of their hosts (*Pytilia*) and their mimics. *Animal Behaviour*, 1973b, *21*, 762–771.

Payne, R. B. Clutch size, egg size, and the consequences of single vs. multiple parasitism in parasitic finches. *Ecology*, 1977a, *58*, 500–513.

Payne, R. B. The ecology of brood parasitism in birds. *Annual Review of Ecology and Systematics*, 1977b, *8*, 1–28.

Payne, R. B. Local dialects in the wingflaps of flappet larks *Mirafra rufocinnamomea*. Ibis, 1978a, *120*, 204–207.

Payne, R. B. Microgeographic variation in songs of splendid sunbirds *Nectarinia coccinigaster*: Population phenetics, habitats, and song dialects. *Behaviour*, 1978b, *65*, 282–308.

Payne, R. B. Sexual selection and intersexual differences in variance of mating success. *American Naturalist*, 1979a, *114*, 447–452.

Payne, R. B. Song structure, behaviour, and sequence of song types in a population of village indigobirds, *Vidua chalybeata*. *Animal Behaviour*, 1979b, *27*, 997–1013.

Payne, R. B. Behavior and songs in hybrid parasitic finches. *Auk*, 1980a, *97*, 118–134.

Payne, R. B. Behavior, songs, and populations of parasitic finches. In National Geographic Society, *Research Reports*, 1980b, *12*, 541–550.

Payne, R. B. Population structure and social behavior: Models for testing the ecological significance of song dialects in birds. In R. D. Alexander and D. W. Tinkle (Eds.), *Natural selection and social behavior*. New York: Chiron Press, 1981a.

Payne, R. B. Song learning and social interaction in indigo buntings. *Animal Behaviour*, 1981b, 29, 688–697.

Payne, R. B. Ecological consequences of song matching: Breeding success and intraspecific song mimicry in indigo buntings. *Ecology*, 1982, *63*, 401–411.

Payne, R. B. Whydah. In B. Campbell and E. Lack (Eds.), *A new dictionary of birds* (Rev. ed.). In press.

Payne, R. B., and Payne, K. Social organization and mating success in local song populations of village indigobirds, *Vidua chalybeata*. *Zeitschrift für Tierpsychologie*, 1977, *45*, 113–173.

Payne, R. B., Thompson, W. L., Fiala, K. L., and Sweany, L. L. Local song traditions in indigo buntings: Cultural transmission of behavior patterns across generations. *Behaviour*, 1981, *77*, 199–221.

Richards, D. G. Recognition of neighbors by associative learning in rufous-sided towhees. *Auk*, 1979, *96*, 688–693.

Shiovitz, K. A., and Thompson, W. L. Geographic variation in song composition of the indigo bunting, *Passerina cyanea. Animal Behaviour*, 1970, *18*, 151–158.

Smith, D. G., and Reid, F. A. Roles of the song repertoire in red-winged blackbirds. *Behavioral Ecology and Sociobiology*, 1979, *5*, 279–290.

Sokal, R. R., and Rohlf, F. J. *Biometry*. San Francisco: Freeman, 1969.

Sullivan, G. A. Song of the finch *Lagonosticta senegala*: Interspecific mimicry by its brood-parasite *Vidua chalybeata* and the role of song in the host's social context. *Animal Behaviour*, 1976, *24*, 880–888.

Thielcke, G. Geographic variation in bird vocalizations. In R. A. Hinde (Ed.), *Bird vocalizations*. London/New York: Cambridge Univ. Press, 1969.

Thielcke, G. Formen der Programmierung des Vogelgesanges. *Vogelwarte, Sonderheft*, 1977, *29*, 153–159.

Thompson, W. L. Song variation in a population of indigo buntings. *Auk*, 1970, *87*, 58–71.

Thompson, W. L. Singing behavior of the indigo bunting, *Passerina cyanea. Zeitschrift für Tierpsychologie*, 1972, *31*, 39–59.

Thompson, W. L., and Rice, J. O. Calls of the indigo bunting, *Passerina cyanea. Zeitschrift für Tierpsychologie*, 1970, *27*, 37–46.

Thornhill, R. Competitive, charming males and choosy females: Was Darwin correct? *Florida Entomologist*, 1980, *63*, 5–30.

Verner, J. Complex song repertoire of male long-billed marsh wrens in eastern Washington. *Living Bird*, 1976, *14*, 263–300.

Wade, M. J., and Arnold, S. J. The intensity of sexual selection in relation to male sexual behaviour, female choice, and sperm precedence. *Animal Behaviour*, 1980, *28*, 446–461.

Wiley, R. H., and Richards, D. G. Physical constraints on acoustic communication in the atmosphere: Implications for the evolution of animal vocalizations. *Behavioral Ecology and Sociobiology*, 1978, *3*, 69–94.

Williams, G. C. Sex and evolution. *Princeton Monographs in Population Biology*, 1975, *8*.

5

Female Manipulation of Male Avoidance of Cuckoldry Behavior in the Ring Dove*

SUSAN LUMPKIN

I. Introduction

In any species in which males invest more than just their gametes in the offspring of their mates, there is a possibility of males' being cuckolded. This is so because it is to the benefit of a male to inseminate another male's mate if there is a chance that the latter male will invest in the former's young. This is, however, extremely costly to the cuckolded male, who may inadvertently increase the reproductive success of a male competitor and,

*This chapter is based in part on the author's doctoral dissertation, conducted at Duke University with the support of U.S.P.H.S. Research Grant HD04482 and Training Grant T32MH14259-03. During preparation of the chapter I was supported by a Smithsonian Institution Postdoctoral Fellowship.

SOCIAL BEHAVIOR OF
FEMALE VERTEBRATES

ISBN 0-12-735950-8

simultaneously, decrease his own reproductive success. Thus, the reproductive strategies of males in species exhibiting paternal care may include mechanisms to avoid cuckoldry. These behavioral and physiological adaptations to prevent competing males from fertilizing the eggs of their mates have received much attention (e.g., Barash, 1976; Beecher and Beecher, 1979; Power and Doner, 1980). They are extremely interesting because they reveal the complex and subtle effects of the sexual selection pressure of male–male competition on the evolution of male reproductive behavior. Females also play a significant role, however, in shaping the evolution of these male adaptations—a role often ignored despite Parker's (1970, p. 559) warning against treating females "as an inert environment in and around which this form of adaptation evolves." Females are often viewed as passive spectators in this subtle form of combat between males; at best, the passive recipients of indirect benefits derived from the males' behavior, and, at worst, suffering ill effects from it. To take this view, however, is to underestimate female animals, and to ignore the possibility that the reproductive interests of females may differ from those of males. In this chapter I suggest that in at least one species, the ring dove (*Streptopelia risoria*), the females exhibit a pattern of active manipulation of the male doves' mechanisms to avoid cuckoldry. Female ring doves, far from being "inert environments," may actually use this male behavior to their own benefit.

II. Female Determinants of Male Tactics to Avoid Cuckoldry

A. Female Behavior and the Possibility of Cuckoldry

Zenone, Lumpkin, and Erickson (1978) and Zenone (1979) have shown that the male ring dove displays a complex of behavioral adaptations apparently designed to avoid investing in the offspring of another male. This finding is consistent with Trivers' (1972) prediction that when male parental investment is high, males should evolve mechanisms to ensure that the offspring they care for are their own genetic offspring. Because the male ring dove makes a significant contribution to his young in the form of nest-building, incubation of eggs, and feeding of squabs, mechanisms to avoid cuckoldry were expected in this species. High paternal investment, however, is insufficient to predict the particular form of mechanisms that will appear. Rather, the type of mechanism seen will depend largely on the behavioral and physiological characteristics of females.

For example, one of the tactics employed by male ring doves is to behave aggressively toward a potential mate who has been recently courted (Erickson and Zenone, 1976). This male aggression may delay the ovulation of the female so that his sperm, and not that of the previous male, will fertilize the eggs of the female (Zenone, 1979). This type of tactic is necessary for the male to avoid cuckoldry for these reasons: (*a*) female doves can store viable sperm from a previous insemination for up to 7 days before oviposition or 6 days before ovulation (Zenone *et al*, 1979); (*b*) their ovarian development, which leads to ovulation, continues in the absence of the male who initially inseminated them (Lehrman *et al.*, 1961); and (*c*) female doves remain sexually responsive (i.e., willing to copulate) to other males after an initial mating (P. G. Zenone, personal communication, 1979; Lumpkin, 1980). In other words, these female traits greatly increase the necessity for a mechanism of cuckoldry avoidance to be employed by males at mate selection.

After pair formation, another set of tactics may be implemented by the male ring dove—surveillance and guarding (Lumpkin, 1980; Lumpkin *et al.*, 1982). These tactics in birds are analogous to the use by male insects of postcopulatory passive phases to avoid sperm competition (Parker, 1970) and to the copulatory plugs of many reptiles and mammals (e.g., *Thamnophis* spp., Devine, 1975). Males may keep their mates under fairly constant surveillance during the fertile period and guard them from other males in an attempt to detect and forestall the insemination and potential fertilization of the female by a competing male conspecific. Surveillance and guarding as tactics to prevent cuckoldry have been observed in several species of birds (e.g., *Pica pica*, Birkhead, 1979; *Riparia riparia*, Beecher and Beecher, 1979; *Sialia currucoides*, Power and Doner, 1980), but the issue is more complex than is commonly assumed. Such devices are necessary only if females exhibit features making them so, and need be implemented only when the behavior of the female is such that the genetic paternity of her mate's offspring can be threatened. Power and Doner (1980), for instance, noted that mechanisms of cuckoldry avoidance will not appear if females refuse to copulate with intruder males.

Female ring doves display several features that seem to make it necessary for their mates to use guarding and surveillance to avoid cuckoldry. Two of these were noted above. The ability of the female dove to store sperm makes an insemination by another male a threat to the genetic paternity of her mate's offspring, and her continued sexual responsiveness after an initial mating allows for the possibility of her insemination by an intruder. Most significantly, female ring doves will solicit copulation from other males if they are briefly separated from their mates. In one study* in which 26 females were briefly separated from their mates (after 3 or 4 days of being paired) and exposed to another male for 4 min, 7 (26.9%) of the females solicited copulation from the "intruder," and one copulation between a mated female and an intruder was observed (Lumpkin, 1980). Zenone (1979) reported similar observations of mated female doves' soliciting and copulating with intruders when the females' mates were briefly absent.

Given these features of the female dove, sexual responsiveness itself in a mate seems to constitute a threat to the male dove's genetic paternity, since a sexually responsive female is able and willing to allow insemination by a competing male even after copulation with her mate has occurred. To avoid cuckoldry, a male dove must attempt to prevent this by keeping his mate under surveillance and guarding her from other males whenever she displays sexual responsiveness. A female could aid her mate in his attempts to avoid cuckoldry by limiting her sexual responsiveness to a brief period of time. Instead, the female dove appears to have a pattern of prolonged sexual responsiveness that actually increases the time during which her mate may be cuckolded and thus extends the time during which guarding and surveillance must be employed.

Female sexual responsiveness, as evidenced by the display of copulation solicitation, was examined in some detail as part of a larger study of reproductive behavior in ring doves (Lumpkin *et al.*, n.d.). Eight pairs of doves breeding in large (89 cm^3) individual cages were observed continuously during the daylight hours throughout a complete breeding cycle, from first introduction to fledging of young. For all pairs this cycle included a second courtship period in preparation for another clutch. The results indicate that as a consequence of the females' pattern of solicitation, male doves are in danger of cuckoldry during a significant portion of the breeding cycle.

First, we found that females begin to solicit copulation quite soon after pairing, usually on the first day, and continue to do so until their first eggs are laid 6–9 days later. Females resume soliciting shortly after squabs are hatched, which in some cases is as long as 19 days before the second clutch of eggs is laid (Table I). Second, female doves were found to solicit frequently (Table II) and to do so with no highly regular diurnal

*Unless otherwise noted, the doves used in the experiments described here were sexually mature, reproductively experienced (i.e., had mated and successfully reared at least one offspring), and had been visually isolated for 2–4 weeks prior to the beginning of the experiment. In all cases, pair mates were assigned; that is, the birds were given no choice of mates.

TABLE I
Onset of Female Copulation Solicitation

Pair	Day of first solicitation[a]	Day of first oviposition[a]	Day(s) of hatching	Day of first solicitation[b]	Day of first oviposition[b]
1	1	6	21, 22	28	38
2	2	9	25	32	37
3	1	7	22, 23	38	45
4	1	7	23	30	47+[c]
5	1	7	22, 23	37	47+[c]
6	1	6	21, 22	34	40
7	1	7	22, 23	29	47+[c]
8	1	6	22	37	43

[a]First courtship period.
[b]Second courtship period.
[c]These females had not laid their first eggs of the second clutches at the end of observations; however, all subsequently laid within a few days. Precise days are not given since the females were moved to different cages, which might have affected oviposition.

TABLE II
Daily Frequency of Female Copulation Solicitation

	First courtship period			Second courtship period		
		Solicitations per day			Solicitations per day	
Pair	Days[a]	Median	Range	Days[a]	Median	Range
1	6	15.5	8–18	11	12	2–21
2	8	11	4–16	6	9.5	1–12
3	7	5	2–8	8	2.5	0–4
4	7	5	3–9	18	1	0–6
5	7	6	3–9	10	3	1–5
6	6	9.5	6–11	7	3	1–7
7	7	9	2–23	19	8	1–27
8	6	7	6–26	7	2	1–6

[a]Number from first solicitation to first oviposition.

periodicity. Soliciting was seen at all times of day, although it was more common in the afternoon and evening hours. Third, female doves exhibit no significant refractory period after solicitation or copulation. In most females the minimum interval between successive occurrences of either of these was less than 5 min. Thus, there is no time of day during which the female dove will not solicit copulation and could not be inseminated by another male. All of these findings together suggest that from the moment a female mate begins to solicit copulation until she lays her eggs there is little time when there is no chance of her being inseminated by another male. Therefore, there seems to be little time during which a mated male may safely neglect surveillance and guarding to forestall this insemination.

B. Female Sexual Responsiveness and Surveillance by Males

This pattern of female sexual responsiveness should compel a female dove's mate to guard her and to keep her under surveillance during a major part of their breeding cycle. Does the male dove actually do this? A variety of studies concerned with surveillance indicates that he does. Lumpkin *et al.*, (1982) found that male ring doves seem to try to maintain visual contact with their mates by searching and calling for them if they disappear. In this experiment, the responses of males and females to the temporary absence of their partners on the third or fourth day of pairing were compared in 15 pairs of doves. Each pair was breeding in a three-chambered cage (each chamber measured $43 \times 43 \times 15$ cm) designed so that individuals could move through the chambers and in and out of visual contact freely. For the test, one individual was removed for 15 min, and the behavior of the remaining animal was observed. Specifically, the time the remaining animal took to search all three chambers of the cage was measured, and the number of perch-coos, a vocalization thought to be used to maintain contact between the sexes, was counted. The results (Table III) indicate that males respond much more strongly than females to their partners' absence. Males were quicker and more likely to search for the females than the reverse, and males emitted the perch-coo much more often than females.

In another study (Lumpkin *et al.*, 1982), I directly measured the time that individual members of a pair spent in visual contact, which, as shown, is maintained more assiduously by males. The activity of 5 pairs of birds (housed in the cages described previously) was monitored continuously using time-lapse videotape recording, from 3 days before the first oviposition until 3 days after the second egg of the two-egg clutch was laid.

TABLE III
Responses of Males and Females to the Absence of Partners

	Animals (N)	% visiting all compartments	Mean time to reach 3rd compartment	% Perch cooing	Mean number perch-coos (responding animals only)
Males	15	80	209 sec	93	61.1
Females	15	40	360 sec	40	25.0
		$\chi^2 = 3.47$	$U = 51.5$[a]	$\chi^2 = 7.35$	$U = 15$[a]
		$p < .05$	$p < .1$	$p < .005$	$p < .025$

[a]Mann–Whitney *U*-test, one-tailed.

The results showed that prior to egg laying, time in visual contact averaged above 70% of the total time. After egg laying—when there is no longer a threat of cuckoldry to the male—time in visual contact decreased to an average of less than 30% of the time. Also, prior to oviposition, most separations were of 4 sec or less, and on only three occasions were partners out of visual contact for more than 15 min. After oviposition, males and females were often out of contact for several hours at a time.

Although data on female copulation solicitation were not collected in these two studies, all other data indicate that almost certainly the females were soliciting before egg laying and had stopped after oviposition. Thus, males were keeping their mates under surveillance during a high proportion of the time when the females were sexually responsive and were no longer doing so when sexual responsiveness ended with oviposition. A third study (Lumpkin, 1980) was designed to determine the effect of female solicitation on the initiation of surveillance behavior. In this experiment, time in visual contact was measured, using time-lapse video monitoring, for 1 or 2 days after a male and female dove were first introduced, and surveillance behavior was compared before and after the first female copulation solicitation. Five pairs of doves were tested. The results (Table IV) show that prior to a female's first copulation solicitation, the time the male spent in visual contact with her was fairly low, but the time in visual contact increased significantly after the female solicited for the first time. Further, the average duration of periods out of visual contact decreased after the first copulation solicitation, although the differences were not statistically significant. These results support the hypothesis that the display of female sexual responsiveness signals to the mated male the possibility of a threat to the genetic paternity of his offspring. This display and its attendant threat to the mated male apparently then compel him to

TABLE IV
Time in Visual Contact before and after the First Female Solicitation

Pair	Hours of observation		Percentage of time in visual contact		Mean duration of periods out of visual contact[a] (min)		Interval between first solicitation and first oviposition (days)
	Before	After	Before	After	Before	After	
1	6	21	57.5	89.7	14.40	.88	13
2	8	6	47.6	87.6	9.03	1.54	15
3	14	14	28.5	82.2	37.86	6.34	15
4	8	20	57.6	79.6	2.25	1.25	10
5	10	18	75.1	64.4	1.22	1.74	15
Mean			53.26[b]	80.7	12.95	2.35	

[a]Only episodes greater than 7.65 sec (1 sec tape time) were used to compute this mean.
[b]This difference is significant at the $p < .05$ level using the t-test for related measures.

begin keeping his mate under surveillance as soon as the display appears, and to continue to watch her fairly constantly until she lays her eggs.

Studies of other monogamous bird species support the trends observed in the ring doves. In bank swallows, *Riparia riparia,* for example, Beecher and Beecher (1979) found that males pursue their mates on every flight from the burrow for a 7–8-day period after pair formation. They interpreted these chases as mate guarding to avoid cuckoldry and supported this by showing that when other males join the chase, the mated male tries to drive them off, and that males do attempt and sometimes succeed in copulating with mated females. Most relevant here, however, is the time during which these chases occur. Chases begin 3–5 days before the first egg is laid and end by the day the fourth egg is laid, when incubation is begun—a period which probably corresponds to the female's fertile period. Moreover, the onset of chasing is coincident with pair formation, which event Beecher and Beecher believe is triggered by the first copulations. Similarly, Birkhead (1978) reported that mate guarding in magpies (*Pica pica*) was confined to a period beginning 3 days before the first egg is laid and ending with the laying of the penultimate egg of the clutch. He assumed this to be the fertile period and thus the time during which there is a danger of cuckoldry. Unfortunately, since copulations are rarely seen in magpies, it is not known whether this also corresponds to the copulation period. Burger (1981) also found that mate defense in black skinners (*Rynchops niger*) was more pronounced prior to and during egg laying than during incubation and after hatching.

In contrast, Power and Doner (1980) found that in mountain bluebirds (*Sialia currucoides*) males follow and escort their mates starting at pair formation (undefined) and continue until hatching. They suggest that male bluebirds may use hatching to signal the end of the cuckoldry risk because females may begin incubation before their clutches are complete. This creates ambiguity about the end of laying and, therefore, about the termination of the fertile period.

C. Do Females Deceive Males about Their Fertility?

Data collected in the aforementioned study (Lumpkin *et al.*, 1982) also showed that the females began to solicit copulation 10 or more days before their first oviposition (Table IV). Similarly, a previously cited study (Lumpkin *et al.*, n.d.) revealed that in the second courtship period, 4 of 8 females began to solicit more than 10 days before they laid their first eggs (Table I). This was surprising because the fertile period (i.e., the time during which viable sperm from an insemination can be stored for fertilization) of the female dove lasts only about the 7 days before oviposition (ovulation occurs about 40 hr before oviposition) (Zenone *et al.*, 1979). These females were soliciting copulation before any insemination was likely to lead to fertilization! That this is a fairly common occurrence is shown by the observation that of another 26 females tested, 19 (73.09%) solicited copulation within 15 min after being introduced to an unfamiliar male, and of these, 13 (68.4%) did not lay their first eggs until more than 7 days later.

While the precise length of the fertile period of female doves is probably somewhat variable, its maximum duration is unlikely to exceed 7 or 8 days. In the Zenone *et al.* (1979) study, only 1 of 3 females laid a fertile egg 7.5 days after insemination, and her second egg (laid about 40 hr after the first) was infertile. Moreover, 2 females laid infertile eggs 6.5 and 10.5 days after insemination, respectively, and a third female's first egg was fertile when it was laid 6.5 days after insemination, but the second egg was not. These findings confirm those of an early study (Riddle and Behre, 1921) of the fertile period in hybrid ring doves. Riddle and Behre tested a large number of females and found only two cases of fertile eggs laid 8 days after insemination of the females. However, 3 other eggs laid 8 days after insemination were infertile, and an additional 15 eggs laid from 8.7–17 days after insemination were all infertile.

Since females in the experiments described here were soliciting more than 10 days before oviposition, it is possible that female ring doves have moved from manipulation to deceit in trying to increase the duration of male surveillance. If female sexual responsiveness is a signal of fertility at least part of the time, then the female dove could deceptively signal fertility by soliciting during infertile periods. Her mate might then be deceived into premature, and, from his point of view, unnecessary surveillance and guarding. The results of the experiment just described suggest that the male might be so deceived. In all cases for which time in visual contact increased after the first female solicitation, it increased and remained high before sperm from an insemination would be likely to survive to fertilize eggs.

Is such deception possible in the ring dove? Dawkins and Krebs (1978) pointed out that successful deceit depends primarily on two conditions, both of which obtain in the ring dove. First, the occurrence of deceit—in this case, the display of sexual responsiveness during infertility—must constitute a relatively small fraction of genuine behavior, so that it pays the recipient (i.e., the male dove) to react the way he does. This is certainly true in the ring dove. For at least 7 days before egg laying, sexual responsiveness does signal fertility, and it is to the advantage of the male to prevent another male from inseminating a fertile mate.

Second, the male dove must be unable to distinguish between the false and the true signals of fertility. Since the signals (i.e., the copulation solicitation) are the same in both cases, it is difficult to see how a male could accurately discriminate between fertility and infertility. A male dove might use female behaviors that are most frequently performed fewer than 7 days before oviposition, such as nest soliciting or nest building (Cheng, 1973), to determine with certainty whether his mate is fertile. Similarly, Cheng (1973, 1979) has suggested that the behavioral interactions preceding solicitation may change as ovulation approaches. She found that female doves were less likely to engage in allopreening, billing, and begging before soliciting in the 3–4 days before oviposition than earlier during pairing. Although this finding could not be confirmed in our laboratory (Lumpkin *et al.*, n.d.), if there is such a transition, the male dove might use it as an indication of the fertility of his mate. But because these behaviors may not appear until 3 or 4 days before laying—after the female's fertile period has begun—their absence is not a perfect predictor of female infertility. Therefore, according to the conditions imposed by Dawkins and Krebs (1978), successful female deception of males seems possible in ring doves. Furthermore, the data indicate that the deception may be successful—male doves do initiate surveillance prematurely if their mates begin to solicit copulation early.

D. Female Sexual Responsiveness and Male Guarding Behavior

The studies just described deal exclusively with the male ring dove's maintaining surveillance over his mate and indicate that surveillance is employed only when the female's behavior presents a threat to the male's paternity of offspring. However, guarding—the aggressive expulsion of conspecific males from proximity to mates—seems not to follow a similar pattern. Male ring doves do not respond differentially to the presence of another male near their mates, whatever the sexual responsiveness of the female. This was found in an experiment designed to compare the responses to male intruders of male doves mated to unresponsive females (i.e. females who had not begun to solicit copulation) and those of males mated to sexually responsive females (Lumpkin, 1980).* Food-deprived males and their female mates were given access to food in a large cage (89 cm^3) and then confronted by another male. The "intruder" males were

*The birds in this experiment were sexually mature but were not reproductively experienced.

confined within a smaller cage so that mated males were minimally affected by the intruder's behavior; the mated males were food-deprived in order to impose a cost on their responding to the intruder in that it would prevent feeding. During the 10-min confrontation, no significant differences were found between the two groups of males on any measure of male response nor was there a difference between groups in the time that the males spent feeding. In both groups, however, males spent little time feeding (median = 12 of 60 intervals), and much more time in trying to attack the intruder male (median = 26.5 of 60 intervals). Thus, female sexual responsiveness did not affect the degree of mate guarding by males.

It is not really surprising that aggressive exclusion of males is not solely contingent on female sexual responsiveness and the threat of cuckoldry. Male ring doves in reproductive condition are always likely to respond aggressively to the presence of another male. Males may be in competition for resources other than females, such as food and nesting sites. Another male may also steal a male's mate altogether. It is probable that there is no specific "mate-guarding" response that is triggered only by an intruder near a fertile female mate. Instead, the proximity of a rival to any resource may elicit an aggressive response by a male dove. Zenone (1979) found that male doves displayed similar amounts of aggression toward intruder males near their mates and nest sites before and after oviposition (i.e., when there was a danger of cuckoldry and when there was not). She suggested that aggression toward male intruders might serve multiple functions, including, but not confined to, guarding a fertile mate to avoid cuckoldry before oviposition and protecting nest and eggs during incubation. Although not tested here, a male should not respond to an unresponsive female in the same way. An unresponsive female, even if a possible rival for some resource such as food, is also a potential mate and as such should not be aggressively excluded from his proximity. Similarly, Power and Doner (1980) found in mountain bluebirds that though the behaviors most directly related to the avoidance of cuckoldry (e.g., escorting and following) were no longer seen after hatching, a mate's defense of his nest site and territory (and incidentally of his mate) against male (but not female) intruders was independent of the stage of reproduction.

If the presence of a rival near any resource triggers an aggressive response by a male dove, then a female, regardless of her state of fertility or sexual responsiveness, may be relieved by her mate from defending herself (and shared resources) from an intruder. However, the studies of surveillance suggest that only when a female dove is sexually responsive will her mate consistently be available to provide this service. When a male intruder appears, a nonresponsive female may be protected from him if her mate is present; but, by conveying a threat of cuckoldry through her sexual responsiveness, the female dove increases the probability that her mate will be present whenever an intruder arrives. This may be particularly significant in that it occurs at a time when the female is putting a great deal of energy into egg production.

III. Do Females Benefit from Surveillance and Guarding?

Is there any benefit to the female dove in extending the period of surveillance and thereby the period of being guarded by her mate? It has been suggested that one result of male guarding is that mated females are protected from the unwanted courtship of other males (Parker, 1974). For the female dove, such an effect could have an impact on her reproductive success. The courtship of male doves has a large aggressive component: females are usually chased and pecked by courting males, especially on first meeting. While aggressive courtship rarely results in physical injury to the female, it may be costly in terms of wasted time and energy. A female dove must try to escape from the aggressive courtship of the male, or, conversely, participate in sexual interactions with him and risk aggression from her mate as a result (Zenone, 1979). Moreover, solicitation itself does not necessarily reduce the courting male's aggression: male doves may ignore, mount, or even attack a soliciting female. In either case the female will be prevented from carrying out other activities, such as feeding and resting. Over a period of several days before egg laying, this is likely to have a negative effect on the reproductive success of the female. A female frequently disturbed when feeding, for instance, will probably be less well prepared nutritionally to meet the high costs of egg production than a female that did not suffer the disruptive effects of unwanted male courtship.

To test whether being guarded does protect female doves from the disruptive effects of the courtship of other males, 26 mated and food-deprived females were individually given access to food in a large cage and 1 min later exposed to a second male (the intruder) either alone or with their mates present. During the 4-min encounter, aggressive courtship toward the female and female feeding behavior were recorded. Episodes of male feeding behavior and of fighting between mated and intruder males were also noted. In addition, females were tested alone and with their mates, but without the introduction of an intruder, to determine baseline feeding activity. Encounters were only 4 min long because preliminary experiments showed that most food-deprived doves will eat continuously for about 5 min; after 5 min, continued feeding behavior was highly variable between individuals.

The results of this experiment (Table V) showed, first, that females receive less aggressive courtship from intruder males if their mates are present than if they are alone with the intruders. This is the case because mated males almost invariably fight with the intruders, preventing the latter from courting or attacking the females. Second, although an intruder always reduces a female's ability to feed, the negative effect is significantly diminished by her mate's presence, probably because she receives less aggressive courtship from the intruder than does a female alone. The latter hypothesis is supported by the significant inverse correlation ($r = -.60$, $p < .01$) between intervals of female feeding and intervals of aggressive courtship by the intruder. In summary, this experiment indicated that guarded females are the object of less aggression and thus are able to feed more than unguarded females.

This study suggests one way by which female doves may benefit from prolonging the duration of male surveillance and guarding. The extent to which this benefit contributes to the reproductive success of the female, however, will depend on the frequency of encounters with intruders and how costly (in terms of time and energy) these encounters are to an unguarded female. For instance, if female doves are courted by other males only very rarely, then the costs to the females may be negligible, as would the benefits of being constantly guarded. Conversely, a very high rate of encounters with intruders would impose high costs on unguarded females,

TABLE V
Behavior During the Food Access Period[a]

	Female alone	Female and intruder	Female and mate	Female, mate, and intruder	p^b
Female feeding	36	0	39	16	n.s.[c], <.05[d]
Aggressive courtship[e]	—	22	—	4	<.01
Mate feeding	—	—	37	2.5	<.01
Mate–intruder fight	—	—	—	28.5	

[a]Median number of 5-sec intervals out of 48.
[b]Wilcoxin Sign Test.
[c]Comparison of female alone group with female and mate group.
[d]Comparison of female and intruder group with female, mate, and intruder group.
[e]Includes bow-cooing, cackling, chasing and pecking.

and being guarded might significantly increase female reproductive success by offsetting these costs.

In addition, there are other possible benefits that may be associated with the pattern of sexual responsiveness shown by the female ring dove. For example, constant surveillance by her mate may reduce the risks of predation, since the vigilance of a male in attempting to detect intruders should also result in quicker detection of potential predators. A second possible advantage is that forcing a male, especially an unfamiliar one, to display his prowess in aggressive encounters with other males early in pairing may enable a female rapidly to evaluate his future ability to protect offspring and other resources, and then to reject him as a mate if necessary. The female dove's continued sexual responsiveness after copulation makes this a viable tactic. Third, by keeping her mate in constant attendance, a female dove may prevent him from deserting during the prelaying period of the reproductive cycle. Finally, it is also possible that a female dove may use the pattern of sexual responsiveness to secure matings from other males so as to increase the genetic diversity of her offspring (Williams, 1975). This explanation does not, however, account for prefertile-period female solicitation, nor does it explain why a female dove should so blatantly and constantly display her responsiveness to her mate. A female interested in copulating for the purpose of fertilization with a male other than her mate should be highly secretive about her solicitation behavior to prevent her mate from staying with her constantly.

A major problem, however, in evaluating the significance of the conclusions reached here is the fact that the research was conducted in the laboratory on a species whose ecology and natural history are little known. Thus, for instance, without information about breeding densities, breeding synchrony, and adult sex ratios, it is difficult to determine whether "intruder" males are abundant enough in natural populations to pose a significant threat to the reproductive success of either male or female ring doves. The very fact, however, that male doves possess a suite of behavioral responses, all of which in some way seem to reduce the risk of cuckoldry, suggests that at some time in the species' evolutionary history rivals were a threat to the genetic paternity of the offspring of mated males. Similarly, the behavioral and physiological features of the female dove that have been discussed here suggest that at some time it was to her advantage to extend the period during which her mate is in jeopardy of cuckoldry.

Experimental studies of captive animals also have the advantage of bringing into relief behavioral patterns and outcomes that might be difficult to detect in the field. It is possible to create situations that essentially challenge the animal to reveal its behavioral capabilities. Ecological factors, such as breeding density and sex ratios, are likely to differ not only between populations but also from year to year in long-lived species like doves, and the "optimal strategies" of individuals will also change in response to differing conditions. Thus, for example, the benefits discussed here to the female dove of the constant presence of her mate might be evident in the wild only in those populations and/or years in which the breeding demography was such that encounters with intruders were frequent. Moreover, even if these conditions were relatively rare, the fitness of females that received these benefits would be relatively greater than the fitness of females that did not. Experimental studies, then, can reveal the animals' potential for adaptive responses to changing conditions.

Information on the natural history of other Columbids can also be used to buttress the results of studies on ring doves. The family Columbidae, pigeons and doves, is remarkably conservative, and few significant differences have been observed in the social and reproductive behavior of columbid species even though they are found in a wide range of habitats (Goodwin, 1977).

Goodwin (1977), for instance, described behavioral interactions between mated pigeons (*Columba livia*) in natural populations that may support some of the findings reported here. Male pigeons frequently perform a "driving" behavior, which they use to move their mates out of the proximity of other males. It is usually seen when birds are gathered at a common feeding ground. Driving begins a week or more before egg laying and coincides with the advent of female sexual responsiveness. Goodwin also noted that during the driving period male pigeons follow their mates everywhere. Following and driving are probably comparable to the surveillance and guarding behavior of male doves, in that they apparently serve to prevent cuckoldry. Thus, the prolonged period of sexual responsiveness of female pigeons may also extend the period of time during which their mates must employ this behavior. Descriptions of pigeon behavior, however, give no indication of how this might benefit the females.

Dawkins and Krebs (1978, p. 309) maintained that "natural selection favors individuals who successfully manipulate the behavior of other individuals, whether or not this is to the advantage of the manipulated individuals." The studies described here suggest that the female ring dove may successfully manipulate, to her own advantage, the avoidance of cuckoldry behavior of her mate by her pattern of prolonged sexual responsiveness. *It is primarily the females' behavior that both makes cuckoldry a possibility for male doves and determines when and what kind of avoidance tactics the male must employ to prevent it. Given this essential control over their mates, females may use it so that their own best interests are served.* This may not, however, be entirely to the male's disadvantage. Since the reproductive success of a male is largely dependent on that of his mate, any behavior that contributes to the success of the female without imposing an extremely high cost on the male will be beneficial to males as well as females. Thus, for instance, if females produce healthier offspring as a result of improved feeding during the period of egg development, this will be advantageous to their mates. Moreover, a male dove may retaliate if he finds it possible that his mate is participating in his cuckoldry. Zenone (1979) found that a male will severely attack his mate if he finds her with another male prior to egg laying.

IV. Female Sexual Responsiveness in Other Birds

Female ring doves are not unique in displaying prolonged periods of sexual responsiveness prior to egg laying, and the explanation developed here for this phenomenon may apply to some other species. For example, in the purple martin, *Progne subis,* Allen and Nice (1952) identified two contexts in which fighting between males occurs. First, male purple martins defend a small area around their nest boxes by fighting with males who try to enter the area; fighting in this context was seen throughout the breeding season. Second, male martins were observed to fight with other males anywhere their female mates are found. The latter may be analogous to mate guarding in ring doves and other species. Fighting in this context, however, begins only about 2 weeks before the eggs are laid, which is also the time at which copulation begins to occur; that is, when the females become sexually responsive. After egg laying, male–male fighting in this latter context is no longer seen. Like the female ring dove, the female purple martin may be using the early onset of sexual responsiveness to prolong the time during which she is protected from other males by her mate.

Female mallards (*Anas platyrhynchos*) also displayed prolonged periods of sexual responsiveness (Hailman, 1978). Mallard pairs may form months before egg laying, and although mates may change several times before breeding, females are sexually responsive throughout this period. Female mallards are severely harassed by males other than their mates, and in crowded populations, promiscuous rape attempts are common enough to be a significant cause of female mortality during the breeding season (Titman and Lowther, 1975). Males try to protect their mates from the rape attempts of other males although, should a rape occur, the male will, in turn, rape his mate; these behaviors have been interpreted as mechanisms for the avoidance of cuckoldry (McKinney, 1975; Barash, 1978). Clearly, in this dangerous social environment, the female mallard will benefit from protection by her mate against rapists, and her early and prolonged sexual responsiveness may ensure that she has a mate available to do this. Further, it may be to the advantage of the female to test the guarding ability of a potential mate early in the season.

In two other species of ducks, females are also sexually responsive for months prior to egg laying, and it is suggestive that in both cases, observers report a benefit to the females due to the constant attendance of their mates. Dwyer (1974) suggested that male gadwells (*Anas strepera*) protect their mates from other males so that females can forage without disruption. Similarly, Milne (1974) found that the constant attentiveness of the male European eider (*Somateria mollisima*) was a significant factor in allowing his mate to increase her food intake prior to egg laying.

Clearly it is unlikely that a unitary explanation for the evolution of prolonged female sexual responsiveness will apply to all of the avian species in which it is displayed. The females of species with highly diverse mating and social systems have been observed to begin copulating weeks and months before egg laying (Table VI). What is likely, however, is that in all species the pattern of female sexual responsiveness is one that has evolved because of its selective advantage to the females that exhibit it. It is hoped that the ideas discussed here will stimulate further research into this previously ignored facet of avian sexual behavior in particular and contribute to the growing awareness that the behavior of female animals is as complex and as interesting as that of males.

V. Summary

The extent to which the behavior of mated female animals affects both the necessity for and the types of male tactics to avoid cuckoldry has been underestimated. The behavior of the female ring dove seems not only to determine the avoidance of cuckoldry tactics of her mate but also to manipulate these to her own advantage. Such manipulation, which may include deception, is effected by, among other factors, the pattern of prolonged and frequent displays of sexual responsiveness exhibited by the female dove. The female's pattern of sexual responsiveness, combined with her mate's large investment in parental care, makes cuckoldry a risk for him. This forces the male to maintain surveillance over his mate to

TABLE VI
Partial List of Species in Which the Females Display Prolonged Periods of Sexual Responsiveness

Species	Duration of responsiveness prior to first egg[a]	Reference
Ring dove *Streptopelia risoria*	7–19 days	Lumpkin, 1980
Pigeon *Columba livia*	7 or more days	Goodwin, 1977
Hairy woodpecker *Dendrocopus villosus*	Several months	Kilham, 1974
Orange-rumped honey guide *Indicator xanthonatus*	up to 17 days	Cronin and Sherman, 1976
American kestrel *Falco sparverius*	6–7 weeks	Balgooyer, 1976
White-fronted bee-eater *Merops bulockoides*	1–3 months	Emlen, 1978
North Atlantic gannet *Sula bassana bassana*	6–8 weeks	Nelson, 1978
Mountain plover *Eupoda montana*	16–25 days	Graul, 1973
Gentoo penguin *Pygoscelis papua*	77 days	Richdale, 1951
Waved albatross *Diomedea irrorata*	27 days	Harris, 1973
Flightless cormorant *Nannopterum harrisi*	4 weeks	Harris, 1979
Shag *Phalacrocorax aristotelii*	21–40 days	Snow, 1963
Black-headed gull *Larus ridibundus*	30 days	Moynihan, 1955
Common guillemot *Uria aalge*	6 months	Birkhead, 1978
California quail *Lophortyx californica*	30 days	Raitt, 1960
Gadwell *Anas strepera*	Several months	Dwyer, 1974
Mallard *Anas platyrhynchos*	Several months	Hailman, 1978
African black duck *Anas sparsa*	Year-long	McKinney *et al.*, 1978
Purple martin *Progne subis*	3–4 weeks	Allen and Nice, 1952
Song sparrow *Melospiza melodia*	14–18 days	Nice, 1943
Common grackle *Quiscalus quiscalus*	2 weeks	Ficken, 1963

[a]In most cases, this was inferred from data on copulations.

prevent her from being inseminated by another male. By this strategy, the female dove may benefit from the protection her mate provides against the disruptive courtship of conspecific males. The relationship between female sexual responsiveness and male avoidance of cuckoldry behavior in the ring dove may also obtain in other species; this possibility, as well as other factors involved in the evolution of patterns of sexual responsiveness in female birds, should be explored further.

Acknowledgments

I wish to thank Carl J. Erickson and Patricia G. Zenone for their criticism and discussions of this chapter, and James Malcolm, Sam Wasser, and two anonymous reviewers for helpful comments on the manuscript. I also thank Virginia S. Garber for typing the drafts.

References

Allen, R. W., and Nice, M. M. A study of the breeding biology of the purple martin (*Progne subsis*). *American Midland Naturalist*, 1952, *47*, 606–665.

Balgooyer, T. G. Behavior and ecology of the American kestrel (*Falco sparverius* L.) in the Sierra Nevada of California. *University of California Publications in Zoology*, 1976, *103*, 1–83.

Barash, D. P. Male response to apparent female adultery in the mountain bluebird (*Sialia curroides*): An evolutionary interpretation. *American Naturalist*, 1976, *110*, 1097–1100.

Barash, D. P. Rape among mallards. *Science*, 1978, *201*, 282.

Beecher, M. D., and Beecher, I. M. Sociobiology of bank swallows: Reproductive strategy of the male. *Science*, 1979, *205*, 1282–1285.

Birkhead, T. R. Behavioural adaptations to high density nesting in the common guillemot *Uria aalge*. *Animal Behaviour*, 1978, *26*, 321–331.

Birkhead, T. R. Mate guarding in the magpie *Pica pica*. *Animal Behaviour*, 1979, *27*, 866–874.

Burger, J. Sexual differences in parental activities of breeding black skimmers. *American Naturalist*, 1981, *117*, 975–984.

Cheng, M.-F. Effect of estrogen on behavior of ovariectomized ring doves (*Streptopelia risoria*). *Journal of Comparative and Physiological Psychology*, 1973, *83*, 234–239.

Cheng, M.-F. Progress and prospects in ring dove research: A personal view. In J. S. Rosenblatt, R. A. Hinde, C. Beer, & M.-C. Bushnel (eds.), *Advances in the study of behavior* (Vol. 9). New York: Academic Press, 1979.

Cronin, E. W., and Sherman, P. W. A resource-based mating system: The orange-rumped honey guide. *Living Bird*, 1976, *15*, 5–32.

Dawkins, R., and Krebs, J. R. Animal signals: Information or manipulation. In J. R. Krebs and N. B. Davies (Eds.), *Behavioural ecology: An evolutionary approach*. Sunderland, Mass.: Sinauer Associates, 1978.

Devine, M. C. Copulatory plugs in snakes: Enforced chastity. *Science*, 1975, *187*, 844.

Dwyer, T. J. Social behavior of breeding gadwells in North Dakota. *Auk*, 1974, *91*, 375.

Emlen, S. T. The evolution of cooperative breeding in birds. In J. R. Krebs and N. B. Davies

(Eds.), *Behavioural ecology: An evolutionary approach.* Sunderland, Mass.: Sinauer Associates, 1978.

Erickson, C. J., and Zenone, P. G. Courtship differences in male ring doves: Avoidance of cuckoldry?. *Science*, 1976, *192*, 1353–1354.

Ficken, R. W. Courtship and agonistic behavior of the common grackle, *Quiscalus quiscalus. Auk*, 1963, *80*, 52–72.

Goodwin, D. Pigeons and doves of the world. Ithaca, N.Y.: Cornell Univ. Press, 1977.

Graul, W. D. Adaptive aspects of the mountain plover social system. *Living Bird*, 1973, *12*, 69–94.

Hailman, J. P. Rape in mallards. *Science*, 1978, *201*, 280–281.

Harris, M. P. The biology of the waved albatross *Diomedea irrorata* of Hood Island, Galapagos. *Ibis*, 1973, *115*, 483–510.

Harris, M. P. Population dynamics of the flightless cormorant *Nannopterum harrisi. Ibis*, 1979, *121*, 135–146.

Kilham, L. Copulatory behavior of downy woodpeckers. *Wilson Bulletin*, 1974, *86*, 23–34.

Lehrman, D. S., Wortis, R. P., and Brody, P. Gonadotropin secretion in response to external stimuli of varying duration in the ring dove (*Streptopelia risoria*). *Proceedings of the Society for Experimental Biology and Medicine*, 1961, *106*, 298–300.

Lumpkin, S. *Sexual solicitation behavior in the female ring dove* (*Streptopelia risoria*). Unpublished doctoral dissertation, Duke University, 1980.

Lumpkin, S., Kessel, K., Zenone, P. G., and Erickson, C. J. Proximity between the sexes in ring doves: Social bonds or surveillance? *Animal Behaviour*, 1982, *30*, 506–513.

Lumpkin, S., Zenone, P. G., and Erickson, C. J. Diurnal rhythms in the behavior of the ring dove. Manuscript in preparation, n.d.

McKinney, F. The evolution of duck displays. In G. Baerends, C. Beer and A. Manning (Eds.), *Function and evolution in behaviour.* London/New York: Oxford Univ. Press (Clarendon), 1975.

McKinney, F., Siegfried, W. R., Ball, I. J., and Frost, P. G. H. Behavioral specializations for river life in the African black duck (*Anas sparsa* Eyton). *Zeitschrift für Tierpsychologie*, 1978, *48*, 349–400.

Milne, H. Breeding numbers and reproductive rates at the Sands of Forvie National Nature Reserve, Scotland. *Ibis*, 1974, *116*, 135–152.

Moynihan, M. *Some aspects of reproductive behavior in the black-headed gull* (*Larus ridibundus ridibundus L.*), *and related species.* Leiden: E. J. Brill, 1955.

Nelson, J. B. *The Sulidae.* London/New York: Oxford Univ. Press, 1978.

Nice, M. M. Studies in the life history of the song sparrow. II. *Transactions of the Linnaean Society, New York*, 1943, *6*, viii–329.

Parker, G. A. Sperm competition and its evolutionary consequences in the insects. *Biological Review*, 1970, *45*, 525–567.

Parker, G. A. Courtship persistence and female-guarding as male time investment strategies. *Behaviour*, 1974, *48*, 157–184.

Power, H. W., and Doner, C. G. P. Experiments on cuckoldry in the mountain bluebird. *American Naturalist*, 1980, *116*, 689–704.

Raitt, R. J. Breeding biology in a population of California quail. *Condor*, 1960, *62*, 284–292.

Richdale, L. E. *Sexual behavior in penguins.* Lawrence: Univ. of Kansas Press, 1951.

Riddle, O., and Behre, E. H. Studies on the physiology of reproduction in birds. IX. On the relations of stale sperm to fertility and sex in ring-doves. *American Journal of Physiology*, 1921, *57*, 228–249.

Snow, B. K. The behaviour of the shag. *British Birds*, 1963, *56*, 77–103.

Titman, R. D., and Lowther, J. K. The breeding behavior of a crowded population of mallards. *Canadian Journal of Zoology*, 1975, *53*, 1270–1283.

Trivers, R. L. Parental investment and sexual selection. In B. Campbell (Ed.), *Sexual selection and the descent of man 1871–1971*. Chicago: Aldine, 1972.
Williams, G. C. *Sex and evolution*. Princeton, N.J.: Princeton Univ. Press, 1975.
Zenone, P. G. *Protection of genetic paternity in the ring dove* (*Streptopelia risoria*). Unpublished doctoral dissertation, Duke University, 1979.
Zenone, P.G., Lumpkin, S., and Erickson, C. J. Does the male ring dove protect his genetic paternity? Paper presented at the annual meeting of the Animal Behavior Society, Seattle, 1978.
Zenone, P. G., Sims, E. M., and Erickson, C. J. Male ring dove behavior and the defense of genetic paternity. *American Naturalist*, 1979, *114*, 615–626.

6

The Evolution of Polyandry in Shorebirds: An Evaluation of Hypotheses*

W. JAMES ERCKMANN

*My research was supported financially by a Doctoral Dissertation Grant from the National Science Foundation, BNS 76-17667; a grant from the Frank M. Chapman Fund of the American Museum of Natural History; the National Audubon Society; and my personal fortune.

SOCIAL BEHAVIOR OF
FEMALE VERTEBRATES

ISBN 0-12-735950-8

I. Introduction

Although polyandry is a rare mating system, an understanding of its origins is crucial to our understanding of female reproductive strategies in general. In polyandrous birds the usual roles of the sexes are reversed, with males performing virtually all parental care and females competing aggressively for mating opportunities (Jenni, 1974; Ridley, 1978). Polyandry has been discovered in more species of shorebirds (i.e., the sandpipers, plovers, and their relatives) than in all other birds combined. Shorebirds also have a greater diversity of mating systems than any other avian order (Johnsgard, 1981), but, most interestingly, polyandry has evolved in more families and more species of shorebirds than has polygyny. In this chapter I take advantage of this remarkable pattern of reproductive strategies in shorebirds to evaluate a variety of hypotheses for the evolution of polyandry and to consider characteristics of shorebirds that may have led to the unusually frequent evolution of polyandry in this group.

Polyandrous shorebirds not only occur in a wide variety of habitats from the Tropics to the High Arctic, but the precise form that polyandry takes also varies considerably among shorebirds, both with regard to the manner by which females acquire mates and the extent of female participation in parental care. In some species, females monopolize males by controlling breeding resources; in others, females defend males directly (Emlen and Oring, 1977). A number of shorebirds have a unique system of double-clutching that frequently has elements of both polyandry and polygyny (Graul, 1974; Hildén, 1975). In double-clutching species, competition for mates can occur in both sexes, and each female lays two clutches, the first incubated by a male and the second by the female.

Considering the substantial variation in the biology of polyandrous shorebirds, it is not surprising that a variety of models for the evolution of polyandry have been developed by investigators studying species with quite different mating systems and in vastly different environments (Graul, 1974; Jenni, 1974; Graul *et al.*, 1977; Emlen and Oring, 1977; Oring, 1982). In this chapter, I critically analyze the evidence for five major hypotheses, or models, for the evolution of polyandry in shorebirds by evaluating both the underlying assumptions and a number of predictions for each one. I also discuss a number of other factors that may have led to the frequent evolution of polyandry in shorebirds, including environmental variables and biological characteristics of shorebirds themselves. The first four hypotheses I discuss are ecological in nature and differ in several fundamental respects. Two of the four hypotheses are based on the proposition that exclusive parental care by males preceded the evolution of female competition for mates and the evolution of polyandry. A third hypothesis assumes that males must benefit by female emancipation for polyandry to evolve, and the fourth proposes that sex-role reversal can evolve in the absence of polyandry.

In the *replacement clutch hypothesis* it is proposed that, when nesting failure is frequent, monogamous males may benefit by assuming all parental care to allow their mates more time to forage for the production of replacement clutches; this creates conditions that can favor mate desertion and polyandry (Jenni, 1974; Emlen and Oring, 1977; Oring, 1982). The *stressed female hypothesis* argues that extreme food scarcity renders females incapable of sharing incubation after their energy reserves have been depleted during laying. This condition leads to the assumption of all parental care by males and the evolution of sex-role reversal without polyandry (Graul *et al.*, 1977).

The *differential parental capacity hypothesis* argues that egg laying may not stress females so much that they cannot participate in incubation but may leave them less able to assume *all* parental care if deserted. Under some environmental conditions males, but not females, might be able to assume all parental care, and males could be subject to desertion (Maynard Smith, 1977; Erckmann, 1981). The *fluctuating food hypothesis* (Parmelee and Payne, 1973; Graul, 1974; Graul *et al.*, 1977), proposes that females evolved multiple-clutch reproductive strategies to capitalize on temporary superabundance of food in highly variable environments. The *stepping-stone hypothesis* (Pienkowski and Greenwood, 1979; Ridley, 1980) is a nonecological model postulating that true polyandry and sex-role reversal in shorebirds evolved from monogamy via the intermediate condition of double-clutching, a mating system in which males have already evolved the tendency to rear broods without female assistance.

The following analysis of these five hypotheses is motivated by my recent experimental studies of arctic shorebirds, which provide critical tests of key elements of the first three hypotheses identified above (Erckmann, 1981). I have also reviewed the rapidly accumulating shorebird literature to provide more comprehensive tests of all five hypotheses. From the analysis I draw a number of general conclusions:

1. No single model is sufficient to explain all cases of polyandry in shorebirds, but several models in modified form may apply under particular circumstances.

2. Polyandry evolved to the benefit of females, and males need not have benefited from female emancipation for polyandry to evolve.

3. Sex-role reversal and true polyandry probably evolved from a condition of monogamy and biparental incubation at one nest rather than from double-clutching.

4. Shorebirds have several biological characteristics that made the evolution of polyandry probable. Most important among these are the tendency of monogamous males to take a large share in all aspects of parental care, the ability of a single parent to hatch and rear a brood successfully when foraging conditions are good, the ability of females to lay successive clutches very rapidly because clutch size is small, and the tendency for females to be larger than males.

II. A Summary of Multiple-Clutch Mating Systems in Shorebirds

A. Biology and Taxonomy of Shorebirds

The shorebirds, a suborder (Charadrii) of the Charadriiformes, consist of 14 families comprising about 200 species. Most species are migratory. The majority of species breed in the North Temperate Zone and Arctic (Johnsgard, 1981), but a number of relatively sedentary species breed in the Tropics and in the South Temperate Zone.

Most species are in two families, the sandpipers (Scolopacidae: 84 species) and the plovers (Charadriidae: 61 species). The highly migratory Scolopacidae have the greatest diversity in feeding morphology and mating systems. Small invertebrates comprise the primary food of most shorebirds, and most species are essentially insectivorous when breeding (Johnsgard, 1981).

The young of all shorebirds are precocial and in about 85% of species feed themselves from hatching. Average clutch size decreases from the Artic to the Tropics, and no species lay more than the four-egg clutch typical of all arctic species (Maclean, 1972)—a limit probably determined by the ability of incubating adults to warm the clutch effectively (Hills, 1980). The primary components of parental care are incubation (3–5 weeks), brooding of small chicks to prevent hypothermia (1–5 weeks), leading young to feeding areas (parental feeding in some species), and protecting eggs and young from predators. The young usually become independent of their parents when they can fly, at 2–5 weeks of age.

Polyandry occurs in five families of shorebirds, double-clutching in two, and polygyny in only one (Table I). In all polygynous species, only

TABLE I
Synopsis of Mating Systems and Parental Care in Shorebirds

Family	Number of species	Mating system and parental care[a]				
		Monogamy	Polygyny	Double-clutch	Polyandry	Cooperative
Jacanidae	8	1? [B/B]			5–7 [M/M]	
Rostratulidae	2	1? [B/?]			1 [M/M]	
Dromadidae	1	1 [B/B+]				
Haematopodidae	6	all [B/B+]				
Ibidorhynchidae	1	1 [B/B]				
Recurvirostridae	6	all [B/B]				
Burhinidae	9	all? [B/B(+)]				
Glareolidae	17	all [B/B(+)]				
Charadriidae	61	>50 [B/B]		1 [MF/MF]	1 [M(F)/M]	1 [B/B]
Scolopacidae	84	>50 [B/M or B]	9–15 [F/F]	3 [MF/MF]	4–6 [M or M(F)/M]	
Thinocoridae	4	all [F/B]				
Chionididae	2	all [B/B+]				
Pluvianellidae	1	1 [B/B+]				
Pedionomidae	1				1 [M/M]	

[a]Parental care shown in brackets, incubation before the slash, care of young after. M, male only; F, female only; B, both sexes; MF, male one brood, female one brood; M(F), male, sometimes aided by female; +, parents feed young; (+), parents of same species feed young.

females perform parental care; males compete for females on leks, nesting territories, or in display groups (Pitelka *et al.*, 1974; Johnsgard, 1981). More than ¾ of all shorebirds have monogamous mating systems, which is believed to be the ancestral condition from which polyandrous and polygynous species evolved (Jenni, 1974; Pitelka *et al.*, 1974). In nearly all monogamous shorebirds, both parents share in incubation; the young are usually reared by both parents, but in many cases by the male only.

B. Multiple-Clutch Mating Systems in Shorebirds

Two types of multiple-clutch breeding strategies can be identified in the shorebirds, *true polyandry* and *double-clutching*. True polyandry is associated with sex-role reversal and is characterized by four features: (*a*) exclusive parental care by some or all males; (*b*) little or no parental care by females; (*c*) competition for mates primarily among females, which are considerably larger and in some cases more vividly colored than males; and (*d*) greater variance in annual gametic contribution among females than males, with a greater average gametic contribution per breeding individual in females than males.

Polyandry may also occur in double-clutching species, but the following features distinguish double-clutching from true polyandry. First, the average annual gametic contribution per breeding individual is about the same for males and females. Second, although females occasionally compete for males, most intrasexual competition occurs among males, and only males defend nesting territories. Sexual dimorphism in double-clutching species is similar to that in closely related monogamous shorebirds. Third, the contribution of parental care by the sexes is nearly the same, as each bird incubates at one nest. In no shorebirds with true polyandry do females incubate alone at a nest. Finally, mate switching between clutches (polygamy) is not a consistent feature among double-clutching species (Kochanow, 1973; Graul, 1974).

1. *True Polyandry*

a. Types of Polyandry

Polyandrous shorebirds may be categorized by the nature of the pair bond or by the manner in which females acquire mates (Jenni, 1974; Emlem and Oring, 1977). Pair bonds in some species are brief, and females desert one male to mate with another (*sequential polyandry*). In others, females may have several mates at the same time, and pair bonds may be prolonged (*simultaneous polyandry*). Females may acquire mates either by controlling breeding territories (*resource defense polyandry*) or by defending males directly (*female access polyandry*). The literature on polyandry in birds has been reviewed elsewhere (Jenni, 1974; Emlen and Oring, 1977; Ridley, 1978; Wittenberger, 1979, 1981; Faaborg and Patterson, 1981; Oring, 1982). Table II summarizes the major features of the breeding systems of shorebirds in which polyandry is known to occur.

b. Polyandrous Shorebirds

The phalaropes (*Phalaropus* spp.) and the dotterel (*Charadrius morinellus*) have female access and sequential polyandry, and females court males in groups of variable size. Polyandry in these species is uncommon, and females have no more than two mates in a season (Hildén and Vuolanto, 1972; Nethersole-Thompson, 1973; Howe, 1975a; Schamel and Tracy, 1977). Female phalaropes perform no parental care, but some female dotterels share in incubation (Pulliainen, 1970).

Jacanas (Jacanidae), the spotted sandpiper (*Actitis macularia*), and possibly the old-world painted snipe (*Rostratula benghalensis*) all have resource defense polyandry. Female jacanas may have as many as three, and possibly more, mates in a season. Polyandry in jacanas is common and may be simultaneous or sequential (Hoffmann, 1949; Jenni, 1974). In spotted sandpipers, polyandry is common and usually sequential (Hays, 1972; Oring and Maxson, 1978; Oring, 1982), and females may have as many as three mates in a season. Female spotted sandpipers sometimes assist their final mates in incubation (Hays, 1972), but females take a smaller share in incubation than males (Maxson and Oring, 1980). Female jacanas contribute little or no parental care and do not incubate (Jenni and Betts, 1978). In both jacanas and spotted sandpipers, males usually defend small nesting territories.

Details of the breeding biology of other known polyandrous shorebirds listed in Table II are too poorly known to discuss. Polyandry and sex-role reversal may also be present in two tringine sandpipers other than the spotted sandpiper (Oring and Knudson, 1972; Raner, 1972), in two other jacanas (Johnsgard, 1981), and possibly in the South American painted snipe (*Nycticryphes semicollaris*).

2. *Double-Clutching*

Double-clutching is known to occur in three calidrine sandpipers and one plover (Table II), and it is likely that a few other species in these families have similar mating systems. In all double-clutching species, each

TABLE II
Shorebirds with Multiple-Clutch Breeding Systems

Species	Mating system[a]	Breeding		References[b]
		Latitude	Habitat	
Scolopacidae				
Wilson's phalarope	FAP	Temperate	Small ponds	7, 9
Northern phalarope	FAP	Arctic–Subarctic	Tundra ponds	5
Red phalarope	FAP	Arctic	Tundra ponds	11, 17
Spotted sandpiper	RDP	Temperate	Ponds, rivers	3, 14
Sanderling	DC	High Arctic	Tundra	16
Temminck's stint	DC	Subarctic	Tundra	4
Little stint	DC	Subarctic	Tundra	4
Charadriidae				
Dotterel	FAP	Subarctic–Arctic	Montane tundra	12
Mountain plover	DC	Temperate	Short-grass prairie	2
Rostratulidae				
Painted snipe (old-world)	RDP(?)	Tropics–Subtropics	Swamps, marshes	1, 12
Jacanidae				
Northern jacana	RDP	Tropics–Subtropics	Swamps, marshes	8
Wattled jacana	RDP	Tropics–Subtropics	Swamps, marshes	15
African jacana	RDP	Tropics–Subtropics	Swamps, marshes	18
Bronze-winged jacana	RDP	Tropics–Subtropics	Swamps, marshes	10
Pheasant-tailed jacana	RDP	Tropics–Subtropics	Swamps, marshes	6
Pedionomidae				
Plains wanderer	?	Subtropics	Open plains	13

[a]FAP, female access polyandry; RDP, resource defense polyandry; DC, double-clutch.

[b]1, Ali and Ripley (1969); 2, Graul (1974); 3, Hays (1972); 4, Hildén (1975, 1978a); 5, Hildén and Vuolanto (1972); 6, Hoffmann (1949); 7, Howe (1975a, 1975b); 8, Jenni (1974); 9, Kagarise (1979); 10, Mathew (1964); 11, Mayfield (1978); 12, Nethersole-Thompson (1973); 13, Olson and Steadman (1981); 14, Oring and Knudson (1972); 15, Osborne and Bourne (1977); 15, Parmelee and Payne (1973); 17, Schamel and Tracy (1977); 18, Vernon (1973).

female lays one clutch that is tended by a male, followed by a second that she tends herself. In some, but not all, cases females switch mates between clutches (Kochanow, 1973; Graul, 1974; Hildén, 1975). Males compete for mates and defend nesting territories. Competition among females occurs less frequently (Graul, 1974; Hildén, 1975), but females of two species perform aerial song displays (Parmelee, 1970; Hildén, 1978a).

III. A Critical Evaluation of Hypotheses for the Evolution of Polyandry in Shorebirds

Models for the evolution of polyandry should be viewed in the general context of polygamous strategies. Because all polygamous mating systems in shorebirds evolved from monogamy (Jenni, 1974; Pitelka *et al.*, 1974), the constraints of selection for monogamous strategies in both sexes must have been relaxed in species that are now polygamous. These constraints can be grouped into two categories (Lack, 1968; Emlen and Oring, 1977; Wittenberger and Tilson, 1980): (*a*) the necessity or advantage of having both parents share in care of eggs and/or young, and (*b*) limitations on the potential for and ability of individuals of one sex to monopolize individuals of the other. Models for the evolution of polyandry must, in addition, provide solutions to two specific problems. First, if one parent is capable of incubating and rearing a brood unassisted, when should selection favor mate desertion by and/or polygamy among females rather than males? Second, under what conditions should a male accept exclusive parental care if his mate terminates parental care or deserts? Whether polyandry or polygyny evolves will depend on the options open to individuals of both sexes, which will be determined by characteristics of both the environment and the species in question.

When monogamous males both defend nesting territories and share in all aspects of parental care, as in shorebirds, females can potentially acquire additional mates by deserting after the first clutch is completed and then competing for males on other territories. Females that desert may also benefit by reducing reproductive effort to increase the chance of survival in the event additional males are not available, and sex-role reversal can potentially evolve either when the chances of polyandry are small, or even in the absence of polyandry.

A. The Stressed Female Hypothesis for the Evolution of Sex-Role Reversal

Graul *et al.*(1977) argued that when food is extremely scarce the energy reserves of females might be so depleted following egg laying that further parental investment would entail a substantially increased risk of mortality. Under such conditions, females should desert their mates. Males should then assume all parental care, as long as the physiological cost of uniparental care is not prohibitively high. Male parental care should eventually lead to competition for mates among females. The *stressed female hypothesis* is thus a model for the evolution of sex-role reversal without polyandry. Polyandry would only be possible when food availability subsequently increased, allowing females to recover reserves rapidly to lay more than one clutch (Graul *et al.*, 1977; discussed in Section III, D).

In the *stressed female hypothesis*, food scarcity is the critical environmental variable that starts the sequence of selection for male incubation, sex-role reversal, and, in some cases, polyandry. The hypothesis assumes that food scarcity precludes female incubation yet still allows males to perform all parental care, This implies that uniparental incubation by males of monogamous species should be feasible. Since, according to this hypothesis, females both lay eggs *and* share in incubation, monogamous species must experience more favorable feeding conditions than the conditions selecting for desertion by females.

1. *Prediction: Males of monogamous shorebirds should be able to incubate unassisted if deserted.* I tested this prediction with an Alaskan population of western sandpipers (*Calidris mauri*), a monogamous species in which incubation is shared by both parents, but the young are reared almost exclusively by the male parent (Erckmann, 1981). Twelve males were experimentally deserted during incubation by the removal of their mates. Experimentally deserted males that continued incubating lost a significant amount of body weight, while paired control males lost no weight. All experimental males abandoned their nests before any eggs hatched, although some persisted as long as 15 days after experimental desertion. Because they were found to incubate more and forage less than control males, experimental males were under energetic stress. The males in the best initial physiological condition, as measured by a ratio of body weight to bill length, persisted longest, but no males were able to incubate unassisted for a full incubation period.

In trying to compensate for the loss of their mates, experimentally deserted male western sandpipers were unable to maintain a nonnegative daily energy balance, even though they spent an average of only 61% of each day incubating compared to 51% for males of control pairs and a 92% combined nest attendance for control pairs. In the western sandpiper, food availability appears to preclude uniparental incubation by males.

There are few comparable data for other monogamous shorebirds that are useful for testing this prediction. Failure of male sandpipers to hatch clutches when their mates disappeared has been reported (Jehl, 1973). However, males of several monogamous plovers have been reported to hatch clutches after their mates had deserted them part way through incubation; successful clutches of deserted male Wilson's plovers (*Charadrius wilsonia*) were incubated an average of 64% of each day compared to 76% for nests of pairs (Bergstrom, 1981). No body weight data were taken in this study but these observations suggest that male incubation may be feasible, at least for part of the incubation period, in some monogamous shorebirds. Whether uniparental incubation in such species is physiologically costly or whether uniparental incubators are less successful than pairs needs to be determined.

2. *Prediction: Species characterized by sex-role reversal and infrequent polyandry should experience poorer feeding conditions than monogamous species.* Among shorebirds with sex-role reversal, polyandry is least frequent in phalaropes and dotterels (Ridlye, 1978). For example, only 8% of female northern phalaropes (*Phalaropus lobatus*) were biandrous in a seven-year study (Hildén and Vuolanto, 1972). Yet time- and energy-budget studies indicate that the feeding conditions experienced by phalaropes and dotterels during the breeding season in the Arctic are relatively *better* than conditions experienced by monogamous arctic shorebirds. In Alaska, I found nest attentiveness (percentage of day spent incubating) of male red phalaropes (*P. fulicarius*) (72%) and northern phalaropes (77%) to be significantly higher than attentiveness in experimentally deserted western sandpipers (61%), yet male phalaropes lost no body weight during incubation (Erckmann, 1981). Nest attentiveness appears to be even higher for male dotterels than for male phalaropes (Nethersole-Thompson, 1973; Wilkie, 1981).

Other observations of nest attentiveness in arctic and temperate shorebirds with sex-role reversal and exclusive male incubation are comparably high (Rubinstein, 1973; Maxson and Oring, 1980), and time-budget studies have shown that the percentage of time that incubating males spend foraging is generally less in polyandrous shorebirds with uniparental incubation than in monogamous species with biparental incubation (Table III). These observations do not indicate that any shorebirds with sex-role reversal, including those in which polyandry is least frequent, experience relatively poor feeding conditions as predicted by the *stressed female hypothesis.*

TABLE III

Percentage of Day Spent Foraging by Incubating Males of Polyandrous Shorebirds with Uniparental Incubation and Males of Monogamous Species with Biparental Incubation

Species	Breeding latitude[a]	% of day male spent foraging	References
Polyandrous			
Red phalarope	A	21	Erckmann, 1981
Northern phalarope	A	17	Erckmann, 1981
Spotted sandpiper	t	15	Maxson and Oring, 1980
Bronze-winged jacana	T	21	Mathew, 1964
Monogamous			
Western sandpiper	A	45	Erckmann, 1981
Semipalmated sandpiper	A	35	Ashkenazie and Safriel, 1979
American avocet	t	24	Gibson, 1978

[a]A, arctic; t, Temperate; T, Tropics.

3. *Conclusions.* Although the *stressed female hypothesis* provides a plausible reason for female desertion and exclusive male parental care, the hypothesis does not address the question of how uniparental incubation by males is possible under conditions of food scarcity. Food scarcity should affect the energy budgets of both sexes. Experimental evidence indicates that the feeding conditions experienced by some monogamous shorebirds are inadequate for males to meet nutritional needs while incubating unassisted, even when females are able to take a large share in incubation (Erckmann, 1981). It follows that if feeding conditions were so poor that females could spend *no* time incubating because they needed to forage all the time to recover energy lost during laying, then the energetic stress on males incubating alone would be even greater.

Regardless of the possibility that uniparental incubation by males may be energetically feasible in some monogamous shorebirds, the demonstration that food scarcity renders uniparental incubation by males infeasible in other shorebirds strongly indicates that food scarcity generally selects for biparental incubation. Thus, extreme food scarcity can select for desertion by females only if nest attentiveness requirements are particularly low. Among polyandrous shorebirds, measured nest attentiveness is lowest for the tropical jacanas (Table IV); yet male bronze-winged jacanas (*Metopidius indicus*) apparently need to forage only about a third of their off-duty time (Table III), indicating that food availability is relatively good. Attentiveness requirements may be low in jacanas because of the contribution of solar heating to incubation (Jenni and Betts, 1978) and the absence of desiccation problems in the humid tropical environment. The relatively

TABLE IV
A Comparison of Nest Attentiveness among Polyandrous Shorebirds with Uniparental Incubation by Males

Species	Average nest attentiveness[a]	References
Red phalarope	72	Erckmann, 1981
Northern phalarope	77–85	Erckmann, 1981; Rubinstein (1973)
Dotterel	87	Wilkie, 1981
American jacana	49	Jenni and Betts, 1978
Bronze-winged jacana	35	Mathew, 1964
Pheasant-tailed jacana	61	Hoffmann, 1949

[a]Percentage of day spent incubating.

high attentiveness values and relatively low foraging requirements of other polyandrous shorebirds (Tables III and IV) indicate that most, if not all, shorebirds with sex-role reversal experience relatively good foraging conditions rather than food scarcity as required by the *stressed female hypothesis*.

Evidence reviewed in Section III, B, also suggests both that females do not need to forage continuously after laying eggs to recover energy losses and that whenever feeding conditions are adequate to support uniparental incubation by males, egg laying cannot stress female shorebirds to the degree that participation in incubation is precluded. Finally, it does not appear that sex-role reversal evolved in any shorebird in the absence of polyandry. Although polyandry may be infrequent in several species, there is no evidence that food scarcity generally prevents females from laying successive clutches. I conclude that food scarcity cannot have selected for female desertion in shorebirds and reject the *stressed female hypothesis* for the evolution of sex-role reversal in shorebirds. Food scarcity appears to select for biparental incubation rather than exclusive male parental care.

B. The Differential Parental Capacity Hypothesis for the Evolution of Polyandry

Although it is unlikely that the stress of egg laying prohibits female participation in incubation as argued in the *stressed female hypothesis*, laying may still constitute a stress on females that can constrain their future reproductive options (Graul, 1974; Maynard Smith, 1977; Drent and Daan,

1980; Erckmann, 1981). When laying is more costly than the preincubation activities of males, females should be in poorer physiological condition than males. Though able to participate in incubation, females may be less able than males to *increase* investment in parental care if deserted. Thus, under some environmental conditions, males may have a greater chance than females of successful reproduction as single parents; females may be more likely to abandon their nests if deserted and may be at greater risk of mortality if they remain with their clutch.

The *differential parental capacity hypothesis* thus assumes that better feeding conditions are required to sustain uniparental care, especially incubation, by females than by males. Given that relatively poor feeding conditions favor biparental incubation and monogamy as argued above, then somewhat better feeding conditions could favor desertion by females but not males. Females deserting or terminating parental care could profit by seeking other males and/or by leaving the breeding grounds early, as long as males were willing and able to rear a brood unassisted. In contrast, males would not profit by deserting under such conditions, since females would be unable to rear a brood unassisted and would be likely to abandon their nest in response to desertion. There would be no advantage to a male that abandoned in response to desertion, as long as feeding conditions were adequate to meet nutritional needs while incubating, and as long as the male could successfully rear his brood unassisted.

1. *Prediction: Females of monogamous shorebirds should be less able than males to incubate unassisted if deserted.* I tested this prediction experimentally by collecting an equal number of male and female western sandpipers (12 each) during incubation (Erckmann, 1981). All experimentally deserted sandpipers of both sexes abandoned their nests before any eggs hatched, and several results suggest that females were under more stress than males: (*a*) the mean time from experimental desertion to nest abandonment was significantly shorter for females (4.2 days) than males (7.8 days); (*b*) initial physiological condition, as measured by the ratio of body weight to bill length, was a much better predictor of persistence for females than males; (*c*) although mean nest attentiveness was the same for experimental males and females, females lost weight more rapidly than males; and (*d*) the increase in nest attentiveness between experimental birds and paired controls was greater for females (20%) than males (10%), and deserted females experienced a greater reduction in time spent foraging.

The observation that female seed snipe (Maclean, 1969) and females of several monogamous plovers regularly perform nearly all incubation (Rubinstein, 1973; Phillips, 1980) indicates that both females and males of some monogamous species may be able to incubate alone and may be subject to desertion. Indeed, both occasional simultaneous polygyny and sequential polyandry have been reported in otherwise monogamous shorebirds (Wilson, 1967; Nethersole-Thompson, 1979; Bergstrom, 1981). These observations, however, do not preclude the possibility that females of some monogamous species may be less able than males to incubate unassisted, as indicated by my experimental results with western sandpipers. Rather, it seems probable that factors other than the inability of single parents to hatch a clutch may select against mate desertion and polygamy in some monogamous shorebirds. For example, individuals who desert may have difficulty in acquiring second mates (Wittenberger and Tilson, 1980), and in larger species of shorebirds, in which parents actively and more effectively defend their young, the presence of both parents may be more critical to successful protection of young (Erckmann, n.d.)

2. *Prediction: Females of monogamous species should have higher cumulative reproductive effort than males at the onset of incubation.* Reproductive effort is defined as investment that negatively influences the ability of individuals to make further reproductive investment (Trivers, 1972). Therefore, to show that the reproductive effort of females exceeds that of males, it is desirable to show not only that the energy expenditure of females is the greater, but also that this difference in expenditure results in a differential effect on physiological capacity. One indication of a reduced physiological capacity in females would be a depletion of energy or nutrient reserves as a consequence of laying (Graul *et al.*, 1977; Wittenberger, 1981).

Time- and energy-budget studies of semipalmated sandpipers (*Calidris pusilla*) and western sandpipers (Ashkenazie and Safriel, 1979; Erckmann, 1981) have shown that the maximum daily energy expenditure for females during egg laying is considerably (approximately 40%) greater than the maximum expenditure for males defending territories. The estimated sex difference in preincubation energy expenditure is even larger in spotted sandpipers (Maxson and Oring, 1980).

In spite of this empirical difference in energy *expenditure*, a review of available evidence indicates that laying does not produce large reductions in energy *reserves* in shorebirds, at least among arctic species (Erckmann, 1981). The erroneous conclusion reached by Ashkenazie and Safriel (1979) that female semipalmated sandpipers experience a huge energy deficit during breeding stems from an error in their calculations; the postulated energy deficit would result in an impossible loss of several times a female's initial body weight. A review of data on body weight and composition for both monogamous sandpipers and polyandrous phalaropes indicates that the weight of females after laying is rarely much less than weight just prior to the onset of rapid follicular growth, and that laying itself has little effect on levels of body lipids or protein (Holmes,

1966, 1972; MacLean, 1969; Yarbrough, 1970; Erckmann, 1981). Apparently females acquire energy and nutrients for laying primarily during the laying period and do not deplete nutrient reserves appreciably during laying. Some monogamous female sandpipers may have minimal lipid reserves at the completion of laying, but lipid reserves of males may be equally low (Yarbrough, 1970). It is possible that females may be more sensitive than males to stresses (such as food shortage) during incubation if they are less able to mobilize nutrient reserves, but I know of no data that can be used to evaluate this possibility.

3. *Discussion and Conclusions.* The results of tests of two predictions from the *differential parental capacity hypothesis* are conflicting. Experiments with western sandpipers provide support for the prediction that monogamous females should more readily abandon their nests if deserted than males. But, although the energetic expenditure for egg laying substantially exceeds the expenditure for territorial defense by males, the available evidence for both polyandrous and monogamous shorebirds suggests that females may not deplete body protein or lipids to any appreciable extent as a result of laying. If this is true, any reduction in a female's ability to withstand stress after laying must therefore stem from a more subtle degradation of physiological capacity, possibly from a reduced ability to mobilize body nutrients in response to stresses, such as temporary food shortage, or from a deficiency of specific nutrients, such as calcium (MacLean, 1974). The evidence linking egg laying to a decrease in the resistance to stress and an increase in mortality is limited to studies of precocial game birds in which, unlike the apparent situation in shorebirds, females heavily utilize protein and lipids stored in body organs, both for laying and during incubation (Ricklefs, 1974).

A particularly important test of the *differential parental capacity hypothesis* would be experimental comparisons of male and female ability to incubate unassisted for polyandrous species, such as the spotted sandpiper or the dotterel in which females regularly share in incubation. There are indications that females of both these species may accept sole care of a clutch if their mates disappear (Nethersole-Thompson, 1973; Oring, 1982), but there are no data on comparative performance of the sexes as single parents.

The difference in male and female capacity to be single parents is probably small. Since females can avoid some of the negative physiological consequences of uniparental incubation by reducing nest attentiveness to increase foraging time, the *differential parental capacity hypothesis* is most likely to apply in environments in which incremental reductions in nest attentiveness or time spent tending young may have the greatest negative effect on breeding success. The model might also apply in environments where periods of temporary stress, such as food shortage, occur during incubation. Nest attentiveness requirements are probably least stringent and food most dependable on a daily basis in wet tropical areas, but in the Arctic and in hot, dry habitats, adequate attendance of eggs may be more critical and food supply more erratic (Maclean, 1967; Norton, 1972; Hildén, 1979b). It therefore seems unlikely that the model can explain the evolution of polyandry in tropical jacanas or painted snipe, but the potentially superior ability of males as single parents could have contributed to selection for female desertion and sex-role reversal in arctic dotterels and phalaropes. However, the rigors of the arctic climate have not precluded the evolution of exclusive incubation by females. Of 24 species in the arctic sandpiper subfamily Calidridinae, which includes the western sandpiper, females incubate unassisted in nine species (Pitelka *et al.*, 1974). That sex-role reversal may have evolved in one of these species at most (Flint, 1973) suggests that the differential parental capacity of the sexes may not have been of overriding importance in the evolution of reproductive strategies, even in arctic shorebirds. Furthermore, the annual mortality of double-clutching female Temminck's stints is no higher than female mortality in related species (Soikkeli, 1970; Hildén, 1978b), indicating that the cost of uniparental incubation (even after multiple clutches) can be small.

It may be significant that in the arctic-breeding and polygynous pectoral sandpiper (*Calidris melanotos*), females carry unusual amounts of fat during incubation and apparently nest only when body fat levels are high (MacLean, 1969). Whether this applies to other polygynous calidridines is unknown, but in the polygynous american woodcock (*Scolopax minor*), females have relatively high body weight while incubating and caring for young (Sheldon, 1967). The possession of such nutrient reserves could very well mitigate the effects of temporary food shortage for deserted females.

C. The Replacement Clutch Hypothesis for the Evolution of Polyandry

Opportunities for replacing clutches destroyed by predators or other environmental forces can influence the optimal division of parental responsibilities in monogamous species. As the frequency of nesting failure increases, the ability of females to lay replacement clutches takes on increasing importance for both members of a pair. Insofar as involvement in incubation reduces female ability to produce eggs by restricting available foraging time, frequent nesting failure might select for emancipation of females as long as foraging conditions were adequate for uniparental incubation by males (Jenni, 1974; Emlen and Oring, 1977). Once males began performing most or all incubation, females might then maximize reproductive success not only by replacing clutches but also by competing to lay for other available males, and polyandry could evolve. The *replacement clutch hypothesis* asumes that males should benefit by female emancipation unless deserted by their mates.

1. *Prediction: Polyandrous species should experience relatively high rates of replaceable clutch loss.* I compared nesting failure rates between polyandrous shorebirds and other species using data reported in the literature, and included only studies giving data for a minimum of 10 nests for a species. I combined data reported for several years at one location but considered data from different locations and studies as separate samples (i.e., as location samples). Average rates of both overall nesting success and replaceable nesting failure were computed for each species by averaging all available location samples. Replaceable nesting failure was defined to include losses to predators or weather-related causes (e.g., flooding, hail damage, freezing) and exclude losses due to human disturbance, infertility, and nest abandonment for unknown reasons, since the latter often do not result in renesting. The following results, expressed as percentage of nests producing or not producing any young, are based on a preliminary analysis of data from more than 70 studies and 53 species to be reported in more detail elsewhere. Several studies (Kagarise, 1979; Howe, 1982) were excluded because the authors felt that their activities may have caused excessive nest predation. Inclusion of these studies does not affect the statistical conclusions drawn.

A comparison of polyandrous species with other shorebirds by *t*-test revealed no significant differences in either mean nesting success or mean rate of replaceable nest loss (Table V). I also analyzed the data for a latitudinal trend by linear regression. As expected from other studies (Ricklefs, 1969; Skutch, 1976; Koenig, 1982), nesting success was found to increase and replaceable nest losses correspondingly to decrease with latitude (Fig. 1). In both cases the trends are weak but significant. Species means for polyandrous and double-clutch species do not differ systematically from values predicted by the regressions (Fig. 1). Results of studies of arctic shorebird communities (Mickelson, 1979; Erckmann, 1981) reinforce the conclusion that polyandrous phalaropes do not differ from other arctic shorebirds with respect to either average overall nesting success or nest predation.

There is no striking differences in nesting success between polyandrous shorebirds as a whole and other shorebirds. In particular, rates of replaceable nesting failure in arctic polyandrous species are relatively low

TABLE V
Comparisons of Replaceable Nesting Losses and Overall Nesting Success between Polyandrous and Other Shorebirds[a]

	Replaceable nest loss		Overall nest success	
	No. of species	Mean ± SD (%)	No. of species	Mean ± SD (%)
Computed using species averages				
Polyandrous	6	24.8 ± 13.0	6	65.3 ± 14.4
All other[b]	43	27.7 ± 24.2	45	68.4 ± 20.5
Computed using location samples	No. of samples		No. of samples	
Polyandrous	20	24.1 ± 20.6	19	70.2 ± 23.7
All other[b]	78	25.7 ± 21.3	87	67.8 ± 20.8

[a]See text for explanation of calculations.
[b]Includes monogamous, polygynous, and double-clutching species.

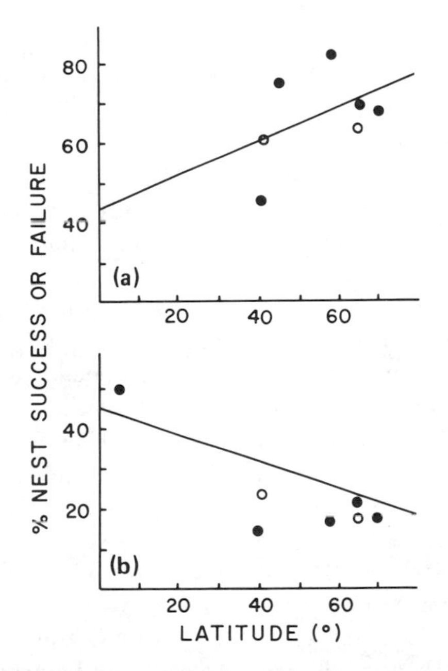

Fig. 1 The relationship between latitude and nesting success in shorebirds. (a) percentage of clutches hatching any eggs. (b) percentage of clutches lost that are potentially replaceable. Regression was calculated for species with monogamous or polygynous mating systems. Polyandrous (●) and double-clutching (○) species are plotted for comparison but were not included in the regression. In (a), $r = +222$, $n = 102$, $p = .02$; in (b), $r = -.203$, $n = 86$, $p = .03$.

(Tables V and VI, Fig. 1). Because nest predation rates are apparently high in the Tropics (Fig. 1), especially the Humid Tropics (Skutch, 1976), it is reasonable to conclude that nesting failure may be relatively high in polyandrous tropical jacanas and painted snipe, although few data are available for comparison. A survey of a number of studies of the polyandrous spotted sandpiper, which breeds in temperate areas, indicates that replaceable nest losses on the whole are about average for shorebirds (Tables V and VI), although Oring (1982) indicates that nest predation can be extreme in some years. The regular occurrence of extremes in nest predation might select for female emancipation and replacement laying (Oring, 1982), although extreme fluctuations in nesting success may be typical for monogamous as well as polyandrous shorebirds (Harris, 1967; Soikkeli, 1970; Mickelson, 1979 and included references; Mayfield, 1978; Erckmann, 1981).

2. *Prediction: Renesting should be relatively frequent in polyandrous species.* Available information is insufficient to permit a quantitative between-group comparison of the frequency of renesting, but a review of the literature indicates several trends among monogamous shorebirds. First, renesting is known to be infrequent in some arctic species, but is regular in others (Norton, 1973; Mickelson, 1979; Erckmann, 1981). Second, renesting is very common among subarctic and temperate shorebirds following nest losses that are early in the nesting cycle and breeding season (e.g., Laven, 1940; Klomp, 1951; Wilcox, 1959; Lind, 1961; Soikkeli, 1967; Nethersole-Thompson, 1979; Parr, 1980). Finally, pairs of most, if not all, species that have long breeding seasons typically renest more than once (e.g., Summers and Hockey, 1980).

Among polyandrous species there is a latitudinal trend in frequency of renesting that essentially parallels the trend in monogamous species and the trend in the frequency of nesting failure. Renesting in phalaropes and dotterels is uncommon (Hildén and Vuolanto, 1972; Nethersole-Thompson, 1973; Kagarise, 1979; Erckmann, 1981). Less than one in four male northern phalaropes and male dotterels losing clutches apparently renest, and renesting has been reported in phalaropes primarily when clutches are lost very early in the nesting cycle (Schamel and Tracy, 1977; Mickelson, 1979). Male phalaropes have been observed to refrain from renesting even after pairing again with females apparently capable of laying (Hildén and Vuolanto, 1972; Kagarise, 1979; Erckmann, 1981). It is not surprising then that female phalaropes typically terminate contact with their mates within several days of clutch completion (Howe, 1975a; Kistchinski, 1975; Mayfield, 1978), even in years of heavy nest predation (Erckmann, 1981). The frequency of renesting in both phalaropes and the

TABLE VI
Nesting Success and Replaceable Nest Losses among Polyandrous Shorebirds[a]

Species	Nest success (%)			Replaceable loss (%)			References[c]
	Mean	Range[b]	No. of samples	Mean	Range[b]	No. of samples	
Red phalarope	68	(50–100)	6	18	(0–43)	5	1, 8, 9, 14
Northern phalarope	70	(18–92)	4	22	(8–64)	5	1, 3, 5, 9
Spotted sandpiper	75	(37–96)	4	25	(4–67)	5	2, 7, 10, 12
Dotterel	83	(60–100)	3	17	(0–40)[d]	3	11, 13
American jacana	50[e]		1	50[e]		1	6
Pheasant-tailed jacana	46[f]		1	15		1	4

[a]See text for explanation of calculations.

[b]Range of averages for area or study; annual variation is much greater.

[c]1, Erckmann (1981); 2, Hays (1972); 3, Hildén and Vuolanto (1972); 4, Hoffmann (1949); 5, Höhn (1968); 6, Jenni (1974); 7, Maxson and Oring (1978); 8, Mayfield (1978); 9, Mickelson (1979); 10, Miller and Miller (1948); 11, Nethersole-Thompson (1973, pp. 149, 266); 12, Oring and Knudson (1972); 13, Pulliainen (1970, 1971); 14, Ridley (1980).

[d]Maximum values based on hatching success.

[e]Approximate.

[f]39% of nests robbed by humans.

dotterel is apparently less than the frequency for most monogamous shorebirds at similar latitudes.

In other polyandrous shorebirds replacement laying is a regular feature of breeding. Between 40 and 50% of lost clutches are replaced in temperate-breeding spotted sandpipers, and about one of four males renests at least once per season (Oring and Knudson, 1972; Maxson and Oring, 1980). Individual males have renested as many as three times with the same female. Although renesting is frequent in spotted sandpipers, many monogamous shorebirds, from the Subarctic to the Tropics, have also been reported to renest from two to four times in a season (Klomp, 1951; Bunni, 1959; Shibnev, 1973; Prater, 1974; Nethersole-Thompson, 1979; Summers and Hockey, 1980). Pairs of some monogamous species almost invariably renest immediately after clutch destruction (Laven, 1940; Soikkeli, 1967; Gibson, 1971). Among polyandrous shorebirds, renesting is most frequent among tropical species. Female jacanas have been reported to replace as many as three successive clutches for their males (Hoffmann, 1949; Mathew, 1964; Jenni and Collier, 1972; Osborne and Bourne, 1977), and some probably lay more. The long breeding seasons of many populations probably increase the opportunity for renesting.

3. *Prediction: Females emancipated from incubation should be able to replace lost clutches with less delay than females sharing in incubation.* This prediction is really an assumption of the *replacement clutch hypothesis,* from which derives the proposition that males benefit by female emancipation. The best test of this prediction would be a comparison of the interval between nest loss and commencement of replacement laying (i.e., replacement interval) between females of a polyandrous species that either were or were not incubating at the time of clutch loss. Limited data presented in Oring and Knudson (1972) indicate that freedom from incubation does not improve a female spotted sandpiper's ability to renest more rapidly. For three females that were sharing in incubation, the mean replacement interval was 4.7 days. For two females not incubating, the replacement interval averaged 4.3 days.

Another test of this prediction can be made by comparing typical or mean replacement intervals for polyandrous species with male incubation, and monogamous species with biparental incubation. Mean replacement intervals for polyandrous species do not differ from intervals in monogamous species (Table VII).

This prediction derives from the assumption that because females store nutrients in advance of laying and need to recover from nutrient depletion after laying, they need to forage nearly full-time after laying one clutch to prepare for laying another. As I argued in Section III, B, most female shorebirds probably do not suffer substantial energy or nutrient

TABLE VII

Comparison of Intervals between Clutch Loss and Commencement of Replacement Laying for Polyandrous and Monogamous Shorebirds[a]

	Replacement interval (days)					
	Polyandrous			Monogamous		
	No. of species	Mean ± SD	Range[b]	No. of species	Mean ± SD	Range[b]
Small Scolopacidae (<100 g)	3	4.3 ± 1.0	(4–5.5)	7	4.6 ± 1.2	(3–6.4)
All shorebirds	7	6.7 ± 2.3	(4–8.7)	16	6.0 ± 2.3	(3–12)

[a]References: Bannerman (1961), Bunni (1959), Gibson (1971), Harris (1967), Heldt (1966), Hildén and Vuolanto (1972), Hoffmann (1949), Howe (1982), Jayakar and Spurway (1968), Mathew (1964), Miller (1979a), Nethersole-Thompson (1973, 1979), Norton (1973), Oring and Knudson (1972), Osborne and Bourne (1977), Parr (1980), Schamel and Tracy (1977), and Soikkeli (1967).

[b]Range of typical or mean values among species. Shortest interval recorded for a polyandrous species is 2 days and for a monogamous species, 1 day.

depletion after laying and do not need to recoup afterwards. The rapidity with which polyandrous females are able to lay successive clutches (Hildén, 1975; Osborne and Bourne, 1977; Schamel and Tracy, 1977; Oring, 1982) attests to the validity of this conclusion. Furthermore, time-budget studies of polyandrous and monogamous shorebirds indicate that females spend a considerable proportion of time prior to and following laying in nonforaging activities, such as loafing, that could be eliminated in favor of parental care. For example, polyandrous female spotted sandpipers spend no more than 57% of their time foraging during any phase of the nesting cycle, including laying (Maxson and Oring, 1980), indicating they could prepare for and actually lay eggs while sharing in incubation. Monogamous killdeers (*Charadrius vociferus*) are often double-brooded, and females may devote as much as 45% of their time to caring for a first brood while laying a second clutch (Lenington, 1980). No more than 50% of a female killdeer's time is required for foraging during laying (Mundahl, 1977). In general, the percentage of time that laying females need to forage correlates positively with the percentage of time incubating males of the same species forage during incubation (Fig. 2). If polyandrous species experience relatively good foraging conditions (Section III, A), then females can find food for laying by foraging about half of each day. These observations indicate that the ability of female shorebirds to replace eggs should depend far more on feeding conditions at the time of clutch loss than on history of parental investment.

4. *Discussion and conclusions*. There is no indication that recent production of a clutch substantially impairs a female shorebird's ability to lay again immediately under normal conditions or that emancipation from incubation allows females to replace clutches more rapidly. The ability of females, even of monogamous species, to lay many eggs in short order is clearly demonstrated by records of a killdeer that produced a clutch of 20 eggs (Mundahl *et al.*, 1981), a Temminick's stint that laid 12 eggs in 16 days (Hildén, 1975), and a spotted sandpiper that laid 20 eggs in 42 days (Maxson and Oring, 1980).

Rather than benefiting males, female emancipation can be costly to males. In cold climates, males are less able to maintain incubation constancy in bad weather without help from their mates, and reduced attentiveness may increase risk of predation and decrease egg hatchability (Norton, 1972; Maxson and Oring, 1980; Erckmann, 1981). Male spotted sandpipers lose all parental assistance from their mates if unmated males arrive, and the tendency of females to share in incubation until the opportunity for polyandry arises (Maxson and Oring, 1980) indicates that females do not refrain from incubation to enhance replacement clutch potential but do so to acquire more mates. It seems likely that males of

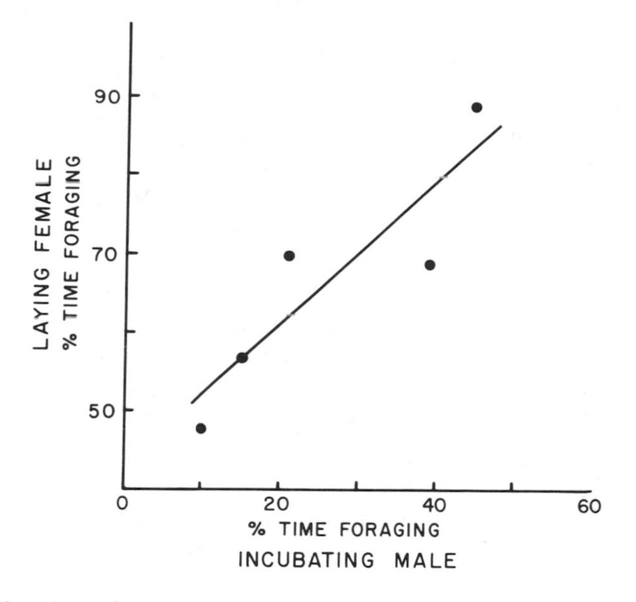

Fig. 2 The relationship between percentage of time spent foraging by incubating males and laying females of the same species for several small shorebirds; $r = +.893$, $p = .02$. References: Mundahl (1977), Ashkenazie and Safriel (1979), Maxson and Oring (1980), and Erckmann (1981).

polyandrous species would "prefer" exclusive access to their mates and assistance in parental care, because in territorially polyandrous species, males try to prevent their females from forming new pair bonds. But males are unable to prevent their mates from deserting or remating, since females are dominant and police the aggressive interactions between their mates and other males (Mathew, 1964; Hays, 1972; Jenni and Collier, 1972; Maxson and Oring, 1980).

The assumption that males must benefit from female emancipation is not critical to the idea that replacement laying may have been an important factor in the evolution of female emancipation. Frequent nesting failure and a willingness of males to renest serve to increase the pool of available males for any female capable of laying multiple clutches (Emlen and Oring, 1977) and should enhance selection for females to terminate parental care in favor of searching for available males whenever their mates can and will incubate without help.

But the conclusion that males of polyandrous species do not benefit by female emancipation raises a problem that will require further analysis and probably experimental studies to resolve: if either parent were able to nest successfully alone, whey were males not selected to desert first? It is possible that the opportunities for polygyny were limited, reducing the benefits of desertion for males, but this remains a speculative proposition. Also, it can be shown mathematically that when nest predation is severe, polygyny may not be advantageous to males if their females are unable to replace their lost clutches, and males may leave more offspring by accepting all parental care when their mates can provide many successive replacements (Erckmann, 1981). But this argument rests on the assumption that uniparental incubation substantially reduces a female shorebird's capacity to renest rapidly, for which I know of no supporting evidence.

The *replacement clutch hypothesis* cannot explain the evolution of polyandry in the phalaropes or dotterels because males renest with very low frequency. The model is more plausible for tropical species, since nest predation rates are apparently high and long breeding seasons allow frequent renesting attempts. More comparative data will be required to evaluate adequately the applicability of the model to temperate polyandrous species, such as the spotted sandpiper. However, available evidence does not indicate that either nesting failure rates or frequency of renesting is unusually high in spotted sandpipers. Replacement laying, however, does play an important role in the reproductive strategies of female spotted sandpipers, as well as females of tropical polyandrous species. When opportunities for replacement laying are frequent, it benefits females to retain contact with and defend their incubating mates. Female jacanas and spotted sandpipers accomplish this by defending the territories of males. In dotterels and phalaropes, females do not usually keep contact with their mates because males only infrequently opt to renest and because females must be mobile to acquire other mates.

D. The Fluctuating Food Hypothesis for the Evolution of Polyandry and Double-Clutching

Graul (1974) and Parmelee and Payne (1973) proposed that double-clutch mating systems evolved to facilitate facultative responses to annual variations in breeding conditions, noting that feeding conditions in the high arctic breeding areas of the sanderling (*Calidris alba*) and the shortgrass prairie habitat of the mountain plover (*Charadrius montanus*) are highly variable both within and between seasons as a consequence of weather fluctuations. The basic argument is that females can capitalize on particularly good conditions by laying two or three clutches and that males can also benefit by fertilizing more than one clutch. Under contrasting (poor) feeding conditions, a female would benefit by refraining from parental care, thereby reducing risk of mortality. Graul *et al.* (1977)

generalized this argument to explain the evolution of polyandry as well as double-clutching, proposing that periods of food scarcity favored assumption of all parental care by males and thus established initial conditions conducive to the evolution of multiple-clutch strategies.

The assumption that food scarcity precludes participation in parental care by females has been discussed above (Section III, A) and judged as invalid, but this assumption is not necessary to the main argument of the *fluctuating food hypothesis.* Temporal variation in food availability could still favor multiple-clutch strategies in females if periods of relative food abundance were required for females to lay multiple clutches.

1. *Prediction: Female shorebirds should be able to lay multiple clutches only when food is particularly abundant.* Little evidence is available to test this prediction directly, but there have been a number of observations of shorebirds laying first and additional clutches under relatively *poor* feeding conditions (Nettleship, 1973; Graul, 1974, 1975; Nethersole-Thompson, 1979; Maxson and Oring, 1980; Ridley, 1980). The general capacity of monogamous females for laying replacement clutches rapidly can also be construed as evidence contrary to the prediction, because many of these species probably do not breed under conditions of relative food abundance (Section III, C). There is no doubt that temporary food abundance should make laying easier for females or that extreme weather-produced food shortages may curtail laying (e.g., Jehl, 1973), but it does not appear that females require particularly abundant food to lay one or successive clutches. The limited evidence indicates that the typical foraging conditions experienced by many, if not most, monogamous and polyandrous shorebirds are adequate for females to lay more than one clutch in a season.

2. *Prediction: Annual variability in the number of clutches produced by females of double-clutching and polyandrous species should be relatively large and correlated with variation in food abundance.* In the double-clutching Temminck's stint (*Calidris temminckii*), which has been studied for a total of 15 years in two different arctic locations, annual variability in clutch production both per breeding female and per breeding pair is slight (Kochanow, 1973; Hildén, 1975, 1978b) and comparable to that typical of other arctic species. There is some indication that female sanderlings may lay only one clutch in some years, but this appears to occur primarily when nesting begins very late and the breeding season is very short, rather than as a result of food shortage per se (Parmelee and Payne, 1973). Graul (1974, 1976) presented evidence that fluctuations in rainfall and probably insect production can be extreme both within and between seasons in the mountain plover habitat, but there is as yet no evidence to indicate either that there is substantial *annual* variability in the average number of clutches

laid per female or that females require food flushes to lay. In fact, a large proportion of clutches were laid during temporary drought periods (Graul, 1974, 1975). In short, although some double-clutching species may breed under conditions of a highly variable food supply, there is little evidence that annual variability results in facultative response by females. Within-season fluctuations in food levels may influence the timing of laying and the interval between successive eggs, but this is probably true for most shorebirds (e.g., Nethersole-Thompson, 1979).

There is also little evidence that annual variability in the number of clutches produced by females of polyandrous species is either extreme or related to variation in food availability. Studies of phalaropes indicate that females are generally able to lay clutches when males are available to accept them. Polyandry in red and northern phalaropes appears to occur primarily when males outnumber females (Hildén and Vuolanto, 1972; Raner, 1972; Schamel and Tracy, 1977), although occasional years of nonbreeding by a fraction of females and males remain unexplained (Hildén and Vuolanto, 1972; Kistchinski, 1975). There is little indication that food flushes are usually necessary for polyandry or replacement laying in spotted sandpipers, and the average quality of foraging conditions appears to be more than adequate for laying successive clutches (Maxson and Oring, 1980). The consistency of intervals between eggs and between clutches in polyandrous jacanas indicates that feeding conditions are not highly variable and rarely impair a female's ability to lay (Hoffmann, 1949; Mathew, 1964; Osborne and Bourne, 1977).

3. *Conclusions*. Two conclusions reached above indicate that the *fluctuating food hypothesis* cannot be accepted as a generally applicable model for the evolution of double-clutching and polyandry in shorebirds. First, most polyandrous species and some double-clutching species do not exhibit sufficient annual variation in the productivity of females to indicate an important influence for variations in food availability. Second, particularly abundant food is evidently not required for females to lay successive clutches. It can be concluded that peaks in food availability should have little influence on selection for desertion by females unless it can be shown that laying multiple clutches increases female mortality and that food abundance mitigates this effect. However, recent population studies of polyandrous and double-cluching shorebirds do not indicate that the production of multiple clutches generally influences female mortality (Hildén, 1978b; Oring, 1982).

If polyandrous and double-clutching shorebirds do experience wide fluctuations in breeding conditions, it might be expected that these species should exhibit a higher frequency of nest abandonment than monogamous shorebirds. Females of double-clutching species, in particular, should be under the greatest stress during food shortages because they incubate unassisted after laying two clutches. Yet a chi-square analysis of nest abandonment reveals no significant differences among polyandrous, double-clutch, and other shorebirds (Table VIII). Hildén (1978b) reports no cases of abandonment by female Temminck's stints as a result of food shortage, but Graul (1975) did observe 6 cases of nest abandonment by mountain plovers, presumably of both sexes, during a temporary drought. Males of polyandrous species typically respond to periods of temporary food shortage by reducing nest attentiveness to levels comparable to those of males of species with biparental incubation (Maxson and Oring, 1980; Erckmann, 1981). During prolonged periods of particularly bad weather, such as storms, incubating Temminck's stints and other arctic shorebirds may even abandon their nests temporarily with resulting decreases in hatching success (Hildén, 1979a). The pattern suggested by the literature is that all shorebirds with uniparental incubation probably suffer greater hatching failure as a consequence of bad weather than do species with biparental incubation (reviewed in Hildén, 1979a). Many polyandrous and double-clutching species certainly breed in variable environments, but probably do not experience more severe periods of food shortage than do the many monogamous species breeding in the same or similar environments.

Arctic phalaropes and some double-clutching species may exploit environmental variability opportunistically by breeding only when and where conditions are favorable (Pitelka *et al.*, 1974), but this would have the effect of exposing these species to good conditions on the average and reducing exposure to poor conditions. Good breeding conditions will favor uniparental incubation, and in that sense the evolution of polyandry and double-clutching, but this argument leaves open the problem of when

TABLE VIII
Rates of Nest Abandonment in Polyandrous and Double-Clutching Shorebirds Compared with Other Species

Mating system	No. of species	No. of nests	% abandoned	References[a]
Polyandry	3	367	3.3	1, 3, 6, 8, 9
Double-clutch	2	362	1.9	2, 4, 5, 7
Monogamy and polygyny	36	3477	2.9	Various

[a]1, Erckmann (1981); 2, Graul (1975); 3, Hays (1972); 4, Hildén (1978a); 5, Kochanow (1973); 6, Maxson and Oring (1978, 1980); 7, Mickelson (1979); 8, Miller and Miller (1948); 9, Oring and Knudson (1972).

such conditions favor these mating systems over polygyny. If the *fluctuating food hypothesis* in its original form does apply to some species of shorebirds, it seems most likely that these would exclude tropical species and include sanderlings, the arctic phalaropes, and mountain plovers. More detailed studies of the effects of variations in food availability in these species would be useful.

There is some evidence that annual variability in food supply is the primary cause of occasional double-clutching in California quail (*Lophortyx californicus*) (Leopold, 1977). It is easier to see how the *fluctuating food hypothesis* could apply to quail than shorebirds. Female quail rely heavily on stored nutrient reserves during laying and incubation, and clutch size in quail is three to four times that of double-clutching shorebirds. Thus, abundant food could substantially increase the female's ability to lay a second clutch, as well as the male's ability to incubate and rear a brood unassisted after experiencing the substantial weight loss typical during the prenesting period.

E. The Stepping-Stone Model: The Evolution of Polyandry from Double-Clutching

The four ecological hypotheses reviewed above were all based on the assumption that true polyandry and sex-role reversal evolved from a condition of monogamy and biparental care, but the double-clutch mating system could conceivably have been the starting point. In this breeding system, each female lays two clutches, the first incubated by a male and the second by herself. Since males regularly care for broods unassisted, it is plausible that polyandry and complete role reversal could evolve if females ceased incubating in favor of seeking and laying for as many males as possible. Thus the double-clutch system could be a "stepping-stone" to sex-role reversal and true polyandry (Jenni, 1974; Pitelka *et al.*, 1974; Pienkowski and Greenwood, 1979). Ridley (1980, p. 224) even concludes that "it is now widely believed that the mating system shown by phalaropes evolved from territorial monogamy and 'soinvestment' through the habit of double-clutching."

The stepping-stone model is appealing since it solves the problem of how males ended up caring for broods unassisted, and it begins with a system in which females often mate with more than one male. However, several lines of reasoning suggest that sex-role reversal did not evolve in this manner.

1. *Arguments Against the Stepping-Stone Model*

1. *The disadvantage to females that give up incubating.* Pienkowski and Greenwood (1979) presented an elegant model demonstrating that females of the double-clutching Temminck's stint switch mates between clutches because they are attempting to lay clutches for more than one male (See Hildén, 1975). Based on the observations that females sometimes compete for males and that some females lay three clutches, Pienkowski and Greenwood developed a second argument that sex-role reversal and true polyandry could evolve from the Temminck's stint mating system if some "male-like" females sacrificed ability as "incubators" in favor of increasing their chances of obtaining more males to tend clutches, primarily by aggressive competition. The critical points in the second argument are that selection could favor females that did not incubate a final clutch over those that did, and that selection for competitive ability in females results in an inability to incubate a clutch.

There are two reasons why I think the process postulated in the second argument is unlikely to occur in a species such as Temminck's stint. First, nesting in arctic sandpipers is usually quite synchronous, and the laying season is relatively short in comparison with the time it takes a female to mate and lay a clutch. Consequently, few unmated males should be available by the time females could be free to seek second mates. Indeed, Hildén (1975) observed that only two of about 115 nesting female Temminck's stints laid three clutches, and in both cases the second clutch remained untended. Under these circumstances any female that incubated a clutch late in the season would always leave more young than a female that spent that period searching for mates, provided that the cost of incubating was not high. Second, there is no convincing reason why an increase in competitive ability should reduce the capacity of females to incubate so much that they could no longer hatch a clutch unassisted. Competitive ability does not limit the male's ability to incubate in double-clutching species, and even in species with sex-role reversal and intense female competition, females may be able to incubate alone (Oring, 1982). Selection for large size may increase the nutritional needs of females, but this could be easily compensated for by a small increase in time devoted to foraging. There is no evidence that there is currently any selection for large female size in any double-clutching shorebird; sexual dimorphism in these species is similar to that in related monogamous species.

2. *The lack of close taxonomic relationships between double-clutching and polyandrous species.* If polyandry evolved from double-clutching, some taxonomic groups should include both mating systems. Yet all three

double-clutching scolopacids are in the subfamily Calidridinae, which includes no polyandrous species. All polyandrous scolopacids are in or closely related to the Tringinae, which includes no double-clutching species. The taxonomic affinities of the polyandrous dotterel are not known (Johnsgard, 1981), but there is no reason to believe that the *stepping-stone model* applies to the Rostratulidae, Jacanidae, or Pedionomidae, since these families have no double-clutching species. It is also significant that the order Galliformes has a number of double-clutching species, but no species with reversed sex-roles. In fact, there are no orders of birds outside of the Charadriiformes that include both double-clutching and polyandrous species.

3. *Shared incubation at one nest in polyandrous species.* Females of some polyandrous species display a tendency to share in incubation with one of their mates (Maxson and Oring, 1978; Pulliainen, 1970), behavior unreported in any double-clutching shorebird. Furthermore, double-clutching has never been observed in any avian species with sex-role reversal. This suggests that polyandry in these species probably evolved from shared incubation at one nest, otherwise it must be explained why shared incubation secondarily evolved in a species that once had exclusive incubation by males. A similar conclusion was reached by Oring (1982).

2. *The Evolution of Double-Clutching*

Double-clutching could have evolved by two routes: (*a*) from the habit of double-brooding in monogamous species, or (*b*) through mate desertion and attempted polygamy by both sexes (Pienkowski and Greenwood, 1979). In the former case, double-clutching can be considered an elaboration of monogamy, proceeding in evolutionary time from a pair rearing two clutches in succession (double-brooding) through brood overlap to a condition in which males are still incubating a first clutch when their mates start a second. I suggest that the mountain plover evolved by the first route, while double-clutching in sandpipers evolved by the second route.

Double-brooding occurs regularly in at least seven temperate and tropical charadriine plovers (Johnsgard, 1981), yet it has never been reported in any scolopacid. Second nests of plovers are usually started before the first young fledge (Conway and Bell, 1968; Lenington, 1980), but are sometimes started while the male is still incubating a first clutch (Laven, 1940). That mountain plovers evolved through the double-brooding sequence is indicated by the fact that females apparently do not attempt polyandry. Graul (1974) presented data demonstrating double-clutching by two marked females. Although the females copulated with more than one male, both laid all clutches on the territory of their original mates, suggesting that they were not attempting to lay clutches for more than one male.

In contrast to female mountain plovers, female Temminck's stints lay their second clutches on territories of males other than their first mate (Hildén, 1975). Whether mate switching occurs in little stints (*Calidris minuta*) or sanderlings is unknown, but, assuming that it does, double-clutching in calidridines may represent a case where *both* parents desert their mates. Both sexes "abandon" the first clutch (males only temporarily) in favor of seeking another mate. As the season progresses, the number of males without clutches on their territory drops, and a female's best strategy is to copulate and lay for herself. From the male's standpoint, it pays to continue advertising for females until most are incubating their own clutches; then males begin incubating themselves. Hildén (1975) reported the hatching of clutches that remained untended as long as 13 days before males began incubating. By the end of the season it is to the advantage of both males and females to rear a brood themselves.

The longer breeding seasons of mountain plovers might be expected to facilitate polygamy in both sexes, but this is apparently not true. The danger of egg mortality from overheating probably deters males from delaying incubation of their first clutches, as do male stints. It is less clear why female mountain plovers apparently do not attempt to lay for several males, if, in fact, they do not.

The environmental conditions leading to the evolution of double-clutching are obscure. Food availability for double-clutching calidridines is apparently good on average as evidenced by an estimated nest attentiveness for sanderlings of 70–80% (Parmelee *et al.*, 1968; Parmelee, 1970). Both the relatively low nest attentiveness (42–58%) and long intervals between the eggs of a clutch in mountain plovers may reflect limited food availability, but attentiveness may be related more to ambient temperature (Graul, 1975). In warm climates, attentiveness may not be maximized since eggs can develop using solar and ambient heat. Moreover, conditions allowing low attentiveness can be equivalent to good foraging conditions in that such conditions favor uniparental incubation. Nest predation in double-clutching species is not substantially different from that in monogamous shorebirds (Fig. 1), and replacement clutches are uncommon in Temminck's stints (Hildén, 1975). To determine whether or not all double-clutching shorebirds experience unusually high annual fluctuations in food availability will require further studies.

It is even less clear under what conditions the polygamous form of double-clutching should evolve in lieu of true polyandry, or for that matter, polygyny. A double-clutching strategy should result in a more consistent reproductive output for a female than a pure polyandrous strategy. The average productivity of female Temminck's stints substantially exceeds the average for female dotterels or phalaropes, and the maximum reproductive output may be similar. A double-clutching strategy would seem superior in environments with very short breeding seasons, unless the cost to females was prohibitive. Lengthy fall migration could conceivably make the care of a late clutch very costly to females, but there are no data that show this. Alternatively, it is possible that, because the evolution of double-clutching requires more complex behavioral adaptations than does the evolution of true polyandry, it is less likely to evolve. Indeed, in birds, double-clutching mating systems are an order of magnitude less frequent than polyandrous mating systems.

F. Other Environmental Influences on the Evolution of Polyandry

In addition to the four major ecological hypotheses for the evolution of sex-role reversal and polyandry, a number of other environmental factors have been implicated in the evolution of polyandry. Although none of these arguments have been suggested as general models for polyandry, the mechanisms postulated enhance selection for female emancipation in important ways.

1. *One parent may have greater breeding success than two: the advantage of uniparental care.* With the exception of the *replacement clutch hypothesis,* all hypotheses for polyandry discussed to this point have assumed that emancipation of females either has a negative influence on a male's chance of successfully fledging his young, or has no effect. An alternative proposition is that one parent may actually benefit by being deserted by the other.

Pitelka *et al.* (1974) proposed this idea as a model for the evolution of polygyny, polyandry, and double-clutching in arctic sandpipers, but the mechanism proposed is of more general importance if valid. They argued that when a species breeds at high density (a) desertion by one parent can improve food availability for the other parent and the young by reducing foraging competition, and (*b*) departure of one parent can reduce conspicuousness of the nest to predators by decreasing the frequency of trips to and from the nest.

Little evidence is available to evaluate the proposition that departure of one parent would improve feeding conditions for the other parent or the young, but two general patterns among the shorebirds suggest that this may not occur in many cases. First, in most northern shorebirds, adults feed away from their nesting territories, often in communal areas (Hildén, 1979b; Miller, 1979b). This is true of polyandrous phalaropes and dotterels, double-clutching species, and some populations of polyandrous spotted sandpipers (Miller and Miller, 1948). Second, territoriality breaks down at hatching (Miller, 1979b). Attending parents take turns guarding the young, and off-duty parents typically feed well away from the brood so that little food competition occurs between parents and young (Hagar, 1966; Gibson, 1971; Parr, 1980). Thus, except in cases where parents and young feed to a large extent on their nesting territory, early desertion of one parent may have little influence on the foraging conditions of the other parent or the young.

Available evidence also indicates that departure of one parent would not usually reduce the risk of nest predation. Recent studies have shown in general that the frequency of daily nest absences is actually higher for shorebirds with uniparental incubation than species with biparental incubation (Norton, 1972; Rubinstein, 1973; Erckmann, 1981), probably due to a greater need for uniparental incubators to forage while incubating. Furthermore, uniparental incubators have lower nest attendance than biparental incubators, and it has also been shown experimentally that untended clutches are at greater risk of detection by predators than nests being incubated (Erckmann, 1981). In spite of these observations, there is no indication that uniparental incubators experience high rates of nest predation (Table IX). Whether nest predation is density-dependent in shorebirds is not known. Studies of nesting success in shorebird communities have generally revealed no consistent relationship between nest predation and nesting density either within or between species (Jehl, 1971; Norton, 1973; Mickelson, 1979; Dyrcz *et al.*, 1981; Erckmann, 1981), but more intensive studies are needed. Whether or not nest predation increases with nest density, some arctic polyandrous and double-clutching species nest at relatively low densities compared to densities of monogamous calidridine sandpipers of similar body size (Fig. 3).

In summary, available evidence suggests that the model proposed by Pitelka *et al.* (1974) cannot explain desertion by one parent in all polygamous species. Deserted parents probably do not benefit by reduced feeding competition, and emancipation of one parent very likely does not reduce the risk of nest predation. The model is least convincing as an explanation for the evolution of double-clutching, which results in a doubling of nest density, increased sexual activity near nests, no change in the number of adults competing for food, and a doubling of the number of young competing for food.

2. *Length of the breeding season.* Hypotheses implicating the length of

TABLE IX

Nest Predation Rates in Shorebirds with Uniparental and Biparental Incubation

	Percentage of nests lost to predators				
	Calculated by species average		Calculated by location samples[a]		
	No. of species	Mean ± S.D. (%)	No. of samples	Mean ± S.D. (%)	References[b]
Uniparental					
Polygynous[c]	5	26.6 ± 15.4	8	28.5 ± 17.7	1, 2, 6, 7, 8, 9
Double-clutch	2	15.5 ± 2.1	3	15.7 ± 16.6	3, 4, 5
Polyandrous	6	25.8 ± 12.6	20	25.4 ± 21.8	10
Biparental					
Monogamous	33	28.3 ± 24.9	57	25.8 ± 21.8	Various

[a]See Section III, C for explanation.

[b]1, Boyd (1962); 2, Dyrcz *et al.* (1981); 3, Graul (1975); 4, Hildén (1978a); 5, Kochanow (1973); 6, Mason and MacDonald (1976); 7, Mendall and Aldous (1943); 8, Norton (1972, 1973); 9, Sheldon (1967); 10, See Table III.

[c]Includes common snipe, in which only females incubate.

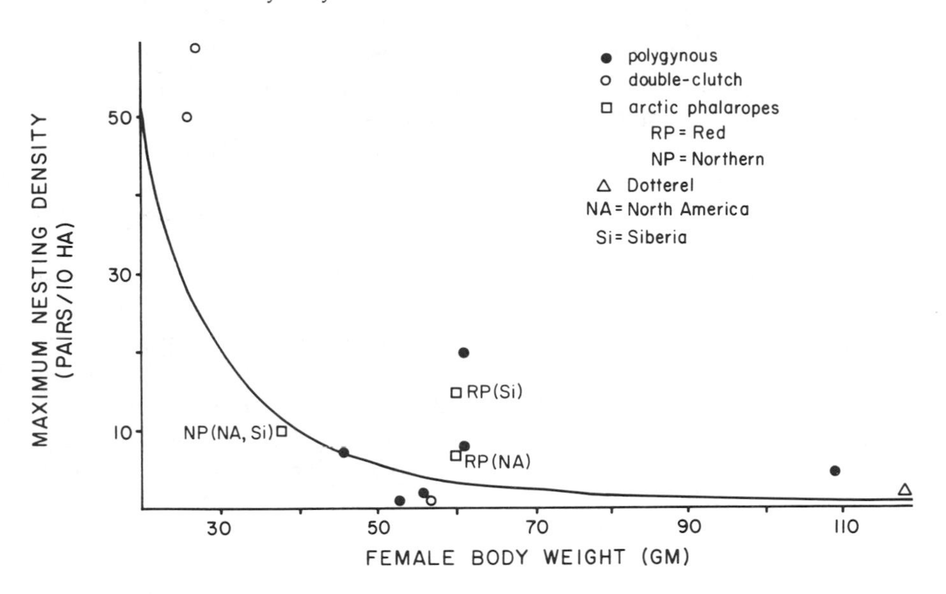

Fig. 3 Relationship between body weight and nesting density for arctic shorebirds with different mating systems. Regression was computed for monogamous calidridine sandpipers. Regression values are $y = 64854x^{-2.378}$, $n = 9$, $p = .001$, $r^2 = .748$. References: Kistchinski and Flint (1973); Nethersole-Thompson (1973); Bergman *et al.* (1977); Hildén (1979b), and Dyrcz *et al.* (1981).

the breeding season and the degree of nesting synchrony in selection for polyandry take two forms. Emlen and Oring (1977) suggested that both long breeding seasons and nesting asynchrony favor polyandry by increasing the opportunities for obtaining multiple males and laying multiple clutches, including replacements. In the extreme, seasons might be so short that females could lay only one clutch. In contrast to this argument, Wittenberger (1979, 1981) suggested that short seasons could favor desertion by females, since males would have little chance for polygyny and would consequently be less likely to abandon their nests in response to desertion.

It is not clear when and why season length should affect female strategies more than those of males. Because females cannot mate and lay clutches as rapidly as males can mate and fertilize clutches, short seasons might be expected to favor desertion by males rather than by females. Indeed, the more frequent occurrence of polygyny than polyandry among high latitude shorebirds (Pitelka *et al.*, 1974) indicates that nesting synchrony does not preclude multiple matings by males. The differential effect of long seasons on multiple mating strategies by the sexes has not been adequately analyzed, and predictions are presently unjustified.

Whatever the selective influence of season length, polyandry and double-clutching have evolved in species experiencing a wide range in length of breeding seasons (Table X). Asynchronous arrival of males in migratory polyandrous species certainly facilitates polyandry (Schamel and Tracy, 1977; Oring 1982), but may have evolved secondarily to polyandry. A thorough analysis of the relationship between nesting synchrony and optimal reproductive strategies in the sexes is warranted.

3. *Sex ratio.* There is no indication that biased sex ratios are either necessary for polyandry to evolve or a consequence of polyandry (Table XI). In only Wilson's phalarope (*Phalaropus tricolor*) does there appear to be a genuine bias in sex ratio on the breeding areas, and females outnumber males. Of interest is the hypothesis that variability in local sex ratios may contribute to selection for facultative polyandry. In phalaropes, local sex ratios can change rapidly. Females often capitalize on an excess of males by

TABLE X

Length of the Egg-Laying Season for Shorebirds with Polyandrous and Double-Clutch Mating Systems

Species	Breeding latitude	Length of laying season (weeks)	References[a]
Polyandry			
Red phalarope	Arctic	2–4	3, 14, 20
Northern phalarope	Arctic–Subarctic	2–4	3, 7
Wilson's phalarope	Temperate	3–5	8, 9
Spotted sandpiper	Temperate	4–7	5, 13, 18
Dotterel	Arctic–Subarctic	1–5	15
American jacana	Tropical	year round[b]	10
Wattled jacana	Tropical	year round[b]	19
Bronze-winged jacana	Tropical	8–10	1
Pheasant-tailed jacana	Tropical–Subtropical	8–25	1
African jacana	Tropical–Subtropical	16–36	11
Painted snipe	Tropical–Subtropical	6–20	11, 12
Double-clutch			
Sanderling	High Arctic	1–3	16, 17
Temminck's stint	Arctic–Subarctic	4	6
Little stint	Arctic–Subarctic	2–3	2
Mountain plover	Temperate	8–9	4

[a]1, Ali and Ripley (1969); 2, Dementev and Gladkov (1969); 3, Erckmann (1981); 4, Graul (1975); 5, Hays (1972); 6, Hildén (1975); 7, Hildén and Vuolanto (1972); 8, Höhn (1967); 9, Howe (1975a); 10, Jenni and Collier (1972); 11, Johnsgard (1981); 12, Lowe (1963); 13, Maxson and Oring (1980); 14, Mayfield (1978); 15, Nethersole-Thompson (1973); 16, Oring and Knudson (1972); 17, Osborne (1980); 18, Parmelee (1970); 19, Pienkowski and Green (1976); 20, Ridley (1980).

[b]Seasonal in some areas.

TABLE XI
Estimated Adult Sex Ratios in Populations of Polyandrous Shorebirds on the Breeding Grounds

Species	Approximate ratio of males to females	References
Red phalarope	0.92	Erckmann, 1981
Northern phalarope	1.05	Hildén and Vuolanto, 1972
Wilson's phalarope	0.61	Erckmann, 1981
Dotterel	1.00	Nethersole-Thompson, 1973
Spotted sandpiper	1.00	Oring, 1982
American jacana	1.33	Jenni and Collier, 1972

laying multiple clutches, and polyandry rarely occurs in the absence of excess males (Hildén and Vuolanto, 1972; Schamel and Tracy, 1977). Ridley (1980) proposed that local sex ratio biases may have selected for male competition and desertion in female phalaropes, both during periods of excess males and when females outnumber males. Alternatively, observed variations in sex ratios might be a consequence of the present mating system. Females move around in groups throughout the breeding season, and males can afford to arrive late because they need not acquire a territory to mate. Both these behaviors contribute to variation in local sex ratios.

An important question is why an excess of males should be necessary for polyandry to occur in phalaropes, as seems to be the case. Females that nest early with one mate should have an equal chance at mating with a late-arriving male. One possible explanation is that females are at a competitive disadvantage after they have laid a clutch. This hypothesis could be tested by experimentally manipulating local sex ratios.

4. *Territoriality and female dominance.* Among shorebirds, only jacanas and spotted sandpipers can presently be characterized as having resource defense polyandry (Emlen and Oring, 1977; Oring, 1982). Three important questions regarding female territoriality need to be addressed: (*a*) In what way does territoriality influence the ability of a female to obtain multiple males? (*b*) What conditions favor territoriality by females? (*c*) How did females evolve to defend territories? The observation that breeding resources may be limited and monopolizable does not necessarily indicate that *females* should control those resources.

a. Correlates of female territoriality. Territorial females can conceivably increase their ability to acquire males by defending territories that are either of large size or high quality. Studies of the production of food on the territories of female spotted sandpipers have revealed no association between the quality of a female's territory and the number of males that she acquires (Maxson and Oring, 1980), but in both spotted sandpipers and American jacanas, territory size correlates positively with mating success (Jenni and Collier, 1972; Oring, 1982). Although no comparisons of nest site quality or nesting success on different territories have been published, it can tentatively be concluded that there is no "polyandry threshold" of territory quality analogous to the polygyny threshold for males (Orians, 1969).

Breeding habitat is limited for American jacanas, and a nonbreeding population of males and females exists. Females are apparently able to defend the territories of more than one male only in productive areas where male territories are small (Jenni and Collier, 1972). In spotted sandpipers, population density varies both spatially and temporally, and exclusion of some females from breeding apparently occurs most regularly at high densities of males (Oring, 1982). In both spotted sandpipers and American jacanas, the local density of nesting males can be extraordinarily high for shorebirds, with as many as eight males per hectare in the latter species and six or more males in the former (Jenni and Collier, 1972; Oring and Knudson, 1972). The ability of territorial females to monopolize males by excluding other females thus depends on the nesting density of males.

b. Conditions favoring territoriality by females. In addition to a relatively high density of nesting males, the defense of territories by females of polyandrous species also appears to be related to two features of breeding biology: (*a*) the dependable occupation of particular areas by nesting males, and (*b*) frequent renesting by males that lose clutches. The former allows females to predict in advance which areas are worth defending, the latter increases the benefit of continuously controlling the nesting area of an incubating male. Females do not monopolize males by territorial defense in migratory polyandrous species characterized by relatively weak nesting site tenacity among males, specifically dotterels and phalaropes (Nethersole-Thompson, 1973; Schamel and Tracy, 1977; Mickelson, 1979; Erckmann, 1981). In all these species, marked annual variation in nesting density occurs, and nesting areas of males may shift. Females do defend territories in species with high male nesting site tenacity (i.e., spotted sandpipers [Maxson and Oring, 1978]) or continuous occupancy of particular areas for long periods (e.g., American jacanas [Jenni and Collier, 1972; Jenni, 1974]). Thus, females defend territories when the location of breeding males is predictable and dependable. In these species, nesting failure is frequent and renesting apparently the typical response to clutch loss, and females defend territories and keep contact with incubating males. In contrast, for dotterels and phalaropes, renesting by males is relatively uncommon, pair bonds are typically short or weak, and females do not acquire mates by territorial defense (Section III, C).

c. The evolution of territorial defense by females. Females largely emancipated from parental care by mate desertion defend territories to increase their opportunities for laying multiple clutches, either for the same or another male. It seems most likely that elaborate female territoriality developed after sequential polyandry had evolved; that is, simultaneous polyandry evolved from sequential polyandry (Jenni, 1974; Hildén, 1975; Graul *et al.*, 1977). Once emancipation from incubation and mate desertion had evolved in females, any females defending good, dependable breeding areas could both retain access to males for replacement laying and gain access to other males. When nesting territories of males are small, simultaneous polyandry can occur; wide spacing of males apparently favors sequential polyandry rather than territorial defense.

d. Preadaptations for territoriality in females. Females of many monogamous shorebirds are larger than males, and, if dominance correlates with size dimorphism, then the step to female control of males and resources may be easily made. Females who desert could prevent males from aggressively interfering with further courtship attempts, as occurs in spotted sandpipers and jacanas. In the former, females are only 20% heavier than males, within the range of dimorphism present in other scolopacids that are monogamous (Erckmann, unpublished data). Females are larger than males in monogamous species of several other families in addition to the Scolopacidae, and size dimorphism may have been present in the ancestors of many polyandrous species.

The reasons for sexual size dimorphism in monogamous shorebirds are not clear, but one hypothesis is that selection for size reduction occurs in males that obtain mates through extensive aerial displays but little fighting (Rohwer and Erckmann, n.d.). Whatever the reason for this dimorphism, the result may be female dominance.

5. *Sources of variation in food availability among species of shorebirds.* I have argued that relatively good foraging conditions are necessary for uniparental care and thus favor the evolution of polyandry and double-clutching. There are several possible reasons why some shorebirds should experience better foraging conditions than others. Many polyandrous shorebirds breed in habitats that are either wet or near water (Johnsgard, 1981). Many of the swampy and marshy habitats utilized by polyandrous shorebirds are known to have a higher productivity of invertebrates than drier habitats at similar latitudes (Whittaker, 1975). Pitelka *et al.* (1974) also argued that polygamous arctic sandpipers experience good feeding conditions not only by feeding in the most productive habitats, but also by

seeking out areas of local food abundance and breeding opportunistically at locally high densities, a hypothesis that may apply to phalaropes. Although the habitat productivity and opportunistic breeding hypotheses provide good explanations for food availability in most polygamous shorebirds, several multiclutching species neither breed in productive habitats nor are locally abundant. Dotterels and mountain plovers breed at low densities in drier, generally less productive habitats (Nethersole-Thompson, 1973; Graul, 1974, 1975), and sanderlings breed at extremely low densities in relatively unproductive high arctic habitats (Parmelee, 1970). Another mechanism is required to explain how these species could experience good foraging conditions.

A point that is frequently neglected in discussions of food availability is that the quality of feeding conditions depends not only on food density and type, but on foraging efficiency. Even if two species feed on the same prey spectrum, foraging efficiency (and realized food availability) could differ as a result of differences in body size, trophic morphology, and foraging techniques. The highly specialized trophic morphology and feeding techniques of many migratory shorebirds are most likely adaptations primarily to foraging conditions in migration and wintering areas, that is, outside the breeding season. During breeding, some species employ entirely different feeding techniques and feed on different substrate types than during the remainder of the year, whereas other species employ the same feeding techniques in all seasons, often feeding in structurally similar habitats. I suggest that species for which the latter is true should be relatively more efficient foragers when breeding than species that change feeding modes or habitats and should thus experience better foraging conditions.

The feeding efficiency hypothesis may help explain the existence of relatively good foraging conditions for a number of migratory, multiple-clutching shorebirds. A great many monogamous shorebirds utilize quite different habitat types and feeding techniques in winter and summer. By contrast, in phalaropes, spotted sandpipers, mountain plovers, dotterels, little stints, and Temminck's stints, there are marked structural similarities in feeding habitats utilized in winter and summer, and similar feeding techniques are employed (Erckmann, 1981; Johnsgard, 1981). That some of the latter species may be more efficient foragers when breeding than some monogamous shorebirds is indicated by reports that both feeding rate and capture efficiency were considerably higher in polyandrous red phalaropes than in monogamous semipalmated sandpipers, even though prey of similar size were being taken (Ashkenazie and Safriel, 1979; Ridley, 1980).

IV. Characteristics of Shorebirds that Favored the Frequent Evolution of Polyandry

A number of important aspects of the breeding biology of shorebirds collectively distinguish them from many other groups of birds. Two major characteristics favor the evolution of polyandry: (*a*) single parents may rear as many young as two parents when foraging conditions are good, and (*b*) the differential capacity of the sexes to capitalize on opportunities for polygamy is minimized, both because males are more likely to assume all parental care if deserted and because females are able to produce many clutches rapidly.

A. Male Parental Care

All polygamous mating systems of shorebirds probably evolved from an initial condition of monogamy with biparental care. More importantly, males are involved in *all* aspects of parental care in all but a few monogamous shorebirds (Johnsgard, 1981). Whereas males participate in incubation in many precocial species in which parents feed their young, males incubate in less than 10% of other precocial birds that do not have parental feeding. In roughly 20% of monogamous shorebirds, males are known to take a larger share of parental care than females. In many cases, females stop tending broods well before males, sometimes even at hatching (Johnsgard, 1981). This historic, extensive involvement of males in parental care could have had important ramifications for the evolution of polyandry whenever breeding conditions secondarily became conducive to uniparental care. A male that is already contributing as much as or more parental care than his mate is less likely to abandon in response to mate desertion than a male that initially was not involved in some aspect of parental care. But why did this unusual pattern evolve in shorebirds?

Male participation in incubation probably evolved in response to food limitations. By aiding in incubation, a male could insure that the clutch was incubated for the maximum time each day while also protecting the physiological condition of his mate. Male participation in tending young could have evolved similarly to maximize the time chicks could be brooded or to better protect the young from predators. The assumption of brood care by males may then have evolved secondarily in cases where one parent was able to accomplish these functions about as well as two, and when females stood to benefit by early abandonment of the brood. Females could potentially benefit more by abandoning the brood if, as a result of laying, they were in poorer physiological condition than males.

Brood abandonment by monogamous females occurs primarily in the most migratory species, probably because females benefit by commencing fall migration early. Myers (1981) has presented convincing evidence that early fall migration has intrinsic advantages and argued that this potential benefit was an important evolutionary determinant of mating systems and patterns of parental care in the arctic calidridine sandpipers. Myers suggested that the benefit of early migration (also the cost of delaying migration) should increase with migration distance and reported a correlation between migration distance and the degree to which one sex is emancipated from parental care, with polygynous species migrating the longest distances. The hypothesis is also consistent with recent observations suggesting that females of some monogamous species may winter further south than males (Page *et al.*, 1972; Spaans, 1976). But in the polygynous species *males* depart first, though evidence suggests that *females* may winter further south than males (Pearson, 1981). This discrepancy requires explanation, since females of these species migrate the longest distances among shorebirds.

Despite unresolved elements of Myers' hypothesis, it seems likely that early termination of brood care by females did evolve to enhance female survival, and this pattern creates a condition that should facilitate the evolution of polyandry. Since females of double-clutching species do not leave the breeding grounds before males, the potential benefit of reduced parental effort is not realized in this mating system, but female phalaropes do depart well before males. The early migration hypothesis cannot help explain the evolution of polyandry in other shorebirds. Although female dotterels do not tend their young, they apparently do not depart on fall migration before males (Nethersole-Thompson, 1973). Many tropical polyandrous species are essentially resident; none are long-distance migrants. In fact, most tropical and south temperate shorebirds are either resident or short-distance migrants, and early abandonment does not occur in females of these species. Nevertheless, males of nearly all these species do contribute heavily to all aspects of parental care.

B. The Need for Only One Parent to Tend Young

Because females leave the rearing of the brood to males in so many species, it can be concluded that one parent should often be almost as successful as two in fledging a brood. This conclusion is supported by published data that reveal no average difference in fledging success between groups of northern sandpipers and plovers in which either one or both parents tend the young (Table XII). In studies of the monogamous killdeer, Lenington (1980) found that young were more closely guarded when both parents shared brood care than when only parent attended the brood, but this does not necessarily mean that biparental tending results in much higher fledging success. Only experimental studies of fledging success will be able to determine what the costs of brood abandonment by one parent actually are. That there is some reduction in fledging success when one parent abandons is suggested by the ubiquity of biparental brood tending in nonmigratory species.

One parent can rear a brood in many shorebirds, since parents do not feed their young and since the young are protected largely by crypsis and predator distraction displays. Because parents actively and more effectively defend young in larger shorebirds (Sordahl, 1981), a second parent may help substantially in protecting young, which may be why polygamy has evolved primarily in smaller species and why, in the larger monogamous species, females rarely abandon their broods before fledging (Erckmann, n.d.).

The time demands of rearing young are generally less than the demands of incubation, and single parents are able to spend more time foraging after the eggs hatch than before (Ashkenazie and Safriel, 1979; Maxson and Oring, 1980; Erckmann, 1981). Consequently, when foraging conditions are adequate to permit uniparental incubation, selection can favor desertion by one parent.

C. Small Clutch Size and the Ability of Females to Lay Successive Clutches

No shorebirds lay clutches of more than four eggs. Clutch size in all but two polyandrous or double-clutching species is four (Maclean, 1972). In all these species, the period for laying the clutch, from first to last egg, is only three to four days and, including the period of rapid follicular growth, is usually less than a week. The small upper limit to clutch size is probably determined by the ability of the parent to warm the eggs adequately (Hills, 1980), because shorebirds have eggs that are extremely large relative to body size (Lack, 1968). Because a clutch is limited to such a small number, females are able to increase reproductive output only by laying multiple clutches. Furthermore, small clutch size allows females to lay successive clutches very rapidly when opportunities for polygamy are present. The rapid production of successive clutches may also be facilitated by a

TABLE XII
A Comparison of Fledging Success between Shorebirds in Which One or Both Parents Tend the Brood.[a]

Parents attending young	Number of species	% of young fledged from eggs hatched Mean ± SD
One	7	54.1 ± 20.0
Both	10	50.6 ± 13.4

[a]Charadriidae and Scolopacidae only. References: Boyd (1962); Glutz von Blotzheim *et al.* (1975, 1977), Graul (1975); Hildén (1978a); Lenington (1980); Nethersole-Thompson (1973, 1979); Parr (1980); Safriel (1975).

substantial reduction in egg size among polyandrous shorebirds (Ross, 1979), but, interestingly, egg size is not significantly reduced in double-clutching species (Erckmann, 1981).

In precocial waterfowl and Galliformes, by comparison, clutch sizes are usually much larger than in shorebirds, and females require two to four times as long to lay a clutch as shorebirds (Johnsgard, 1973; Bellrose, 1976). Clutch size in these groups is also variable, and females can often increase reproductive output by increasing clutch size. Thus, in these two groups there should be less selection for females to desert to increase reproductive output, and females should be less able to capitalize on opportunities for polygamy. It is probably significant that in most polyandrous birds clutch size is either very small, or females lay only a few eggs in one nest shared by other females (Ridley, 1978).

V. Conclusions

Many recent theoretical and review papers have discussed the evolution of reproductive strategies primarily from the standpoint of the male, assuming that such factors as long breeding seasons, habitat productivity and variability, and monopolizable resources favor polygyny. Insufficient attention has been given to the question of why such conditions should favor polygyny and not polyandry, primarily because of the commonly held assumption that the natural assymetry in gamete size (or parental investment) is the most important variable determining which sex competes for mates (Parker *et al.*, 1972). While this assumption is extremely useful for explaining the evolution of reproductive strategies in many animals, it provides little insight into the evolution of diversity in shorebird mating systems. Even when female strategies are discussed, the assumption is frequently made that males must benefit by female emancipation for polyandry to evolve. Yet there is no compelling reason why this must be true, and evidence suggests that male shorebirds do not benefit by female emancipation.

Differential investment in gametes, in a sense, "stacks the deck" in favor of the evolution of male competition and polygamy in males rather than females, both becaus females provide the investment that is the object of competition and because females have less potential to be polygamous. Several important features of shorebird breeding biology reduce these two discrepancies created by gamete dimorphism and consequently favor the evolution of polyandry.

First, paternal care is extensive in shorebirds. Monogamous male shorebirds perform all parental functions and, in many species, even take the larger share of incubation and tend the young with little aid from their mates. Under such circumstances, there should be greater incentive for competition among females, and aggressive competition for mates has been reported in some monogamous females (Howe, 1982). A consequence of the shorebird pattern of parental care is that males are less likely to abandon their clutch in response to desertion for two reasons. Because the young feed themselves, single parents are often capable of successfully rearing a brood. Also, the cost of increased investment of time in parental care that is required of single parents can be mitigated by favorable foraging conditions. Whether "monogamous-type" females are available for a male that abandons in response to desertion is irrelevant if a deserted male can successfully rear his brood with little physiological cost.

Second, the sex differential in potential for polygamy is reduced in shorebirds. The maximum clutch size of four, typical of shorebirds, is relatively small for precocial birds that do not feed their young, and females can potentially produce successive clutches rapidly. In fact, small clutch size is typical of most polyandrous birds (Ridley, 1978). Furthermore, the first deserting females in a population of monogamous shorebirds may be able to locate unmated males more easily than territorial males can acquire additional females. The ability of females to control access to males may be enhanced when they are larger than males, as is the case in a large proportion of monogamous shorebirds.

Polyandry has evolved frequently in shorebirds because both male parental care and small clutch size reduce the "threshold" environmental conditions necessary for desertion by females to be advantageous. A primary problem addressed in this chapter was to determine what environmental conditions do favor polyandry. No single hypothesis evaluated proved to be sufficient as a general model for polyandry in shorebirds, but several are plausible for some species.

An analysis of time- and energy-budgets of breeding shorebirds, in conjunction with data on the frequency of renesting, leads to a number of important conclusions concerning the relationship between food supply and both the capacity of females for laying successive clutches and the capacity of individuals to incubate unassisted. The latter are both necessary conditions for polyandry. First, females of most species basically convert food into eggs at the time of laying and do not appear to deplete nutrient or energy reserves appreciably. Even in monogamous species, which in some cases experience limited food availability, females are able to lay successive clutches. Limited capacity to lay multiple clutches is not likely to have precluded selection for polyandry. Second, whenever foraging conditions are adequate for males to incubate unassisted, females should easily be able to find food to produce eggs, even if they have already laid.

These conclusions have important consequences for three of the hypotheses evaluated. The *stressed female hypothesis* can be rejected because laying does not leave females in an energy-depleted condition and unable to incubate, especially when availability of food is adequate for males to incubate alone. Food scarcity selects for biparental incubation and monogamy. The *fluctuating food hypothesis* can also be tentatively rejected since females apparently do not require unusually abundant food to lay successive clutches. If this hypothesis is valid, it can only explain the evolution of multiple-clutch strategies in a few species breeding in environments with extreme variation in breeding conditions, possibly the High Arctic and short-grass prairies.

Evidence for the *differential parental capacity hypothesis* is inconsistent. There is little indication that females deplete nutrients while laying, but there is some indirect experimental evidence that females have a slightly lower capacity for uniparental care than males. Consequently, the model is potentially useful in explaining the evolution of female desertion only in species breeding in harsh environments in which small differences in parental capacity to attend the eggs regularly may have relatively large effects on egg hatchability. A more general conclusion is that the potential negative effect of egg laying on the physiology of female shorebirds has been overestimated in discussions of reproductive strategies. It is likely that desertion during incubation by female shorebirds is rarely favored because females must recoup losses from laying, and females should usually be able to cope with the rigors of uniparental care nearly as well as males by adjusting the time they spend foraging. However, laying stress may have been important in the evolution of brood abandonment in migratory monogamous species, thus creating conditions favorable to the evolution of sex-role reversal and polyandry.

A comparison of time and energy budgets and intervals for replacement clutches leads to the important conclusion that males do not benefit by female emancipation as assumed in the *replacement clutch hypothesis*, indicating that this is not a prerequisite for the evolution of polyandry. Total emancipation from incubation probably only marginally increases a female's capacity to lay a replacement clutch rapidly. However, high rates of nest predation and frequent renesting (facilitated by long nesting seasons) were probably important factors in the evolution of female strategies in tropical polyandrous shorebirds, because frequent nesting failure increases the pool of males available to a polyandrous female. Resource defense polyandry has evolved only in species in which renesting is common because it benefits females to control access to their incubating males.

The conclusions reached in this chapter suggest one important generalization that shows the relationship between theories for the evolution of polyandry and theories of reproductive strategies in general (e.g., Emlen and Oring, 1977). When the potential for polygamy is similar for both sexes, polyandry may be as likely to evolve as polygyny. Furthermore, when polygamy is secondarily evolved from a system of biparental care in which parental roles are the same, females may benefit by deserting as much as males, and males may be as likely to accept all parental care as females. Females need not benefit by male emancipation for polygyny to evolve, nor must males benefit by female emancipation for polyandry to evolve.

The future development of a general model for the evolution of polyandry seems unlikely, but many of the variables discussed in this chapter may have contributed either singly or in combination to the frequent evolution of polyandry in shorebirds. These include both environmental factors and, more importantly, characteristics of shorebirds themselves. To determine the relative importance of each factor for a particular species will probably require a combination of both experimental and detailed time and energy studies. Many primary questions remain unresolved and many research challenges remain. Our understanding of the role that stress from laying plays in the evolution of female reproductive strategies is particularly poor but should be amenable to experimental study. The potential for other experimental manipulations is also substantial.

The state of our current understanding of the origins of the remarkable diversity in shorebird mating systems has been most eloquently expressed by Desmond Nethersole-Thompson, a superb naturalist who has spent a full half-century chasing shorebirds across the fells of Scotland:

> All these fascinating adaptations are difficult to explain. Is the process of natural selection so rigid that it allows the species only one option? Or does chance help beget the unorthodox? . . . Were all these strangely different adaptations inevitable? Surely other patterns might have been equally successful. To some species, polyandry now offers the means to exploit abundance or to redress sexual imbalance; to others bare survival in the jungle of ecological competition [Nethersole-Thompson, 1973, pp. 54–55].

VI. Summary

Polyandry is a rare mating system in which the usual roles of the sexes are reversed: males perform all parental care and females compete for mates. Polyandry occurs more frequently in shorebirds than in any other group of birds and has evolved in more shorebird families than has polygyny. Polygamy has evolved frequently in shorebirds because one parent is often able to rear a brood unassisted. However, several features of shorebird biology have reduced the usual sex differential in potential for polygamy, resulting in the frequent evolution of polyandry. Monogamous males engage in all aspects of parental care and often take a much larger share than females. Clutch size is small and, since they can lay successive clutches rapidly, females can often benefit by mate desertion. Males are likely to assume all parental care when deserted, because the cost of doing so is small.

An analysis of four models for the evolution of polyandry indicated the environmental conditions most likely to favor polyandry, but no single model proved sufficiently general to explain the diversity of female strategies that has evolved in shorebirds. Several general conclusions were drawn concerning the relationship between food availability and both the capacity of females to lay successive clutches and the capacity of males to incubate unassisted. Even when food is relatively limited, the effect of laying on female physiological capacity is apparently small. Females do not deplete nutrient reserves when laying, but use food available at the time to lay eggs. There is no evidence that polyandrous females desert their mates to recoup energy losses, even when food is scarce. Food scarcity selects for biparental care and monogamy. As a result of laying, some reduction in female capacity as a single parent may occur, making the mate desertion by males less likely than desertion by females. This sex difference in parental capacity probably only influences the evolution of reproductive strategies in harsh environments in which a small difference in the ability of parents to incubate unassisted may result in large differences in egg hatchability.

Females do not require unusually abundant food to lay successive clutches, nor must females forage continuously prior to and during laying. It is thus unlikely that large fluctuations in food availability in general select for polyandry. Furthermore, there is no indication that total emancipation from incubation enhances a female's ability to lay replacement clutches after nesting failure. Therefore, monogamous males do not benefit by total female emancipation when nesting failure is frequent. A high rate of nesting failure, however, increases the availability of males for emancipated females, and can favor the evolution of polyandry when long breeding seasons allow many opportunities for replacement clutches.

When the differential in potential for polygamy is small between the sexes, "threshold" environmental conditions necessary for polyandry to evolve are more frequently met, and polyandry may evolve as often as polygyny. The evidence reviewed in this chapter also indicates that males need not benefit by female emancipation for polyandry to evolve, just as females need not benefit for polygyny to evolve.

Acknowledgments

I extend my deepest appreciation to Lynn Erckmann for assistance at every stage in the preparation of this chapter, from field research to final editing and typing. I also thank Sievert Rohwer, Wendy Hill, Frank Pitelka, and Jim Wittenberger for providing useful comments on an early draft. I thank Dennis Paulson for innumerable stimulating discussions of shorebird biology. Editor Sam Wasser displayed remarkable patience during the many revisions and suggested many improvements. Gordon Orians provided support, much useful advice, and the original inspiration to tackle this difficult problem.

References

Ali, S., and Ripley, S. D. *Handbook of the birds of India and Pakistan* (Vol. 2). London/New York: Oxford Univ. Press, 1969.

Ashkenazie, S., and Safriel, U. N. Time-energy budget of the semipalmated sandpiper *Calidris pusilla* at Barrow, Alaska. *Ecology*, 1979, *60*, 783–799.

Bannerman, D. A. *The birds of the British Isles* (Vol. 9). London: Oliver and Boyd, 1961.

Bellrose, F. C. *Ducks, geese, and swans of North America*. Harrisburg, Penna.: Stackpole, 1976.

Bergman, R. D., Howard, R. L., Abraham, H. F., and Weller, M. W. Water birds and the wetland resources in relation to oil development at Storkersen Point, Alaska. *Fish & Wildlife Service Resource Publication 129*. Washington, D.C.: U.S. Dept. of the Interior, 1977.

Bergstrom, P. W. Male incubation in Wilson's plover (*Charadrius wilsonia*). *Auk*, 1981, *98*, 835–838.

Boyd, H. Mortality and fertility of European Charadrii. *Ibis*, 1962, *104*, 368–387.

Bunni, M. *The Killdeer*, Charadrius v. vociferus *Linnaeus, in the breeding season: Ecology, behavior, and the development of homoiothermism*. Unpublished doctoral dissertation, University of Michigan, 1959.

Conway, W. G., and Bell, J. Observations on the behavior of Kittlitz's sandplovers at the New York Zoological Park. *Living Bird*, 1968, *7*, 57–70.

Dementev, G. P., and Gladkov, N. A. *Birds of the Soviet Union* (Vol. III). Jerusalem: Israel Program for Scientific Translations Ltd., 1969.

Drent, R. H., and Daan, S. The prudent parent: Energetic adjustments in avian breeding. *Ardea*, 1980, *68*, 225–252.

Dyrcz, A., Witkowski, J., and Okulewicz, J. Nesting of "timid" waders in the vicinity of "bold" ones as an antipredator adaptation. *Ibis*, 1981, *123*, 542–545.

Emlen, S. T., and Oring, L. W. Ecology, sexual selection, and the evolution of mating systems. *Science*, 1977, *197*, 215–223.

Erckmann, W. J. *The evolution of sex-role reversal and monogamy in shorebirds.* Unpublished doctoral dissertation, University of Washington, 1981.

Erckmann, W. J. *The relationship between body size and parental care among sandpipers (Scolopacidae).* Manuscript in preparation, n.d.

Faaborg, J., and Patterson, C. B. The characteristics and occurence of cooperative polyandry. *Ibis*, 1981, *123*, 477–484.

Flint, V. E. On the biology of the broad-billed sandpiper, *Limicola falcinellus sibiricus. Fauna and Ecology of Waders*, 1973, *1*, 98–99.

Gibson, F. The breeding biology of the American avocet (*Recurvirostra americana*) in central Oregon. *Condor*, 1971, *73*, 444–454.

Gibson, F. Ecological aspects of the time budget of the American avocet. *American Midland Naturalist*, 1978, *99*, 65–82.

Glutz von Blotzheim, U. N., Bauer, K. M., and Bezzel, E. *Handbuch der Vogel Mitteleuropas* (Parts 6 and 7). Wiesbaden: Akademische Verlagsgesellschaft, 1975, 1977.

Graul, W. D. Adaptive aspects of the mountain plover social system. *Living Bird*, 1974, *12*, 69–74.

Graul, W. D. Breeding biology of the mountain plover. *Wilson Bulletin*, 1975, *87*, 6–31.

Graul, W. D. Food fluctuations and multiple clutches in the mountain plover. *Auk*, 1976, *93*, 166–167.

Graul, W. D., Derrickson, S. R., and Mock, D. W. The evolution of avian polyandry. *American Naturalist*, 1977, *111*, 812–816.

Hagar, J. A. Nesting of the Hudsonian godwit at Churchill, Manitoba. *The Living Bird*, 1966, *5*, 5–43.

Harris, M. P. The biology of oystercatchers *Haematopus ostralegus* on Skokholm Island, S. Wales. *Ibis*, 1967, *109*, 180–193.

Hays, H. Polyandry in the spotted sandpiper. *Living Bird*, 1972, *11*, 43–57.

Heldt, R. Zur Brutbiologie des Alpenstrandläufers, *Calidris alpina schinzii. Corax*, 1966, *1*, 173–188.

Hildén, O. Breeding system of Temminck's stint *Calidris temminckii. Ornis Fennica*, 1975, *52*, 117–146.

Hildén, O. Occurrence and breeding biology of the little stint *Calidris minuta* in Norway. *Anser*, 1978a, Suppl. 3, 96–100.

Hildén, O. Population dynamics in Temminck's stint *Calidris temminckii. Oikos*, 1978b, *30*, 17–28.

Hildén, O. Nesting of Temminck's stints *Calidris temminckii* during an arctic snowstorm. *Ornis Fennica*, 1979a, *56*, 30–32.

Hildén, O. Territoriality and site tenacity of Temminck's stint *Calidris temminckii. Ornis Fennica*, 1979b, *56*, 56–74.

Hildén, O., and Vuolanto, S. Breeding biology of the red-necked phalarope *Phalaropus lobatus* in Finland. *Ornis Fennica*, 1972, *49*, 57–85.

Hills, S. Incubation capacity as a limiting factor of shorebird clutch size. *American Zoologist*, 1980, *20*, 744.

Hoffman, A. Über die Brutflege des polyandrischen Wasserfasans, *Hydrophasianus chirugus* (Scop.). *Zoologische Jahrbucher*, 1949, *78*, 367–403.

Höhn, E. O. Observations on the breeding biology of Wilson's phalarope (*Steganopus tricolor*) in central Alberta. *Auk*, 1967, *84*, 220–244.

Hohn, E. O. Some observations on the breeding of northern phalaropes at Scammon Bay, Alaska. *Auk*, 1968, *85*, 316–317.

Holmes, R. T. Breeding ecology and annual cycle adaptations of the red-backed sandpiper (*Calidris alpina*) in northern Alaska. *Condor*, 1966, *68*, 3–46.

Holmes, R. T. Ecological factors influencing the breeding schedule of western sandpipers (*Calidris mauri*) in subarctic Alaska. *American Midland Naturalist*, 1972, *87*, 472–491.

Howe, M. A. Behavioral aspects of the pairbonds in Wilson's phalarope. *Wilson Bulletin*, 1975a, *87*, 248–270.

Howe, M. A. Social interactions in flocks of courting Wilson's phalaropes (*Phalaropus tricolor*). *Condor*, 1975b, *77*, 24–33.

Howe, M. A. Social organization in a nesting population of eastern willets (*Cataptrophorus semipalmatus*). *Auk*, 1982, *99*, 88–102.

Jayaker, S. D., and Spurway, H. The yellow-wattled lapwing *Vanellus malabaricus* (Boddaert), a tropical dry-season nester. III. Two further seasons' breeding. *Journal of the Bombay Natural History Society*, 1968, *65*, 369–383.

Jehl, J. R., Jr. Patterns of hatching success in subarctic birds. *Ecology*, 1971, *52*, 169–173.

Jehl, J. R., Jr. Breeding biology and systematic relationships of the stilt Sandpiper. *Wilson Bulletin*, 1973, *85*, 115–147.

Jenni, D. A. Evolution of polyandry in birds. *American Zoologist*, 1974, *14*, 129–144.

Jenni, D. A., and Betts, B. J. Sex differences in nest construction, incubation, and parental behavior in the polyandrous American jacana (*Jacana spinosa*). *Animal Behaviour*, 1978, *26*, 207–218.

Jenni, D. A., and Collier, G. Polyandry in the American jacana (*Jacana spinosa*). *Auk*, 1972, *89*, 743–765.

Johnsgard, P. A. *Grouse and quails of North America*. Lincoln: Univ. of Nebraska Press, 1973.

Johnsgard, P. A. *The plovers, sandpipers, and snipes of the world*. Lincoln: Univ. of Nebraska Press, 1981.

Kagarise, C. M. Breeding biology of the Wilson's phalarope in North Dakota. *Bird-Banding*, 1979, *50*, 12–22.

Kistchinski, A. A. Breeding biology and behavior of the grey phalarope *Phalaropus fulicarius* in east Siberia. *Ibis*, 1975, *117*, 285–301.

Kistchinski, A. A., and Flint, V. E The biology of the ruff in the Yana-Indigirka lowlands. *Fauna and Ecology of Waders*, 1973, *1*, 57–60.

Klomp, H. The decline of the lapwing, *Vanellus vanellus* (L.), in Holland and data on its laying mechanism and the egg production capacity. *Ardea*, 1951, *39*, 143–182.

Kochanow, W. D. Studies of the ecology of Temminck's stint in the Kandalakscha Gulf of the White Sea. *Fauna and Ecology of Waders*, 1973, *1*, 66–71.

Koenig, W. D. Ecological and social factors affecting hatchability of eggs. *Auk*, 1982, *99*, 526–536.

Lack, D. *Ecological adaptations for breeding in birds*. London: Chapman & Hall, 1968.

Laven, H. Beitrage zur Biologie des Sandregenpfeifers (*Charadrius hiaticula* L.). *Journal für Ornithologie*, 1940, *88*, 183–187.

Lenington, S. Bi-parental care in killdeer: An adaptive hypothesis. *Wilson Bulletin*, 1980, *92*, 8–20.

Leopold, A. S. *The California quail*. Berkeley: Univ. of California Press, 1977.

Lind, H. *Studies of the behavior of the black-tailed godwit* (Limosa limosa L.). Copenhagen: Munksgaard, 1961.

Lowe, V. T. Observations on the painted snipe. *Emu*, 1963, *62*, 221–237.

Maclean, G. L. The breeding biology and behavior of the double-banded courser *Rhinoptilus africanus* (Temminck). *Ibis*, 1967, *109*, 556–569.

Maclean, G. L. A study of seedsnipe in southern South America. *Living Bird*, 1969, *8*, 33–80.

Maclean, G. L. Clutch size and evolution in the Charadrii. *Auk*, 1972, *89*, 299–324.

MacLean, S. F., Jr. Ecological determinants of species diversity of arctic sandpipers near Barrow, Alaska. Unpublished doctoral dissertation, University of California, 1969.

MacLean, S. F., Jr. Lemming bones as a source of calcium for arctic sandpipers (*Calidris* spp.). *Ibis*, 1974, *116*, 552–557.

Mason, C. F., and Macdonald, S. M. Aspects of the breeding biology of the snipe. *Bird-Study*, 1976, *23*, 33–38.

Mathew, D. N. Observations on the breeding habits of the bronze-winged jacana (*Metopidius indicus* (Latham)). *Journal of the Bombay Natural History Society*, 1964, *61*, 295–301.

Maxson, S. J., and Oring, L. W. Mice as a source of egg loss among ground-nesting birds. *Auk*, 1978, *95*, 582–584.

Maxson, S. J., and Oring, L. W. Breeding season time and energy budgets of the polyandrous spotted sandpiper. *Behaviour*, 1980, *74*, 200–263.

Mayfield, H. F. Red phalaropes breeding on Bathurst Island. *Living Bird*, 1978, *17*, 7–39.

Maynard Smith, J. Parental investment: A prospective analysis. *Animal Behaviour*, 1977, *25*, 1–9.

Mendall, H. L., and Aldous, C. M. *The ecology and management of the American woodcock*. Orono, Maine: Maine Cooperative Wildlife Research Unit, 1943.

Mickelson, P. G. Avian community ecology at two sites on Espenberg Peninsula in Kotzebue Sound, Alaska. *Environmental Assessment of the Alaskan Continental Shelf. Final Reports of Principal Investigators* (Vol. 5), U.S. Dept. of Commerce, OCSEAP, 1979, 289–607.

Miller, E. H. Egg size in the least sandpiper *Calidris minutilla* on Sable Island, Nova Scotia, Canada. *Ornis Scandinavica* 1979a, *10*, 10–16.

Miller, E. H. Functions of display flights by males of the least sandpiper, *Calidris minutilla* (Vieill.), on Sable Island, Nova Scotia. *Canadian Journal of Ecology*, 1979b, *57*, 876–893.

Miller, J. R., and Miller, J. T. Nesting of the spotted sandpiper at Detroit, Michigan. *Auk*, 1948, *65*, 558–567.

Mundahl, J. T. Role specialization in the parental and territorial behavior of the killdeer. Unpublished Master's thesis, Utah State University, 1977.

Mundahl, J. T., Johnson, O. L., and Johnson, M. L. Observations at a twenty-egg killdeer nest. *Condor*, 1981, *83*, 180–182.

Myers, J. P. Cross-seasonal interactions in the evolution of sandpiper social systems. *Behavioral Ecology and Sociobiology*, 1981, *8*, 195–202.

Nethersole-Thompson, D. *The dotterel*. Glasgow: William Collins Sons, 1973.

Nethersole-Thompson, D. *Greenshanks*. Vermillion, S. Dak.: Buteo Books, 1979.

Nettleship, D. N. Breeding ecology of turnstones *Arenaria interpres* at Hazen Camp, Ellesmere Island, N.W.T. *Ibis*, 1973, *115*, 202–217.

Norton, D. W. Incubation schedules of four species of calidrine sandpipers at Barrow, Alaska. *Condor*, 1972, *74*, 164–176.

Norton, D. W. Ecological energetics of calidrine sandpipers breeding in northern Alaska. Unpublished doctoral dissertation, University of Alaska, 1973.

Olson, S. L., and Steadman, D. W. The relationships of the Pedionomidae (Aves: Charadriiformes). *Smithsonian Contributions to Zoology,* 1981, No. 337, 1–25.

Orians, G. H. On the evolution of mating systems in birds and mammals. *American Naturalist,* 1969, *103,* 589–603.

Oring, L. W. Avian mating systems. In D. S. Farner, J. R. King, and K. C. Parkes (Eds.), *Avian biology* (Vol. 6). New York: Academic Press, 1982.

Oring, L. W., and Knudson, M. L. Monogamy and polyandry in the spotted sandpiper. *Living Bird,* 1972, *11,* 59–73.

Oring, L. W., and Maxson, S. J. Instances of simultaneous polyandry by a spotted sandpiper *Actitis macularia. Ibis,* 1978, *120,* 349–353.

Osborne, D. R. *Seasonality of polyandry in the wattled jacana.* Paper presented at the meeting of the American Ornithological Union, Fort Collins, Colorado, August 1980.

Osborne, D. R., and Bourne, G. R. Breeding behavior and food habits of the wattled jacana. *Condor,* 1977, *79,* 98–105.

Page, G. W., Fearis, B., and Jurek, R. M. Age and sex composition of western sandpipers on Bolinas Lagoon. *California Birds,* 1972, *3,* 79–86.

Parker, G. A., Baker, R. R., and Smith, V. G. F. The origin and evolution of gamete dimorphism and the male–female phenomenon. *Journal of Theoretical Biology,* 1972, *36,* 529–553.

Parmelee, D. F. Breeding behavior of the sanderling in the Canadian High Artic. *Living Bird,* 1970, *9,* 97–146.

Parmelee, D. F., Greiner, D. W., and Graul, W. D. Summer schedule and breeding biology of the white-rumped sandpiper in the central Canadian arctic. *Wilson Bulletin,* 1968, *80,* 5–29.

Parmelee, D. F., and Payne, R. B. On multiple broods and the breeding strategy of arctic sanderlings. *Ibis,* 1973, *115,* 218–226.

Parr, R. Population study of golden plover, *Pluvialis apricaria,* using marked birds. *Ornis Scandinavica,* 1980, *11,* 179–189.

Pearson, D. J. The wintering moult of ruffs *Philomachus pugnax* in the Kenyan Rift Valley. *Ibis,* 1981, *123,* 158 182.

Phillips, R. E. Behavior and systematics of New Zealand plovers. *Emu,* 1980, *80,* 177–197.

Pienkowski, M. W., and Green, G. H. Breeding biology of sanderlings in north-east Greenland. *British Birds,* 1976, *69,* 165–177.

Pienkowski, M. W., and Greenwood, J. J. D. Why change mates? *Biological Journal of the Linnaean Society,* 1979, *12,* 85–94.

Pitelka, F. A., Holmes, R. T., and MacLean, S. F., Jr. Ecology and evolution of social organization in arctic sandpipers. *American Zoologist,* 1974, *14,* 185–204.

Prater, A. J. Breeding biology of the ringed plover *Charadrius hiaticula.* In A. J. Prater (Ed.), *Proceedings of the IWRB Wader Symposium.* Warsaw: Warsaw Univ. Press, 1974.

Pulliainen, E. On the breeding biology of the dotterel (*Charadrius morinellus*). *Ornis Fennica,* 1970, *47,* 69–73.

Pulliainen, E. *Breeding behavior of the dotterel, Charadrius morinellus.* Varrio Subarctic Research Station, Report 24, 1971.

Raner, L. Förekommer polyandri has smalnäbbad simsnäppa (*Phalaropus lobatus*) och svartsnäppa (*Tringa erythropus*)? *Fauna och Flora,* 1972, *67,* 135–138.

Ricklefs, R. E. An analysis of nesting mortality in birds. *Smithsonian Contributions to Zoology,* 1969, *9,* 1–48.

Ricklefs, R. E. Energetics of reproduction in birds. In R. A. Paynter (Ed.), *Avian energetics.* Cambridge, Mass.: Nuttall Ornithological Club, 1974.

Ridley, M. Paternal care. *Animal Behaviour,* 1978, *26,* 904–932.

Ridley, M. W. The breeding behavior and feeding ecology of grey phalaropes *Phalaropus fulicarius* in Svalbard. *Ibis,* 1980, *122,* 210–226.

Rohwer, S. A., and Erckmann, W. J. The "efficient male" hypothesis for sexual size dimorphism in sandpipers (Scolopacidae). Manuscript in preparation, n.d.

Ross, H. A. Multiple clutches and shorebird egg and body weight. *American Naturalist,* 1979, *113,* 618–622.

Rubinstein, N. A. Comparative studies of the behavior of certain wader species during the incubation period. *Fauna and Ecology of Waders,* 1973, *1,* 134–137.

Safriel, U. N. On the significance of clutch size in nidifugous birds. *Ecology,* 1975, *56,* 703–708.

Schamel, D., and Tracy, D. Polyandry, replacement clutches, and site tenacity in the red phalarope (*Phalaropus fulicarius*) at Barrow, Alaska. *Bird Banding,* 1977, 48, 314–324.

Sheldon, W. G. *The book of the American woodcock.* Amherst: Univ. of Massachusetts Press, 1967.

Shibnev, B. K. Shore birds of the Bikin River basin. *Fauna and Ecology of Waders,* 1973, *2,* 83–86.

Skutch, A. F. *Parent birds and their young.* Austin: Univ. of Texas Press, 1976.

Soikkeli, M. Breeding cycle and population dynamics in the dunlin (*Calidris alpina*). *Annales Zoologici Fennici,* 1967, *4,* 158–198.

Soikkeli, M. Mortality and reproductive rates in a Finnish population of dunlin *Calidris alpina. Ornis Fennica,* 1970, *47,* 149–158.

Sordahl, T. A. Predator mobbing behaviour in the shorebirds of North America. *Wader Study Group Bulletin,* 1981, *31,* 41–44.

Spaans, A. L. Molt of flight and tail feathers of the least sandpiper in Surinam, South America. *Bird Banding,* 1976, *47,* 359–364.

Summers, R. W., and Hockey, P. A. R. Breeding biology of the white-fronted plover (*Charadrius marginatus*) in the south-western Cape, South Africa. *Journal of Natural History,* 1980, *14,* 433–445.

Trivers, R. L. Parental investment and sexual selection. In B. Campbell (Ed.), *Sexual selection and the descent of man.* Chicago: Aldine, 1972.

Vernon, C. J. Polyandrous *Actophilornis africana. Ostrich,* 1973, *44,* 85.

Whittaker, R. H. *Communities and ecosystems.* New York: Macmillan Co., 1975.

Wilcox, L. A twenty year banding study of the piping plover. *Auk,* 1959, *76,* 129–152.

Wilkie, A. O. M. Incubation rhythm and behaviour of a dotterel *Charadrius morinellus* nesting in Norway. *Ornis fennica,* 1981, *58,* 11–20.

Wilson, J. Trigamy in the lapwing. *British Birds,* 1967, *60,* 217.

Wittenberger, J. F. The evolution of mating systems in birds and mammals. In P. Marler and J. Vandenbergh (Eds.). *Handbook of behavioral neurobiology: social behavior and communication.* New York: Plenum, 1979.

Wittenberger, J. F. *Animal social behavior.* Boston: Duxbury Press, 1981.

Wittenberger, J. F., and Tilson, R. L. The evolution of monogamy: Hypotheses and evidence. *Annual Review of Ecology and Systematics,* 1980, *11,* 197–232.

Yarbrough, C. G. Summer lipid levels of some subarctic birds. *Auk,* 1970, *87,* 100–110.

7

Human Female Reproductive Strategies

WILLIAM IRONS

I. Introduction

This chapter presents a preliminary theoretical discussion of human female reproductive strategies based on a neo-Darwinian perspective.

SOCIAL BEHAVIOR OF
FEMALE VERTEBRATES

ISBN 0-12-735950-8

Since female strategies are partly a response to male strategies, some attention is also given to the latter.

The propositions explored relate to the following four strategies. First, throughout human evolution, female reproductive success has been limited by the assistance women have received from others in nurturing their children (cf. Robinson *et al.*, 1980). The primary strategy women have evolved for increasing such assistance is the reciprocal exchange of various forms of assistance with husbands and with consanguineous kin. The use of reciprocity for this purpose is a particular instance of a generalized, evolved human propensity to use reciprocity as a means of increasing inclusive reproductive success (cf. Darwin, 1871, p. 157; Williams, 1966, pp. 93–96, 203; Trivers, 1971; Fox, 1972, 1979; Axelrod and Hamilton, 1981; Wasser, 1982). Like all human behavior, use of reciprocity for this purpose is highly flexible in response to different environments. Among the features of female reciprocity that have varied are (*a*) the nature of the assistance exchanged; (*b*) the extent to which women have relied on aid gained through exchange and, conversely, the extent to which they have relied on their own resources for child rearing; and (*c*) whether women have developed their closest reciprocal ties with husbands or with sisters, mothers, and other consanguineous kin. The choices women have made concerning reciprocal aid are important determinants of the varied kinship and marriage systems found among human societies.

Second, in addition to attempting to increase resources for parental investment, women have also adjusted the size of their families to the resources available to them for child nurturance. They reduce the number of children born and increase nurturance per child when they can gain genetic representation in future generations by doing so. When they gain by the reverse strategy, they increase the number of children born and reduce parental investment per child. Postpartum taboos on sexual intercourse and suppression of ovulation through lactation are examples of proximate mechanisms related to this strategy. Third, women have attempted to choose mates whose phenotypes indicate a high probability of producing fit offspring, and fourth, women have devised strategies for countering the use of violence by males to coerce female behavior that serves male interests at the expense of female interests. Briefly, these four strategies may be described as: (*a*) using reciprocity to increase parental care, (*b*) family planning, (*c*) choosing fit mates, and (*d*) countering male coercion.

The chapter explores primarily theoretical statements that can be tested with ethnographic data. As a result, more is said about some strategies than others because the ethnographic literature has more to say about some areas of social life than others. In particular, the literature is richest in material relevant to the strategy of increasing resources for parental investment through reciprocity.

The theoretical propositions presented derive from the assumption that human beings behave so as to maximize their inclusive fitness (Hamilton 1963, 1964). In most situations, this assumption works well in the study of animal behavior. In studying human behavior, however, one has to be aware that human environments have been changing at an accelerating rate and that environmental novelty may elicit behaviors that do not maximize inclusive fitness. Therefore, the chapter begins with a discussion of the range of human environments in which the assumption of fitness maximization can be expected to hold.

The chapter is organized as follows. Section II attempts to delineate the range of human societies in which behavior can be expected to maximize inclusive fitness. Section III reviews some ethnographic data concerning human breeding systems. These data serve two purposes: (*a*) to allow inferences concerning some of the constraints influencing reproductive strategies during human evolution; and (*b*) to provide some specific examples of mating and parental investment strategies employed by women in environments similar to those of evolution. Section IV presents the central theoretical argument concerning how and why women in various societies have developed different patterns of reciprocity, different forms of family planning, and different strategies for choosing fit mates and resisting male coercion. The last section is a summary.

II. Neo-Darwinian Theory and Novel Environments

A number of authors have pointed out that the assumption of inclusive fitness maximization often will not hold in novel environments (Alexander, 1975, 1979b, p. 78; Irons, 1977, 1979a, p. 38, 1979c, p. 272; Alcock, 1978, p. 467; Symons, 1979, pp. 31–38; Daly, 1980). Symons has made this point more forcefully than most authors, and has correctly stated that behaviors promoting reproductive success in novel environments should be considered evolved adaptations only if it can be argued that these behaviors had similar effects in the environments of evolution. Symons has also taken an extreme position concerning the limits of ultimate cause explanations in human societies.

A. Agriculture: Artificial Environmental or New Niche?

For Symons (1979), the dividing line between "natural environments," where proximate mechanisms produce adaptive behavior, and "artificial environments," where proximate mechanisms no longer produce adaptive behavior, is the origin of agriculture 10,000 years ago (pp. 35–36).

The expressions "natural environments" and "artificial environments" in this context suggest a parallel with wild animals in such artificial environments as zoos or laboratories. Thinking of human beings in food-producing societies as closely analogous to animals thrust by human captors into such artificial environments is misleading. Human beings themselves created agriculture; they were not—at least initially—thrust into it contrary to their will. Further, one effect of agriculture was a tremendous population growth.

Consider what would happen if an animal behaviorist were to observe the following in a wild population: (*a*) the appearance of a new means of gathering food, (*b*) the spread of this novel food-gathering technique throughout the population, and (*c*) a one hundredfold increase in population as a direct consequence of this new form of food procurement. Our hypothetical observer would probably not describe this new behavior as the creation of an artificial environment rendering most behavior maladaptive.

On the other hand, agriculture and its consequences are in many ways novel for human beings. Initially, domestication was probably merely a means of choosing a method of food acquisition which, compared to earlier methods, was more productive, or more dependable, or both. Eventually, however, it created situations of population density, group size, and other conditions that were indeed novel. The situation is probably not greatly different from that of other populations that have entered new niches. It seems to me reasonable to speak of agriculture as a niche distinct from foraging. The innovative behaviors that represent exploiting a new niche lead initially to a higher rate of reproduction; however, other traits are not finely tuned to the new way of life. Thus, following entry to the new niche, some evolution occurs. In the case of human beings, such evolutionary changes since food production have probably been very minor. Known examples of evolutionary adjustment to food production include the evolution of sickling as a response to malaria, which became endemic in some areas after the establishment of food production (Livingstone, 1958). Another example is the maintenance of lactase in adulthood in human populations with a long history of using the milk of domestic animals as an important food source (Weiss and Mann, 1981, p. 476). These are, however, physiological adjustments to specific environments. Changes in behavioral traits following agriculture seem less probable. This is so, I believe, because human beings before agriculture had evolved to deal with continually changing environments; novelty, in effect, had become the normal condition, and they evolved behavioral mechanisms for coping with a wide range of novelty.

If behavioral changes were to occur, they might be detectable as differences among human populations in behavioral predisposition corresponding to the different time periods of involvement with agriculture. However, attempts to study heritable behavioral differences among human beings point overwhelmingly to the conclusion that such differences occur in small measure between individuals, but not among populations (Spuhler, 1967; Loehlin *et al.*, 1975).

B. How Many Generations for Evolutionary Change?

It is also worth noting that it is hard to defend Symons' (1979) statement that the 10,000 years since the origin of agriculture are too short a time for evolution to occur. The relevant question is the number of generations, not years, during which selection has been able to act. Howell's data on !Kung foragers give the average age of women at childbirth as 29.5 years (Howell, 1979, p. 214). The mean age for becoming a father is probably somewhat higher. This suggests that a century would represent on average roughly three human generations, with the specific number in particular lines of descent varying around this average. Therefore, the average number of generations separating us from each of our ancestors of 10,000 years ago is around 300. Some human populations, thus, have a history of food production going back at least 300 generations. The majority probably have histories of food production going back at least 100 generations. If one really believes that selective pressures have been very different since the origin of domestic food sources, then one would have to conclude that food-producing populations have undergone evolutionary change since adopting agriculture. Cities and the state go back at least 5,000 years or 150 generations in some areas of the world. Again, if the social environments created by cities and states favor very different behavioral predispositions, human populations with long histories of these kinds of environments should have, at least, begun to evolve different behavioral propensities.

C. Adaptation and Maladaptation in Postforaging Societies

It seems to me that in examining any particular behavioral propensity, we should try to determine both its effects on reproduction in foraging societies and its effects in various types of societies that emerged after the origin of food production. Many behavioral predispositions continue to have the same effect on reproductive success that they had in foraging societies. For example, if it is true that women establish and maintain social ties with husbands and other kin with the object of increasing their resources for child nurturance, then I see no need to assume this behavior has effects on reproduction in food-producing societies that are different from its effects in foraging societies. The same can be said of many behavioral propensities, such as the tendency of women to prefer mates with physical characteristics indicating strength and good health, or the propensity of men to seek multiple mates under certain circumstances.

On the other hand, there are behaviors that do have different effects in more recent environments. The most salient of these, in my opinion, are behaviors relating to modern practices of fertility limitation. Intentional fertility limitation can, of course, render other adaptations ineffective. A woman who chooses a mate with vast resources and a healthy phenotype but has herself sterilized as a contraceptive measure renders her mate's choice ineffective as an evolutionary strategy. Nevertheless, this sort of behavior is not typical of female behavior in food-producing societies. Most women since the origin of agriculture have not opted for sterilization. Extensive fertility limitation by contraception is much more recent than food production. The ethnographic record indicates that, in the majority of societies preceding the recent phenomenon known as the demographic transition, both women and men have desired high fertility.

Other good candidates for the status of maladaptive response to a novel environment include certain forms of drug use (Livingstone, 1980) and perhaps overeating and underexercising. Apart from these, the effect of most human behavior, in my opinion, is still to promote the genetic survival of the individual exhibiting the behavior; the only conspicuous exception relevant to female reproductive strategies is modern fertility control. (Stating that modern fertility control is maladaptive biologically in no way implies that it is immoral; one must avoid the naturalistic fallacy of assuming that what selection favors is morally good and what it disfavors is morally bad.)

For this reason, it is useful to examine the behavior of women in all societies not characterized by modern forms of intentional fertility limitation to determine the nature of evolved reproductive strategies. At the same time, in looking at behavior in food-producing societies, one should ask whether its effect on reproduction is different from its effect on foraging societies. In most cases, in my opinion, the answer is no.

III. Human Breeding Systems: Some Ethnographic Examples

The ethnographic record reveals a wide range of behaviors in reference to mating and child-rearing. This statement is true even if one examines only that portion of the ethnographic record dealing with foraging societies, although for the reasons explained above it seems unnecessary to restrict the data examined in that way. The best way to highlight the extent and nature of variation in mating and child-rearing practices among human societies is to provide a few ethnographic sketches. These ethnographic data are offered not as rigorous empirical support for the hypotheses discussed in Section IV, but rather as material that suggests these hypotheses are worth testing and provides some concrete examples of what is meant by the more abstract discussion below. The societies described, the Plateau Tonga, the Tiwi, and the pastoral Yomut, provide a good illustration of the range of variation in female mating strategies among human societies. The data presented below are taken from the following sources: for the Plateau Tonga, Colson (1958); for the Tiwi, Hart and Pilling (1960) and Goodale (1971); and for the pastoral Yomut, Irons (1975).

A. The Plateau Tonga

The Plateau Tonga of Malawi were originally a horticultural people. They practiced shifting cultivation of sorghums and millets, with hoe and axe as their primary tools. They also herded some cattle and did some hunting and fishing. Arable land existed in surplus of demand, and free land was available to anyone who wished to claim some. The basic unit responsible for economic production was a household, the core of which was a husband, wife, and children. This group formed a single work team. The husband and boys worked together to clear two separate fields, one for the husband himself and one for the wife. Herding was done by adolescent males. The wife and girls planted, seeded, hoed, and did most of the harvesting. Produce from each field was placed in separate granaries for the husband and for the wife. Each then made decisions about the use of the grain in his or her granary. The wife was obligated to prepare food for her husband and children, and the husband was obligated to use his produce to buy clothing, tools, and household utensils and in general see to the other economic needs of his wife and the children of the household. The husband did not need to account to his wife for his use of his income; he only needed to meet her and her children's economic needs. She managed the allocation of grain from her granary for home consumption without interference from her husband. She could sell surplus grain from her granary only with his consent and was obligated to share the cash income with him. Her share of the cash in such cases was hers to do with as she saw fit. She was also free to earn money on her own by such activities as contracting to brew beer, making and selling pots and baskets, or acting as a diviner. This income was hers and not subject to her husband's supervision. She could also own cattle and chickens separately from her husband and refuse to use them instead of his to feed the family.

Households commonly also included relatives of either the husband or wife: children by other marriages, siblings, parents, or more distant kin. In fact children, once they were old enough to travel on their own, were free to live with whomever they chose: grandparents, older married siblings, other relatives, or with either parent, should their parents be divorced. Children were a valuable source of labor and were generally welcome wherever they chose to reside.

In Colson's (1958, p. 119) census data, 23% of the men were married polygynously, most with only two wives. Cowives maintained completely separate households and were not obligated to work in each other's fields. Cowives were never closely related and were generally jealous of the favors given by their husbands to other cowives. Men, in contrast, generally preferred polygyny although only a minority succeeded in realizing the preference. Sororal polygyny or polygyny including close cousins as cowives was forbidden.

Lover relationships were common for women before marriage and about 40% had children before marriage. Such "bush" children were usually fathered by the man the mother intended to marry. When "bush" children were born, the mother's lover was obligated to make a payment to the woman's maternal and paternal matrilineage and acknowledge paternity. Though disapproved, clandestine lover relationships after marriage were common. Colson (1958, p. 165) states that "husbands, despite their own affairs—or perhaps because of them—seem considerably exercised about the possibility that their wives have lovers." She also says that the Plateau Tonga considered it "natural" for mature adults to be adulterous at times. Once discovered, such relationships led to little scandal and were not considered to have damaged either partner seriously. Compensation was paid in cattle by lovers to husbands who, though they disliked their wives' infidelity, sometimes kept their wives even though they had caught them in adultery many times over. Divorce was common, however, because of adultery or other grounds, and could be initiated by either wives or husbands.

Children conceived by lovers were considered legal children of the woman's husband. In general, men desired many children and wished to keep children in their households because of the value of their labor. On a man's death his property passed to his matrilineal kin among whom it was divided. His children inherited nothing, although a man had the option of giving wealth to his children while he was still alive, thus circumventing matrilineal inheritance if he wished.

Colson (1958, p. 117) states that in the past, variation in standard of living among households was very small. Accumulated wealth consisted primarily of livestock, which was secondary to land as a source of food and other subsistence products. Inheritance of accumulated wealth was not an important affair. A deceased individual's wealth was dispersed among members of his matrilineage, and his or her land was often abandoned. (Land was, as noted previously, a free good.) In more recent times, land has become scarce and many households convert much of their labor into capital equipment such as ploughs, wagons, and even trucks and tractors. Under these conditions, wives and their children resent seeing property pass to a deceased husband's matriline, and there was, at the time of Colson's (1958, p. 119) observations, a movement underway to replace matrilineal inheritance with the right to dispose of property by will. Thus, the social relations described above were characteristic of only the most conservative Plateau Tonga and were in the process of disappearing at the time of Colson's field research. This social system, however, was viable among the Plateau Tonga so long as their means of exploiting their environment was shifting cultivation of sorghum and millets with limited herding, and so long as land remained a free good. As such it is useful as an example of a society in which the resources a man could offer a woman in return for mating opportunities were of limited value relative to their value in many other societies. As a consequence, the extent of male–female cooperation was limited.

Colson mentions that "informants recall noted amazons who rid themselves of unwanted husbands and succeeded in inducing their daughters to follow the same path. Women then lived together in a separate homestead or hamlet, accepting as lovers men who did not become permanent members of their establishment. They acquired cattle which were herded by their sons" (Colson, 1961, p. 68). She also mentions having met widows and divorcees who claimed to do all of their own horticultural labor (Colson, 1958, p. 138).

Husbands were accorded the right to discipline wives, and this included the right to beat them within well-defined limits. Women, however, were given the right to leave a husband when dissatisfied with his behavior. It was common practice for a woman to pack up her belongings and small children and go to parents or other kin, with no intention of leaving her husband permanently. The husband, if he wanted his wife back (as was usually the case), was forced to appear periodically at the door of his parents-in-law or other affine, begging forgiveness. Men in this situation were frequently the butt of jokes. If the woman's kin sided with her, they had the right to demand that the husband pay damages before they would agree to the wife's return. If a woman were sufficiently dissatisfied, she could demand a divorce, and if her kin supported her she could have one. If not, it was more difficult for a woman to rid herself of a husband, but some women were known to persist, contrary to family wishes, and eventually rid themselves of husbands they did not like. Thus, while the man's greater ability to use violence gave him some advantage in negotiating his relationship with his wife, women could respond with strong countermeasures of their own.

Among the Tonga, it was considered desirable for a woman to marry immediately after completing a set of rituals marking her puberty. Some, however, delayed marriage while having affairs with potential husbands for some years. It was, as noted previously, common for women to have "bush" children during these years. Men generally married later in life, and husbands commonly were older than their wives. It was thought in no way unusual or undesirable for a husband to be as much as 20 years older than his wife.

Marriages were arranged by close kin of the parental generation of the bride and groom. The matrilineage of both spouses' mothers and fathers were involved. Sometimes the wishes of prospective brides and grooms were overridden by the older generation. In these cases, the prospective bridge and groom were sometimes known to resist, sometimes with success, though more often without. Also, occasionally young people eloped or attempted to force a marriage by openly living together. In this case also, the final outcome did not inevitably follow the wishes of either the older or younger generation involved.

When a match was agreed upon, a full commitment to marriage was not made immediately. A series of payments were made by the groom and his kin to the bride's kin and the marriage was thought of as being in a trial stage until the final payment was made. These payments consisted of hoes, spears, and livestock and were not of negligible value. In cases of divorce, a portion of these payments was returned. On the other hand, should the marriage last and the cattle given grow in number, some of the calves born were returned to the groom's kin. Thus, an economic tie was maintained between the lineages of a married pair. Marriages were also commonly arranged between the children of men who were political allies of some sort. However, the Tonga did not have the ideal of reciprocal exchange of women between two lineages over several generations. It was considered improper for close male lineagemates to take wives who were close lineagemates. Thus, a marriage between two lineages of shallow generational depth precluded further marriages between the same groups. This prevented long-term reciprocal exchanges of brides between the same lineages.

During the early years of a marriage, the groom built a hut for himself and his bride, but they were not allowed to cook there, entertain friends there, or conduct certain household rituals there. Each continued to depend on separate households of older kin, usually parents, for these purposes. After one or two children had been born and it was agreed by all concerned that the marriage should continue, the final bridal payments were made and the new couple acquired the right to maintain a hearth for cooking, dedicate their hut to ancestor spirits, and in general assume the status of a fully independent household.

Both men and women desired high fertility. One of the reasons for preferring early marriage for young women was the feeling that their fecundity was a precious, but time-limited, gift that should not be wasted. Abortion was much disapproved of and rare. Infanticide occurred only in the unusual situation of a woman's conceiving before completing her puberty ceremony.

There is, however, evidence of fertility limitation by means of abstaining from intercourse. Before the time of Colson's (1958) observations, intercourse was not resumed for some time after a live birth. However, it was resumed immediately after miscarriage or stillbirth. It was considered shameful to wean children by force, yet necessary to wean a child if its mother became pregnant. The intention of resuming intercourse had to be announced to the midwife who delivered the child, and rituals involving the midwife needed to be performed before resuming sexual relations. Informants told Colson that in the past, birth intervals were long. They further claimed that in these earlier days when raiding was common, a woman did not dare run the risk of having two children who could not run freely. If she did, she would be forced to carry two children while fleeing raiders.

At the time of Colson's visit, these practices had disappeared. Resumption of intercourse was decided on privately by husband and wife and varied from couple to couple. Some felt it appropriate to resume sexual relations after the child was a few months old. Others took the cutting of a first tooth, or the child's ability to stand unaided, as events signalling the appropriate moment. It was still, however, the practice not to wean children early. Colson (1958, p. 156) gives the range of breast feeding as 12–28 months, with 24 as the mode.

Whether birth intervals had diminished in the time period before Colson's observations is not rigorously documented, although it is true that informants had the impression that this had happened. If it did, it is probably relevant that the shortening followed not only the cessation of raiding, but also occurred at a time of increasing economic prosperity. Greater economic prosperity is very likely to be accompanied by improvements in nutrition. Both better nutrition and freedom from the disruptions of raiding would make it easier for women to rear successfully more closely spaced children.

B. The Tiwi

The Tiwi, a group of Australian aborigines, provide another interesting set of both female and male reproductive strategies. Like all Australian aborigines, the Tiwi, before they were altered through contact with Europeans, subsisted exclusively by foraging for wild plant and animal food sources. They followed the division of labor that is universal among foraging peoples, in which women gather the plant foods and men hunt. However, the division of labor was not rigidly adhered to. Women occasionally captured small animals while gathering plant foods and men occasionally gathered plant foods while hunting. The division of labor defined merely the primary activities and responsibilities of each sex. Also, as was common generally among foragers in warmer climates, the plant foods gathered by women provided a larger portion of the diet than the meat procured by the men.

The Tiwi practiced an extreme form of gerontocratic polygyny. The older men—those in their late thirties and older—monopolized females of reproductive age, excluding younger males from legitimate reproduction (Hart and Pilling, 1960, pp. 14–16). This practice was accompanied by the practice of betrothing girls before they were born, and keeping all women married or betrothed throughout their lives.

When a young woman, among the Tiwi, had her first menses, she was put through a puberty ceremony. As part of this ceremony, her as-yet-unconceived daughters were betrothed, as a group, to a man who became her son-in-law. (Note that she already had a husband of her own.) This son-in-law was chosen by the pubescent woman's father, that is, the current husband of her mother. The mean age of men who became new sons-in-law in this way was a little over 30, with a range from 14 to around 65 (Goodale, 1971, pp. 53, 65–66). A son-in-law was obligated to provide his mother-in-law with products of the hunt, as well as other services. For this purpose he usually became a member of his mother-in-law's camp group and moved with that group wherever they went in their nomadic food quest. This was a large camp group consisting of several households each with a separate campfire. The mother-in-law had the right to abrogate the betrothal of her son-in-law to her unborn daughters if she were dissatisfied with his service to her. Thus, she had considerable power over this man who was her senior in years. Should she dismiss him as a son-in-law, the right to betroth her daughters fell again in her father's hands. Later in life the right to betroth her daughters would move to her own husband.

A young girl usually left the circle of people who lived around her parents' campfire well before puberty and went to live with the group around her husband's campfire. Her husband was required at this point to live in the same larger camp group as her parents. This large group, consisting of several campfires, would assemble each evening in the same place to share food from the day's hunting and gathering activities. Thus, although a prepubescent wife was at this point under the supervision of her husband and his older wives if he had any, her parents were also in a position to observe how she was treated. If they were dissatisfied, they could abrogate the marriage and find her a new husband. Usually, however, husbands treated their young wives well, and most women, therefore, lived with their husbands during the latter portion of their childhood. On reaching puberty, a Tiwi woman acquired a son-in-law to serve her while she continued to serve her husband.

Because women were ordinarily much younger than their first husbands, most women became widows at an early age. The right to remarry her fell to her mother's husband during her mother's lifetime. After her mother's death it moved *de jure* to her brothers, and if she had no living brothers, to her sons. Most women exercised considerable *de facto* influence over their own marriages in later years. The men who had rights to bestow women established reciprocal ties with other men. In return for bestowing wives, they demanded wives for themselves and various forms of political and economic aid. Older men with many wives and many daughters to bestow often chose other old men as their allies, reinforcing the older males' near monopoly on wives. Younger men were given unborn girls as wives in return for living near and serving the mothers of their unborn wives. By the time these first wives matured, most men were in their late thirties.

The households of older, more successful men contained a large number of both older and younger wives. These households were economically the most viable ones, according to Hart and Pilling (1960, pp. 33–36). Plant foods were seasonal, and the annual yields of various regions were variable and uncertain. Large households with many wives could cope with this uncertainty by sending foraging parties in several different directions. Older women with much experience chose the directions in which to go and supervised the foraging. With this arrangement they seldom came up empty-handed. Smaller households, however, often did come up empty-handed. Since food could not be stored for more than a few days, such households were occasionally unable to feed themselves. They reacted to this by staying near larger, more successful households, whom they could call on for food when unable to feed themselves. For this they incurred obligations to reciprocate with various forms of aid, thus increasing the power of heads of large households.

Men hunted, and the products of the hunt were valued. However, hunting was, at best, undependable as a *daily* source of food, so that a steady diet depended on the success of the women of a household, not the men.

Hart and Pilling (1960), the first ethnographers of the Tiwi, offer the hypothesis that the Tiwi preferred sororal polygyny because of the economic viability of households with large female labor forces. A large male labor force was, given the specific ecology of the Tiwi, of less value. If this is correct, then in traditional Tiwi society, a woman's most valuable assistance in child-rearing would come from her sister rather than her husband. Sharing a husband with several cowives and changing husbands several times during her life was an acceptable cost to pay for the benefit of coresiding permanently with her sisters, and perhaps also with other close female relatives, such as half sisters by the same father and cousins. She also had the further compensation of having a son-in-law to provide her with the benefits of male economic activities.

In contrast to the Tonga, bestowals of women were assumed to carry the obligation to reciprocate by bestowing women on the wife-giver at some future point. The ideal was to maintain the exchange of women between two matrilineages over several generations, a practice which entailed cross-cousin marriage. Among the Tiwi, of course, there was no accumulated wealth to exchange. Thus, all that males could give in exchange for the scarce resource that women represented was other women or various services. Older men used the right to bestow daughters both to gain more wives and to gain young male clients who performed both economic and political services. It is also important to note that women in their later years acquired considerable power (Goodale, 1971, pp. 228–229). They not only supervised their husbands' younger wives and sons-in-law, but also exerted considerable influence on the marriage that their brothers or sons *de jure* had the right to arrange. Often, two older women would arrange to be married to each other's sons. This gave each woman's son the right to remarry widowed daughters of the other woman. This could then easily be translated into the *de facto* ability of each of these older women to arrange her daughters' remarriages and thus to launch her sons on typical Tiwi marital careers in which the right to bestow would be used to acquire wives (Hart and Pilling, 1960, pp. 19–20). Older women were expected to guard younger cowives against adultery and sometimes were known to use this duty as a pretext for spreading or manufacturing gossip about young men. In this way they could extend their power beyond sons, sons-in-law, and daughters' sons-in-law to other young men of their camp group. Adultery and gossip, and quarrels and feuds stemming from adultery appear to have been common among the Tiwi.

Divorce occurred among the Tiwi. As noted earlier, parents observed carefully both the behavior of their young daughters' husbands and sons-in-law and if dissatisfied, abrogated either arrangement. Because these marriages involved reciprocal obligations among several people, they were likely to cause considerable disruption—especially if older influential husbands were involved. Nevertheless, they did occur. Elopements also occurred, even though the result was often a feud that might last a long time. Thus, whatever one may say about the relative *de jure* rights of men and women in Tiwi society, it is clear that women sought power over both their own lives and those of others. It is also clear that many of them found ways of obtaining the power they sought, even power that *de jure* they were not supposed to have.

The ethnography of the Tiwi says relatively little about practices that might have limited fertility. Abortions did occur among young women who wished to continue sexual relations with lovers. Tiwi do not engage in sex during pregnancy. However, abortion appears not to have been common before the Tiwi began to be transformed by European influence (Goodale, 1971, pp. 145–146). The ethnography also mentions that in traditional Tiwi society, individuals could only gain in status through having children, a fact that would not seem to encourage limitation of fertility.

Information from other foraging societies, however, indicates that both suppression of ovulation through lactation and postpartum taboos on sex contributed to long birth intervals (Howell, 1979, pp. 189–211; Lee, 1979, pp. 325–330). The richest and most widely discussed data on fertility in a foraging society are those on the !Kung of the Kalahari (Howell, 1979; Lee, 1979), where birth intervals average a little over 4 years. Lee (1969, 1972, 1979) has observed that !Kung women often walk many miles in their foraging activities, and that without the long birth intervals that characterize this group, they would often be forced to carry two children on these long foraging trips. Blurton Jones and Sibly (1978) analyzed Lee's (1969, 1972) data and suggested that, given the environment and foraging patterns of the !Kung, birth intervals of less than 4 years would reduce their reproductive success. This would occur, according to their analysis, because a woman's work load would be increased by the necessity both of gathering more food and of carrying two children at once. The resultant stress on the mother during the most difficult season of the year would increase the risks of serious injury and death to the mother. This would not only cut short her reproductive career, but would also increase the risk of inadequate care of children already born. !Kung women who have become sedentary, however, are subject to much less stress and risk if they give birth more frequently, and these !Kung women have shortened their birth intervals and increased their birth rate. Demographic data on foraging societies are not adequate to allow firm empirical support for statements meant to apply to all, or most, foraging societies. Nevertheless, the data on the !Kung at least suggest the hypothesis that human beings evolved both physiological and behavioral mechanisms for adaptively adjusting fertility both upwards and downwards.

Tiwi husbands were not considered to have the right to beat their wives, even adulterous ones, although "a small-scale 'beating' " might be overlooked (Goodale, 1971, p. 132). They were expected instead to direct their violence toward an adulterous wife's lover. Husbands usually resided with their wives' natal groups, and their relationships with their wives were supervised by her close kin. If a husband were to "beat his wife too much," her father or brother could take her away from him (Goodale, 1971, p. 132). The Tiwi data strongly suggest that, in general, uxorilocal residence greatly strengthens a woman's ability to resist coercion by her husband.

Among other evidence of the ability of Tiwi women to resist their husbands' coercion was the widespread practice for women of maintaining lovers with whom they had regular clandestine sexual encounters (Goodale, 1971, pp. 131–32). While husbands never liked the practice and were thought to have a legitimate right to take steps to prevent it, close siblings of the lover and wife were often willing to aid in arranging "bush" meetings of lovers. Lovers were commonly classificatory brothers of the husband (Goodale, 1971, pp. 109–110, 131). This was so because the incest rules, which proscribed most of the other men of a woman's residence group as husbands, were respected in choosing lovers.

C. The Pastoral Yomut

The pastoral Yomut Turkmen, who were the subject of my own ethnographic work (Irons, 1975), offer an interesting contrast to both the Plateau Tonga and the Tiwi. The Yomut herded sheep and goats as their primary means of making a living. Agriculture (using ploughs with horses or camels) was a secondary source of livelihood. Herding required spending long periods of time away from the family campsite, especially in the dry season when young men would travel on their own for days away from the other members of their household. Danger from wolves and thieves, in addition to the hardship of the labor itself, made the task unsuitable for women with young children. Most Turkmen women produce a child every second or third year. Thus, travel for herding would have necessitated carrying infants and toddlers along as well as other younger children. It is not surprising that the care of the livestock was thought of as suitable work only for men. Agricultural labor was also thought of as men's work. Much of it was heavy, and the locations of the small patches of arable land available to the pastoral Yomut were often scattered and far from the family's campsite at the time when work needed to be done.

Women's work consisted of things that could be done near the family tent, and included child care, cooking, sewing, and the manufacture of various textile products, especially pile carpets, for both use and sale. Different households enjoyed different standards of living, and those with greater wealth enjoyed lower death rates and higher fertility (Irons, 1979c, 1980). The Yomut themselves are aware that greater wealth—larger herds and greater amounts of agricultural land—bring them a more comfortable and generally more desirable life. Thus, a woman's own well-being and that of her children is very much affected by her husband's labor. In his herding and agricultural activities, he provides scarce resources that she cannot supply herself. Groups of men are more productive economically than a single adult male, so households are commonly built around groups of men, fathers, and adult sons, or groups of brothers. The more adult males in a household, the greater its ability to acquire wealth. A woman who tried to support herself or to form a coalition of women to support themselves would enjoy little success. Occasionally widows do head households, but such households generally do not prosper and usually depend heavily on the assistance of brothers, or other male kin, of the deceased husband. Thus, a woman's best chance for reproductive success is to acquire a husband who has wealth and many close male kin and to do nothing to disrupt her husband's cooperation with his male kin.

The relationship of husband and wife and the exchange of assistance underlying this relationship, thus, form a clear contrast to that of the Plateau Tonga. Wives always leave their natal families after marriage and reside with their husbands and husband's parents and brothers. A few women are given a dowry in the form of a cow or a few sheep at marriage which is their property and is eventually inherited by their children. This is true, however, only of women from exceptionally wealthy natal families who, as a rule, marry into other wealthy families. The majority of women own no property other than some clothing and jewelry. Property is inherited from father to son. Divorce is extremely rare and results in strong disapproval. It can only be initiated by men, although sometimes a woman is able to persuade male kin to initiate divorce on her behalf. When a man divorces his wife, the bridewealth paid when the marriage occurred is not returned. When a woman's kin seek a divorce on her behalf, they must return the bridewealth. Since bridewealth is substantial, divorce is thus expensive as well as disapproved.

Extramarital liaisons are not tolerated. When young unmarried women become pregnant (which is rare), marriages are sometimes arranged, or alternatively, abortion or child desertion occur. Married women caught in adultery (also rare) are killed either by the husbands or their own brothers. Male adulterers are also killed if they can be caught. Yomut men confer extensive economic benefits on their children and demand a high guarantee of accurate paternity knowledge. Because of their need for male assistance, Yomut women greatly restrict their activities in response to male demands. In their daily activities, Yomut women work with the other women of their household or neighboring households under the supervision of the older women of the group. It is not considered the prerogative of men to interfere in or attempt to supervise these activities as long as the needs of men and children are met. Thus, within their own sphere of activities, which are by and large separate from those of men, Yomut women do enjoy autonomy.

First marriages among the Yomut are arranged by the parents of the bride and groom. Other older relatives offer advice, but the final decision rests with the parents, both the mother and the father. When parents are deceased, older siblings or other older relatives generally make the decision. The bride and groom are generally thought to be too young to make such decisions for themselves, although I was consistently told that young men make their feelings concerning choice of wives known to their mothers. No one ever admitted that daughters did the same, but I suspect they do so very carefully and very privately. Formal ideology has it that daughters should be strictly silent on the issue.

The average age at marriage for women is 15, although generally a woman does not take up permanent coresidence with her husband and his extended family until she is around 18. The average age at first birth is 19. Unpublished data on first menses, gathered in 1974 in a demographic survey of Yomut women by Rogasner, indicate that most women are not fecund until about 18.* In more prosperous, predominantly agricultural Yomut communities, both the age at first menses and the age at first coresidence with their husbands is earlier. At the time of Rogasner's research, this appeared to be a recent trend among the agricultural Yomut.

Thus, Yomut women generally begin their reproductive careers as soon as they are fecund. On the other hand, women who are widowed after they have had children often do not find other mates. When a woman is widowed, the deceased husband's brothers or other close kinsmen in the male line have the right to insist that the children stay with them. Thus, if the widow marries elsewhere, she is separated from her children. This creates a painful situation for all involved, including the brothers of the deceased husband. The situation is generally resolved by having the widow either marry a brother of the deceased husband or by having her stay in the camp group of her deceased husband's brothers without remarriage. In the latter case, she becomes the head of a separate household, closely dependent on the households of her husband's brothers. With either of these latter arrangements, the children remain with both their mother and with their father's brothers.

When a widow does marry one of her husband's brothers, it is sometimes understood that this will be a marriage in name only and that they will continue to reside in separate households. Often this understanding, however, breaks down and the nominal husband and wife, in fact, take up sexual relations.

Men, at first marriage, are generally several years older than their wives. The age at first marriage tends to be older for poorer than for wealthier men. Thus, the difference in age between husband and wife tends to be greater the poorer the household. When widowers remarry, or married men take second or third wives, the husband is often much older than his wife.

*M. Rogasner: personal communication, 1981.

Yomut marriages are contingent on payment of a large bridewealth which is said to be fixed at 10 camels for the bride's father and 1 camel for her mother. The actual payment may be commuted to sheep and goats on the assumption that 1 camel equals 10 small animals, either sheep or goats. At the time of my first research with the pastoral Yomut (1966–1967), the bridal payments were usually made in small stock. The actual value varied with the ages and conditions of the animals given, and the portion of the payment to be made in goats, kids, sheep, and lambs was negotiated by the families of the bride and groom. The bridewealth needed for a first wife was equal in value to 2–4 years' income for a family of median wealth in the pastoral community in which I lived. The value is stated as 2–4 years' income because income was highly variable from year to year.

When a widower marries a previously unmarried woman, he is expected to pay the equivalent of 20 camels to the father of the bride and 1 to the mother. When a married man seeks an additional wife, he is expected to pay the equivalent of 30 camels to the father. Thus, for the Yomut, acquisition of wives requires economic sacrifice.

Polygyny is favored by men for themselves, but is disfavored by women and also by men when they are arranging marriages for their daughters or sisters. Together with the fact that female labor is not of great economic value, this results in a low rate of polygyny. The common pattern is for wealthier men to take a second young wife when their first wife is approaching the end of her reproductive years. Thus, the reproductive history of polygynous men is not very different from that of serially monogamous men; that is, widowers who marry second young wives. About 10% of men who survive past 40 years are sufficiently prosperous to marry polygynously.

Elopements are very rare among the Yomut, but do occur occasionally. When they do, the bride and groom must flee to a community distant from that of the bride's family and must seek protection from local residents. The bride's family pursues the groom with the intent of reclaiming the bride by force and if necessary to do so, killing the groom. Sometimes after the passage of years, neutral parties acting as representatives of the groom negotiate for payment of bridewealth and restoration of peace between the groom and bride's family.

All Yomut desire high fertility. Although at the time of Rogasner's research most Yomut women were aware of contraceptive pills because of government propaganda spread by radio, very few were interested in using contraception. In fact, out of a survey of approximately 350 women, she found only 2 who made effective use of contraceptives. These were both women who had large families already—six children in one case and five in the other—and who had been told by physicians in the nearby city that they were not likely to survive another pregnancy.

Yomut women regularly continue breast-feeding a child until another is born. Birth intervals were fairly long, averaging 33 months. They were much shorter following the birth of children who died in the first year, indicating that lactation suppressed ovulation. For example, among women 20–24 years old, birth intervals averaged 35 months if the first of their two children defining the interval lived past its second birthday. If, however, the first child died before its first birthday, the interval between births averaged 21 months for the same age group. The pattern is similar for other age groups, although the exact numbers are not the same.

Yomut husbands are recognized as having the right to discipline their wives, including the right to strike them. However, they rarely do so. When wives are seriously mistreated, it is considered proper for their kinsmen to intervene on their behalf, but only when the situation is extreme. Usually the bride's father and brothers pay frequent visits to the family she marries into, especially in the early years of the marriage. This allows them some opportunity to see how she is being treated. The opportunity is, however, limited by the fact that the bride is supposed to either stay out of sight or veil herself—not only from parents-in-law but also from her father and brothers when they visit the family she has married into. Also, the bride is forbidden as a topic of conversation between her father and father-in-law or between older members of natal and marital households.

The Yomut in many ways are an example of a society in which a woman's ability to resist her husband's coercion is at a minimum. Nevertheless, even here one can identify means by which women try to counteract male coercion. One feature of Yomut social life that aids in this is the general segregation of the sexes in daily life. Work groups are usually either all male or all female. At mealtime, women eat with women and men with men in separate locations. Times of relaxed conversation, or festive occasions similarly are passed by women in all-female groups. What women do in detail during most of a day's activities is not considered the business of men as long as perceived male needs are met (cf. Waterhouse, this volume).

Also, the usual life cycle reveals some success for women in controlling their own lives and in controlling the lives of others. Normally, a woman's control over her own life is weakest as a new bride, but increases thereafter. Later in life, as mother and mother-in-law, she enjoys considerable influence over adult sons and formal authority over daughters-in-law (cf. Waterhouse, this volume). Her position in later life is ideal for supervising the rearing of her son's children.

IV. Some Preliminary Hypotheses Regarding Female Reproductive Strategies

The societies discussed above are, in my opinion, sufficiently representative of the variation in social environments and reproductive strategies among societies to serve as a basis for preliminary hypotheses. In the first part of Section IV, I discuss several such hypotheses concerning use of reciprocity to increase parental care. These relate to the nature of the reciprocity underlying the husband–wife bond (Section IV, A), trade-offs between cooperation with husbands and with other kin (Section IV, B) and variation in the frequency of polygyny (Section IV, C). Following this, I discuss theoretical topics relating to reciprocity for increasing parental care. These topics include natural selection for phenotypic plasticity in female strategies (Section IV, D), altruism to mates (Section IV, E), and the relationship between scarcities of mates and the sex drive (Section IV, F). At the end of Section IV, I turn more briefly to the other three strategies: family planning (Section IV, G), and choosing fit mates and countering male coercion (Section IV, H).

A. Marriage

From the comparative perspective of animal behavior studies, human marriage is somewhat unusual. Animal species in which mates cooperate in feeding their offspring are not unusual. Human beings, however, carry the reciprocity a step further in that the mates commonly feed each other. A sexual division of economic activities is nearly universal among human societies, and marriage is a means by which both men and women gain access to the goods and services produced by the opposite sex. Human beings are further unusual in maintaining a relatively stable and sexually exclusive bond between mates while also maintaining numerous other bonds of intense reciprocal aid with other individuals (Daly and Wilson, 1978, p. 328).

1. *The Basic Reciprocity Underlying Marriage*

Thus, marriage contains an important element of reciprocal altruism. The things exchanged meet the following conditions: (*a*) each is scarce in the sense of limiting reproduction, for the person receiving it, and (*b*) each is something which the partner receiving it cannot produce or cannot produce as well as the other partner (see Wasser, 1982).

For men, opportunities to copulate are scarce, and they can gain in reproductive success by increasing opportunities to copulate. Past a certain point, women cannot gain in reproductive success by increasing copulation. They can, however, gain by increasing the resources available for child nurturance and by choosing fit mates. These contrasting strategies are predictable from the fact that women invest more parental care in each offspring (Trivers, 1972). Thus, women try to find fit mates and then give copulation in return for assistance in rearing the children. As noted earlier, marriage also entails a sexual division of labor, such that the husband and wife pursue different economic activities and, thus, provide different economic resources for their offspring as well as for each other. Women in many societies also behave in ways that assure their husbands that they are the fathers of their children. This strategy, which can be thought of as giving a husband paternity knowledge or paternity confidence, increases male willingness to invest in children. Also, in prestate societies, men commonly provide their wives with protection from the potential violence of other men. The network of male allies a man can call on is usually an important resource for accomplishing this. The word "allies" is used here to mean other individuals one can call on for aid, whether the basis of their willingness to help is reciprocity, nepotism, or both (see Irons, 1979b).

2. *Variation in Reciprocity among Societies*

The ethnographic examples above, however, illustrate the extent to which both women and men can vary the reciprocally altruistic relationship we call marriage. To a large extent, these variations in the nature of marriage are associated with variation in the nature of other social bonds. For example, in most societies, people have economic allies other than spouses and must adjust relations with spouses to accommodate these other alliances. Among the Tiwi, sisters and other close female kin are apparently economically more important than husbands. To accommodate their ties with sisters, women are willing to share a husband, that is, accept polygyny, as a means of holding their female networks together. Among the Yomut, male–male economic cooperation yields higher payoffs than female–female cooperation. Thus, Yomut men cooperate closely with their fathers and brothers in economic production and demand that wives adjust to their need to keep this network together. Because of their need for male economic aid, Yomut women are willing to pay the cost of leaving their natal families and accepting the initially lonely and disadvantageous position of daughter-in-law in their parents-in-law's household (cf. Waterhouse, this volume). In comparison to the Tiwi, Yomut women are giving more to husbands but are getting something of greater value in return. Value here refers to perceived value, if one is speaking of psychological proximate mechanisms, and refers to resources for reproduction if one is speaking of function. Among the Tiwi, reciprocal obligations among men based on betrothal of women also modify the division of labor among wives. An older Tiwi man, rather than hunting to provide meat for his wives, can call on clients—allies of lower status—to hunt for him. Younger men, in exchange for future rights to women, perform this service for him. There is a high variance among Tiwi men in their ability to demand services from other men, and this variance in male resource underlies polygyny, in addition to the economic advantage of female coalitions (see Section IV, C, 1 below).

Thus, the nature and amount of aid women seek from husbands can vary from one social environment to another. Variance in male ability to give such aid also differs from one social setting to another. The female strategies devised in response to these varying constraints frequently affect other variables, such as the stability of marriage, the degree of tolerance of extramarital sex, preferred residence after marriage, and the frequency of polygyny.

B. Trade-offs between Cooperation with Husbands and Cooperation with Kin

Marriage becomes attenuated when female coalitions are more effective at gaining what is scarce in a particular environment than are either individual men or male coalitions (Irons, 1979b). Given limited resources to engage in reciprocity, the greater payoff of female coalitions can mean less willingness to confer benefits on husbands and less willingness to accept the constraints on their own behavior necessary to confer paternity confidence. It can also mean less unwillingness to leave female kin and take up residence with husband's kin, and less willingness to hold a marriage together once conflict emerges between spouses. Thus, a relative tolerance of extramarital sex, frequent divorce, low paternity confidence, uxorilocal residence, and matrifocal families may frequently emerge as a result of a preference for female allies over male allies. (See also Silk and Boyd, this volume). A preference for female allies may also lead, as noted earlier, to a preference for sororal polygyny.

The exact set of social alliances sought by women in any particular society probably reflects the relative importance of female allies versus male allies among that group. When female allies are more important, male aid still has some value, and this value can vary among those societies in which female allies are more important. Male aid, in smaller amounts than female aid, can be obtained through weaker marital bonds that nevertheless do involve some exchange of benefits. In some cases, aid may also be obtained from brothers. Brothers, who are related to a woman's children by 1/4, can be counted on for less aid, in most cases, than husbands who are confident they are related by 1/2. Nevertheless, when little aid is sought, a brother may be willing to help, and it may cost a woman less to call on a brother than on a husband whose willingness to help depends on constraining her behavior in such a way as to confer paternity confidence. Often, of course, women seek aid from both brothers and husbands.

Attenuation of the husband–wife bond may also occur when women find they can get the assistance they need through a series of relatively weak and shifting social alliances. Given this, they may find that sisters, brothers, parents, or adult children are generally as valuable as, or more valuable than, husbands. It is possible that in some societies a few long-term ties of reciprocity, in which much is invested, represent the optimal strategy, whereas in other societies weaker ties, which are more easily established and broken, represent the optimal strategy. The latter appears to be the case among the Plateau Tonga, where the pattern of close cooperation appears to be highly variable from one individual to another. Thus, some women emphasize one tie, others another, and most shift their alliances several times during their adult life, creating no firm cultural expectation that any tie—that with husband, sister, brother, or other kin—will be invariably a locus of extensive cooperation (see Colson, 1958, p. 61).

1. *Attenuated Marriage and Paternity Confidence*

When divorce and extramarital sex are common, men will often not be the fathers of some of their wives' children; this condition of low paternity confidence can have far-reaching effects on social organization (Alexander, 1974, 1977, 1979b). Many of the things a man does for a wife and children—such as building a house or planting a garden—tend to benefit all of a wife's children, whether his or not. Given this, the male's reaction to attenuated marriage is likely to be an unwillingness to invest heavily in their wives and children. It may cause them to put their resources instead into the pursuit of multiple mates through either promiscuity or polygyny. It may also cause them to prefer forms of investment that can be directed at a particular child rather than all of a woman's children. Thus, they may prefer voluntary gifts to particular children rather than inheritance by all of a wife's children.

Lowered paternity confidence also biases altruism investment toward uterine kin. If paternity misassignment reaches the point where a man is the father of fewer than one-quarter of his wife's children, he will be more related to his sisters' children than to his wife's children; this may explain such widespread institutions as avunculocal residence and the inheritance of property and status from mother's brother to sister's son (Alexander, 1974; Greene, 1978). Even if paternity confidence is well above the one-quarter threshold, the lessening of relatedness through paternal links may bias the flow of assistance toward relatives through female links and away from relatives through male links. It may cause both men and women to prefer aiding sisters' children rather than brothers' children. It may also cause them to prefer forming large solidary groups, such as extended families, lineages, and clans through female rather than male links. The Tonga, who practice both frequent divorce and extramarital sex, have both lineages and clans based on matrilineal descent. The Tiwi situation is similar, except that remarriage has more to do with the age disparity between husbands and wives than divorce. This group also has matrilineages and matriclans. For the Yomut, who have virtually no divorce and draconian reactions to extramarital sex, extended families and descent groups are based on patrilineal descent. Thus, the social ties that women choose to cultivate in order to increase resources for nurturing children appear to have far-reaching consequences for social organization.

Other things being equal, men would prefer to constrain their wives in such a way as to raise paternity confidence. This allows them to invest in children who share one-half of their genes by recent common descent, rather than those of less related individuals. However, males may accept low paternity confidence for two reasons: (*a*) it may be the cost of the pursuit of multiple mates either through promiscuity or polygyny, or (*b*) they may be unable in some social environments to coerce women who prefer to emphasize ties with female kin.

2. *Greater Emphasis on Reciprocity with Lowered Relatedness*

An interesting suggestion made by Richerson and Boyd (1980) in reaction to Alexander's hypothesis is that low paternity confidence, by lowering the relatedness of relatives in general, leads to a greater emphasis on reciprocity and a lesser emphasis on kinship as the basis of cooperation. Richards's (1950) observation that matrilineal societies are unstable may reflect this fact. "Unstable" in her evaluation equates with "less regularly ordered" in terms of formally recognized genealogical ties, as well as less stable marriages. The general freedom among the Tonga of individuals, even adolescents, to reside where they like among a wide range of relatives and even nonkin may reflect this fact as well (Colson, 1958). Among the Tonga, there are a number of practices that would decrease relatedness through all kin links: frequent adultery, frequent divorce, proscription against sororal polygyny, proscription against cowives who are close cousins, and proscription against repeated marital ties among the same matrilineages. On the other hand ,polygyny itself will increase the number of paternal half-siblings and, thus, will tend to raise average relatedness among local groups. Nevertheless, the balance of these various practices affecting relatedness may be to cause relatedness to be low in general compared to other populations. In comparison with the Yomut, for example, the average Tonga would have fewer siblings related, with a high degree of certainty, by one-half. They would have more putative paternal half-siblings, but only some of these would actually be related. Multiple genealogical links owing to repeated marriage between the same lineages would be rare. Given this, people may emphasize reciprocity more than neoptism in choosing their allies, and this could be perceived as instability by an anthropologist who expects social arrangements to reflect genealogical ties and expects social relationships to remain stable over time.

The Tiwi, in contrast, have a different balance of factors affecting relatedness. Frequent sororal polygyny and frequent cross-cousin marriage would cause individuals to be more related on the average to paternal half-siblings, to parallel cousins, and even to their children and full siblings. Frequent changes of husband (owing to vast differences in ages of wives and husbands) and frequent adultery would have a counteracting effect, but the balance would appear to raise the average relatedness of most kin ties in comparison to the Tonga. The average relatedness of siblings, cousins, and other kin varies among societies owing to different mating systems and it is a reasonable hypothesis that the relative importance of reciprocity versus nepotism will vary in response. The extent to which genealogical tie versus complementarity (in Wasser's [1982] sense) determine close social alliance can be used as a test of this.

Alexander's (1974) hypothesis, that paternity uncertainty causes uterine biases in altruism investment, has led to a large body of theoretical literature (see e.g., Greene, 1978; Kurland, 1979; Gaulin and Schlegel, 1980; Flinn, 1981; Hartung, 1981). Empirical testing of these ideas has only begun. However, the tests that have been done recently support the hypothesis (Gaulin and Schlegel, 1980; Flinn, 1981; Hartung, 1981).

All of this suggests that many features of social organization may stem from choices made by women concerning the extent to which they are willing to grant paternity confidence and commitments of long-term marriage to husbands, choices which may in turn reflect the relative values they place on the aid of husbands and the aid of female kin.

C. Polygyny

Women's evolved behavioral propensities appear, from the ethnographic record, to be flexible regarding tolerance of polygyny. While the rules of some societies strictly forbid polygyny, others encourage it. Some ethnographies even report instances in which women prefer polygyny over monogamy (Bohannon, 1954). If human females choose mates with the goal of maximizing the altruism investment given their children, they should prefer to become an additional wife of an already-married man when such a marriage will bring more investment to their children than would the available opportunities to marry monogamously. This can occur in two ways. The most straightforward is in the way suggested by Verner (1964) and Orians (1969). If males vary in the resources they command, some married men may be able to offer a new wife more assistance than can some unmarried men. The evidence available from many societies accords with this, at least to the point of demonstrating that polygyny is confined to higher-status males (see e.g., Cohen, 1961; Malinowski, 1961; Irons, 1975, 1979c, 1980; Chagnon, 1979b). Precise quantitative analyses testing whether women gain in inclusive reproduction success by opting for polygyny over monogamy have not yet appeared. Such studies would have to identify in some way the probable monogamous alternatives to the polygynous marriages. A second reason why human females may opt for polygyny is that suggested previously for the Tiwi: cowives may be an important source of aid for women (see also Wittenberger, 1980).

Another possibility is that polygyny, in some cases, may represent male coercion. In some cases, polygyny would be advantageous to males but not females. This would probably be associated with a tendency for second and later wives of polygynists to be women of lower social status, or women who for some other reason have difficulty attracting desirable mates. It should also be associated with a high level of jealousy among cowives. Polygyny that is advantageous to women because it increases the number of female allies in the form of cowives should be associated with less jealousy. However, since even in these cases cowives may compete for husband's favors, some jealousy is likely but at a lower level, reflecting the fact that cowives are both allies and competitors, not just competitors.

1. *The Distribution of Polygyny among Human Societies*

The majority of human societies allow polygyny, but in most societies that allow polygyny, the majority of men marry monogamously (Murdock, 1967). Some societies exhibit a different pattern, which can be called gerontocratic polygyny. In these societies, men on average marry much later than women; a high proportion of older men are married polygynously, and most younger men are unmarried. This suggests that the resources that women seek from men are unequally distributed and tend to correlate with age. The Tiwi are an example, and in their case, as noted previously, power in the form of the right to bestow women is unequally distributed and correlates with age. (This fact, as noted earlier, may be complementary to the advantage of female coalitions in explaining Tiwi sororal polygyny). This pattern of gerontocratic polygyny is especially common in horticultural societies; that is, societies with the simplest forms of food production (van den Berghe, 1979). It is not especially common among foraging societies outside of Australia; thus, the Tiwi are an exception among foraging societies in exhibiting this pattern.

Agricultural societies—those with more complex forms of food production involving such technology as the use of draft animals and irrigation—exhibit a different pattern of polygyny associated with an unequal distribution of wealth and political power. In these societies, an individual's wealth and political power are largely determined by his or her parents' wealth and power. These societies fit the common pattern in that the majority of married men are monogamous. Polygyny tends to be restricted to the upper strata, and unusually powerful men, such as heads of state, tend to have very large harems (Betzig, in press). The Ottoman sultans, with harems numbering in the hundreds, are examples (Coon, 1958, pp. 274–276).

The distribution of these various patterns of polygyny found among human societies can be explained, I suggest, by two variables: (*a*) the extent to which men have unequal resources, and (*b*) the amount of altruism investment women seek from their husbands. These variables are implied by the Verner (1964) and Orians (1969) polygyny threshold hypothesis. They tend to vary in a consistent way with technology and population density. In most regions of the world, there has been a long-term trend toward greater population density and an accompanying trend toward forms of technology that can feed (and otherwise provide goods for) more people with a given unit of land (Spooner, 1972). These trends have been associated with an increase in competition and warfare among political groups and an increase in the size of political groups. Increasing size and external hostility have been associated—until recent stages of history—with an increase in political and economic inequalities within political groups (Carniero, 1970; Irons, 1979d; Alexander, 1979b,pp. 249–1262; Betzig, in press). The variables noted above—inequalities of resources among men and female demands for male altruism investment—have tended to vary in a consistent way with these long-term trends. In foraging societies, inequalities among men are limited, and therefore it is relatively infrequent that a woman is better off as the second wife of a man with many resources than as the first wife of a man of lesser resources. Horticultural societies are often characterized by larger political units, more intense warfare, and greater political inequalities within groups. The greater inequalities among men make it more common for women to find being a second wife of a man with many resources the better choice. In agricultural societies, inequalities among men increase. This alone might seem to make polygyny more frequent. However, in agricultural (and also pastoral) societies, male labor tends to be of much greater economic value than female labor. This increases female need for the assistance of a male in rearing their children and increases the second variable above. As a result of the greater demands wives place on husbands, fewer men are able to meet the needs of more than one woman. Thus, the proportion of the male population that can be successful polygynists declines. However, a minority of men do have the resources to meet the needs of more than one wife, and within this group there are very large inequalities, with heads of state representing the extreme. I suggest this as a broad trend among societies associated with broad trends in the different relative value of male and female labor and in inequalities among men. These trends are in turn associated with increasing economic intensification and intergroup competition. At the same time, among societies that are similar in terms of their economy and the external hostilities they face, local differences of ecology and history will cause differences in the value of male and female labor and in inequalities among individuals. These will cause additional variation in the frequency of polygyny among societies not associated with these broad historic trends.

2. *The Recent Trend toward Socially Imposed Monogamy*

Alexander *et al.* (1979) have suggested that the more recent historical trend toward socially imposed monogamy is a response to further intensification of intergroup competition and a consequent need to increase internal group cohesion by limiting internal competition and internal inequalities of access to crucial resources. This hypothesis does not contradict the preceding discussion, and it is possible that modern socially imposed monogamy reflects both the forces suggested by Alexander and those suggested above. Seen this way, both women and men of middle and lower status would stand to gain from socially imposed monogamy. Men of higher status would stand to gain from polygyny, but in response to a need for cooperation from other members of their group (all women and most men), they tolerate formal rules against polygyny. Tolerating these rules makes acquisition of simultaneous multiple mates more difficult, but not impossible. Careful examination of the historic emergence of formal rules forbidding polygyny might suggest whether either or both of these forces was, in fact, important in establishing socially imposed monogamy. In addition to increasing competition between groups, an increase in the complexity of the division of labor may also play a role in encouraging socially imposed monogamy.

D. Natural Selection for Phenotypic Plasticity

The various propositions above suggest that the nature of the assistance women seek through social alliances with husbands and kin are important determinants of a number of social institutions. The relative value of aid sought from husbands and from female kin is one important intermediate variable affecting such variables as paternity confidence and uterine biases in overall social organization. The extent of aid sought from husbands and variance in male ability to supply such aid are important variables affecting the occurrence of polygyny. But it is also important to note that the actual set of opportunities and constraints faced by each individual within what anthropologists describe as one culture or one society are different. Different individuals establish different social ties based on reciprocity. The specific social network of any individual is different from that of his or her parents. Usually each individual's network changes over her or his lifetime as well. Given this, I suggest that selection in all human populations has favored individuals with the heritable propensity to modify their behavior in accordance with their changing environment. In other words, selection should have favored genotypes that specify a wide range of possible behavioral phenotypes, along with propensities for individuals to track their environments and behave in the ways that most enhance their fitness, given their current environment. Evolution, of course, has not produced a conscious striving for fitness. Rather, it has produced a conscious striving for intermediate goals—such as a good diet or sexual satisfaction—which in past environments enhanced fitness (see Irons, 1979d, pp. 362–364).

Lumsden and Wilson (1981) have suggested that since different populations experience different selection regimes, selection will favor a tendency to develop more readily the currently favored behavioral phenotype(s) in each population. This could give rise to somewhat different modal norms of reaction in each of these populations. While this is a logical possibility, I would argue that a close look at our best data on human behavior suggests that selection pressures relevant to the traits under discussion have changed rapidly (in evolutionary terms) in all human populations. Evidence also suggests that institutionalized patterns of behavior can change very rapidly and drastically over short periods of time (see e.g., Barth, 1967). Thus, the range of phenotypes specified by a given genotype is probably much larger than that implied by Lumsden and Wilson. This being the case, the range of possible phenotypes available to any given population should be very large and the potential for differences between populations in behavioral norms of reaction is probably trivial. Given similar environmental constraints on different populations, similar patterns of behavior should eventually tend to develop between them. Moreover, as suggested earlier in this chapter, the various mating strategies and other patterns of behavior converged upon should generally, but not always, tend to be adaptive (Irons, 1979a) and predictable, even in novel environments (see also below).

Given our current state of knowledge, both the view I suggest and that suggested by Lumsden and Wilson (1981) are reasonable hypotheses. The contrast between these hypotheses, however, suggests that it is important to record the extent to which individuals within the same society face different constraints and devise different strategies. It is also important to know how rapidly the overall array of constraints faced by a population has changed over time.

E. Altruism to Mates

The networks of allies that both women and men attempt to create and maintain involve many behaviors that are altruistic. Since these networks involve many close kin, inference is strong that kin selection has played an important role in the evolution of the behavioral predispositions guiding the cultivation of these networks of social allies. This view is, for example, supported by recent evidence that individuals prefer close over distant kin in types of societies were kin terms do not distinguish degrees of genetic closeness and distance (see e.g., Chagnon, 1975, 1979a, 1980, 1981; Weisner, 1977; Chagnon and Bugos, 1979; Hames, 1979).

However, it would be a mistake to conclude that nepotism is the only form of altruism in human societies. The assumption that all human altruism is nepotism can be misleading in many ways, and in particular it can be misleading in reference to the relationship between mates. One can find evidence for this misleading assumption in much of the literature on animal behavior. For example, Dawkins (1976) begins a discussion of sexual selection by saying: "If there is conflict of interest between parents and children, who share 50 percent of each others' genes, how much more severe must be the conflict between mates, who are not related to each other?" (p. 151). Sahlins (1976, pp. 26–27) and Freedman (1979; p. 209) apply this logic to human beings by interpreting evidence that human beings are often more altruistic to spouses than to siblings as evidence that human beings do not behave in ways predicted by inclusive fitness theory. The view represents a failure to realize that sexual selection—that is, selection for attributes that increase access to mates—can favor self-sacrificing aid to mates. The issue is somewhat muddled by the confusion of self-sacrificing aid to others with behavior that is technically altruistic in the sense that it lowers one's own fitness and raises another's fitness. A monogamously mated individual, with no chance of obtaining another mate should the current one die, would be serving his or her own reproductive interests to behave in a self-sacrificing way toward his or her mate. The effect of such behavior, however, is not technically altruistic in the sense that it lowers one's own fitness and raises the mate's fitness. Rather, it raises both. It does, however, fit Alexander's (1974) definition of phenotypic altruism. In the extreme condition laid down previously—of no possible alternative mate—reproductive success may hinge on the survival of both mates, not just one's own survival. Under this condition, behavior that lowers one's own chances of survival, but raises the chances that both mates survive, will be adaptive. Such classical nonhuman examples as giving a warning call after detecting a predator could easily have this effect; that is, decrease the chance that the call giver will survive, increase the chance that the individual warned will survive, and increase the chance that both will survive (which is the product of the probabilities of each individual's surviving). (Exceptions might occur if most offspring are already born and one parent can successfully finish rearing them). Much human behavior toward mates, and even toward potential mates, fits this pattern and is both phenotypically altruistic in Alexander's (1974) sense and self-sacrificing in the everyday sense.

It may be useful to observe that mating can be "cooperative" in the technical sense (Hamilton, 1964) in that it simultaneously raises the fitness of both individuals (assuming the mating does not exclude better alternate mating opportunities for either partner). When, in some human cases, a female offers a male immediate opportunities to reproduce in return for a promise of future aid in rearing children, the exchange is technically one of reciprocal altruism since the benefit to one party is delayed in time and may never occur. Thus, reciprocity theory in general (Williams, 1966; Trivers, 1971; Axelrod and Hamilton, 1981; Wasser, 1982) is relevant to understanding human male–female cooperation in child-rearing and self-maintenance. Each partner offers benefits to the other with an expectation of reciprocation, either immediate or delayed. At the same time, once children are produced, altruism to a spouse can be interpreted as nepotism since it amounts to preserving resources useful to one's offspring and hence increasing their probable reproductive success.

Self-sacrifice for mates should, in general, be higher the scarcer alternative mates and the greater the fidelity of mates. Also, greater self-sacrifice for mates would become more adaptive as the value of a mate's assistance in rearing offspring increases. This third variable—value of mate's aid in rearing offspring—should covary strongly with scarcity of mates. Mates that contribute little parental investment are not likely to be scarce (Trivers, 1972). The extent to which self-sacrifice is adaptive may be different for the two members of a mated pair. For example, alternate opportunities to mate may be scarcer for one mate than for the other. Thus, one member may be more willing to offer benefits than the other. In general, mates are scarcer, in most preindustrial societies, for younger males. Females in such societies have mates chosen for them by their kin before puberty. Thus, they face no scarcity initially, but at the same time, limited opportunity to exercise their own choice. Later in their reproductive years, human females may well face scarcities of mates in a serially monogamous or polygynous breeding system and, ironically, at the same time, they may have greater opportunities to exercise choice. Widows in their thirties often do not remarry among the Yomut and Tonga. Among the Tiwi, there is evidence that men with many wives neglect older wives as sex partners (Goodale, 1971, pp. 226–227). Customs involving changes in residence during the course of a marriage or delayed bridal payments, as among the Tonga, may be related to this.

In general, the social relationship between mated human males and females may be the most difficult to track of all social relationships, since under some conditions great self-sacrifice for one's partner may be favored by natural selection. At the same time, the relationship is rife with opportunities for cheating—such as cuckolding or desertion. It is, therefore, not surprising that human emotions and motivations regarding this kind of relationship reach the highest level of subtlety and complexity (cf. Symons, 1979 for a discussion of the relationship between emotional complexity and careful tracking of subtle features of the environment).

The conditions that make altruism to mates adaptive may also promote adaptive altruism to potential mates and altruism to relatives' mates, that is, affines. Thus, the fact that human societies are characterized by important exchanges with affines and potential mates is not prima facie evidence against the proposition that human beings behave so as to maximize inclusive fitness (pace Sahlins).

F. Human Motivation and Scarcity of Mates

Kinsey (1953, pp. 714–717) observed that the male sex drive is strongest at the beginning of a man's fecund years, whereas the female sex drive peaks around the middle of a woman's fecund years; this may derive from a different timing of scarcities of mates in the environments of evolution. In general, selection should favor a stronger drive the greater the obstacles to fulfilling the goal toward which the drive is directed (Symons, 1979, 270–271). At the same time, it should favor greater consciousness of the drive the greater the need to shift carefully and weigh possible alternative courses of action and the greater the need to gather information relative to these choices (cf. Dawkins, 1976, pp. 60–64). Given this and the probability that in human evolution, males have usually faced a scarcity of mates at the beginning of their fecund years and human females later in their fecund years, it is not surprising that their conscious interest in mating opportunities and their motivation to seek them should peak when they do.

Because of polygyny, serial monogamy, and the larger number of fecund years for males, there have probably been more males seeking mates than females during most of human evolution. The usual expression of this in the ethnographic record is a later age at marriage for males than females. Thus, throughout human evolution, most human males have been excluded from mates during their early fecund years. A high sex drive during these years may have been adaptive in that it led to a striving both to acquire legitimate mates and to take advantage of opportunities for clandestine matings with the spouses of older men. During these years, females, in contrast, have had mates that were in most cases chosen for them. By their thirties, however, if the Yomut, Tonga, and Tiwi are good examples of the conditions of earlier evolution, they have occasionally faced a shortage of mates. Also, some women among all of these groups may be competing with younger cowives for their husbands' attention during these years. The possibility of divorce or desertion by a husband would also be greater for older than for younger wives. Since males are often more attracted to women with higher reproductive value, that is, younger women, older cowives may have been more successful if they actively sought their husbands' attention and if they were motivated to make sacrifices to hold their attention.

After menopause, a woman may still have need for male aid, but at that time the possibility of holding males through sex may be too low to make further increases in the sex drive adaptive. Thus, the sex drive should begin to decline again as a woman approaches and eventually reaches menopause.

Strong believers in cultural relativism may suggest that Kinsey's data on the ages at which interest in sex and mate acquisition peak are drawn from a limited range of cultural settings and that in different societies the situation may be different. Indeed it may. The preceding hypothesis, however, can be tested and is worth testing. It could easily incorporate the assumption that some facultative adjustment of the sex drive is possible as the scarcity of mates varies from one environment to another.

G. Fertility Regulation

The idea that infanticide is widely practiced as a means of population regulation, in the Carr-Saunders (1922) and Wynne-Edwards (1962) sense of a means of preventing long-term environmental degradation, has been popular among anthropologists (Polgar, 1972; Divale and Harris, 1976), but empirical support for this idea is very weak (Bates and Less, 1979; Chagnon *et al.*, 1979; Howell, 1979, p. 135).

The possibility that human beings have practiced fertility regulation as a means of maximizing the number of adult progeny produced (Lack, 1954, 1966, 1968) is a more promising hypothesis, in my opinion, but one which has not been explored for the most part (see also Dublin, Silk and Boyd, Wasser, this volume). The study by Blurton Jones and Sibly (1978) discussed earlier is an exception. Data mentioned above suggest that both lactation and postpartum taboos on intercourse regulated birth intervals among the Tonga in a way paralleling what Blurton Jones and Sibly (1978) suggest for the !Kung. Among the Yomut there is good evidence that lactation plays a role in lengthening birth intervals, but no evidence of a role for postpartum taboos. Such adaptive lengthening of birth intervals, however, is not as dramatic in its effects on fertility as modern contraception.

As noted earlier, modern fertility appears to be the most salient example of a maladaptive behavior induced by a novel environment, and as such, it can best be discussed in terms of proximate mechanisms that produced adaptive effects in past environments. One relatively straightforward hypothesis of this sort states that one proximate mechanism regulating fertility was the sex drive, which often led to reproduction even when the burdens of child rearing were unwelcome (cf. Burley, 1979). Contraception by allowing sexual outlets of the most preferred form, heterosexual intercourse, without the unwelcome later births, in a simple way leads to novel maladaptive effects. (Contraceptive technology is here seen as a novel environmental element, and the regular use of it is seen as a behavioral phenotype.) This hypothesis is worth testing and I believe may eventually provide a partial explanation of modern low fertility. I believe, however, that it works somewhat better as an explanation of male, rather than female, motivations for fertility limitation.

On the other hand, a prospective desire for children is a common—though not universal—motive. This, I have the impression, is more common among women than men, although by no means absent in men. The number of children desired also does seem to be sensitive to what some economists refer to as the quality of children as well as the quantity of children (Becker, 1975). This means that people often prefer a few children on whom they lavish more nurturance and who achieve high status, rather than many children of low status (Davis and Blake, 1956; Irons, 1977; Barkow, 1978, p. 16). A preference of this sort has the appearance, in many ways, of a set of motivations for optimizing the trade-off between child-bearing and child-caring in accord with Lack's (1954, 1966, 1968) principle. On the other hand, in most societies preceding the demographic transition, people appear to have a conscious desire for higher fertility than do people in postdemographic transition societies. The fertility limitation before the demographic transition appears to be accomplished largely by physiological mechanisms rather than behavioral ones. On the other hand, postpartum taboos may represent a conscious attempt to space children in order to have greater resources per child, as well as to limit stress on the mother (cf. Whiting, 1964). It may also be the case that most ethnographically described societies, like the Tonga and the sedentary !Kung were described at a time when resources for population growth had recently increased, so that we do not find descriptions of earlier means of family planning.

Also it can be noted that Yomut women who are widowed sometimes face a choice between leaving existing children in order to remarry and have more children, or staying with existing children and foregoing further child-bearing. Polygynously married women who were neglected by their husbands for younger wives in many traditional societies may have faced a similar choice, that is, leave a husband with many resources in order to seek one with less resources who will copulate with them more regularly and provide probable higher future fertility. It is possible that in human evolution, choices of this sort occurred frequently enough to favor the evolution of a facultative motivation to limit fertility in order to improve the quality of life for existing children. This would be another situation, in addition to that suggested by Blurton Jones and Sibly (1978), in which fertility would be adaptively limited in earlier human societies. Further, it is possible that modern environments, by providing novel opportunities to do this, cause this evolved motivation to produce novel maladaptive behaviors.

In foraging societies and in the simpler food-producing societies, children were an economic burden until about age 12–15. At about that age, they began to become economic producers who aided parents in rearing younger children (see, however, Draper [1976] for evidence that child labor was unimportant among the !Kung; the issue these data raise is how the cost of child-rearing in foraging societies compares to the cost for the modern middle class). Although spacing children too closely may have been disadvantageous for both children and parents as noted above, after a certain interval, birth of additional children was directly beneficial to existing children. This was so because adults, in these societies, found it advantageous to have a large number of siblings and other close kin. Although these societies were characterized by free access to basic economic resources, such as arable land or wild food sources, and by lack of heritable formal political office, there were significant inequalities in access to mates for males, access to preferred mates for females, and in the number of allies one could call on for support in disputes; those with large numbers of kin were favored in the unequal distribution of these things (Chagnon, 1979a, 1979b). This would mean that with sufficient birth spacing, additional children would improve the long-term quality of life for existing children. This is dramatically different from the situation in modern environments.

I suggest as a working hypothesis that human beings evolved in social environments similar to those observed in ethnographically described foraging societies and preindustrial food-producing societies, and that in these environments, occasional conscious attempts to space births or limit additional births were adaptive. The proximate mechanisms governing these decisions included a conscious trade-off between several desired goals: (*a*) maintaining a high qualty of existence for onself, (*b*) maintaining a high quality existence for existing children, and (*c*) having additional children who could also enjoy a high quality of existence by the standards of those societies. The proximate mechanisms also included the ability to understand one's social environment and make good judgments concerning the long-term effects of additional children on oneself and existing children. A set of relative values placed on one's own welfare, existing children's welfare, and the birth of additional children, which would lead to optimal fertility in these social environments, leads to low fertility in modern environments. This occurs because children who are to enjoy a high quality of life by current standards place much greater burdens on parents and because additional children have a long-term detrimental rather than beneficial effect on existing children, even if very widely spaced in age. Quantitative data on the long-term effects of number of siblings and birth spacing on an individual's life conditions and reproductive success in foraging and simple food-producing societies, combined with comparative data for modern societies, should allow testing of this hypothesis. It is important to keep in mind that the constraints affecting fertility in modern middle-class America did not influence much of the world's human population until a few centuries ago.

H. Other Strategies: Choosing Good Genes and Countering Male Coercion

Ethnographic data on choice of mates with phenotypes indicating fit genes are scanty. Occasionally, ethnographies mention briefly what physical characteristics are thought to be desirable in a mate, but generally it has not been thought to be a question of great importance. Yomut men—with whom I could talk freely about the subject, in contrast to women—are very much interested in the physical appearance of women even though local standards of modest dress limit what they can see. They are always attentive to female attractiveness even when decorum prevents them from discussing what they are observing. They consider women who are tall and robust most attractive. In seeking mates, they are also attentive to the fertility of a young woman's family. Women from families with high fertility are preferred.

The reasons people give for falling in love might also help in answering this question. Ethnographies tell us little about this but do reveal that young people fall in love even when the practice of arranging marriages makes falling in love largely irrelevant and a nuisance. Falling in love is the usual reason for elopement, for example, and elopement is a subject that also is generally not treated at length by ethnographers. (For one of the few treatments of this subject, see Bates [1974].)

The means by which women try to counter male coercion is another subject that receives little attention in ethnographies. However, as noted in the descriptions of the Tiwi, Tonga, and Yomut, women do call on brothers and fathers to counter attempts by husbands to use their greater physical strength for coercion. I would predict that this practice is very widespread. In all social relations the individuals involved have different interests (identical twins aside). This implies that all parties will occasionally try to renegotiate their social contracts and that men, in dealing with women, will occasionally use the threat of violence (Alexander, 1979b, pp. 159, 183). However, fathers, brothers, and adult sons are at least one resource wives can call on to counter this tendency. It is also worth noting that men also use threats of violence against other men, and similarly, men call on male allies to counter such strategies. How much those with greater ability to wield violence are able to skew social rules and expectations in their direction is an interesting question. Generally, the outcome is not a simple dominance of the physically stronger over the physically weaker. If it were, young men would generally dominate older men, but the reverse is the case.

The pastoral Yomut also provide other data which can yield insight into female strategies for countering male coercion. As noted previously, the Yomut are a sexually segregated society. In a Yomut extended family, husband and wife are not in contact during the day. Rules of proper behavior also forbid husband and wife from conversing in the presence of senior members of the husband's extended family. Thus, Yomut women live largely in a social world of other women. A woman's striving to assert and protect her perceived self-interest is largely carried out in a world of women (cf. Waterhouse, this volume).

V. Summary

Neo-Darwinian theory and the ethnographic record taken together suggest a number of preliminary hypotheses, which if tested, may increase our understanding of women's reproductive strategies.

Those who attempt to explain women's behavior as the outcome of evolved behavioral predisposition must, however, be attentive to the possibility that novel environments may produce maladaptive responses. The argument that particular behaviors are evolved adaptations is reasonable only if it can be reasonably argued that these behaviors promoted reproductive success in the environments of human evolution.

The patterns of mating and child-rearing in three traditional human societies—the Plateau Tonga, the Tiwi, and the pastoral Yomut—are described. These examples highlight the extent to which women face different constraints in different social environments and illustrate their flexibility in adjusting to these variable constraints. They also provide concrete examples, which can aid in the formation of preliminary hypotheses.

Women are hypothesized to exhibit the following four reproductive strategies:

1. They attempt to increase their resources for parental investment by establishing a network of social relations based on reciprocity. Usually the relations are established with mates and close kin, but the underlying strategy is not simply nepotism, but rather a mixture of reciprocity and nepotism. The means of earning a livelihood and other features of particular societies cause different types of coalitions—male–male or female–female coalitions—to yield different payoffs in terms of resource gathering. Women respond to these different features of various social environments by emphasizing different alliances in different environments. In particular, the trade-off between cultivating a close alliance with a husband and cultivating a close alliance with close female kin varies among societies and appears to be a distal cause of the emergence of certain other features of social organization, such as matrilineal versus patrilineal descent. Women also vary in their willingness to accept polygyny, a factor that influences their ability to acquire resources from husbands. The variables hypothesized as relevant for animal polygyny (Verner, 1964; Orians, 1969) appear relevant for human polygyny. Additional variables may also be relevant in the human case: sororal polygyny may be a means of holding together female coalitions in some human societies.

2. Women sometimes space births and may under some circumstances cease reproducing before menopause as a means of optimizing the trade-off between number of children and resources per child. The evolved motivational predisposition that optimized fertility in earlier societies may cause modern low fertility, which is evolutionarily suboptimal.

3. Women attempt to choose mates whose phenotypes indicate they will probably produce fit offspring. Falling in love and consequent elopement or other attempts by women and men to circumvent parental control of their mate choices may be expressions of this strategy in conjunction with conflict of reproductive interests between parents and children.

4. Women are faced with the fact that husbands and male kin have superior ability to exploit violence in situations of conflict of interest. To protect their interests in these situations, they cultivate different male allies, husbands and brothers for example, who can be played off against one another in these situations. They may also counter male coercion in some societies by encouraging segregation of the sexes (which males may desire for somewhat different reasons). Segregation of the sexes during certain activities can limit male knowledge of, and attempts to control, female behavior.

The discussion is intended as a contribution to the formulation of testable predictions which, if tested in sufficient number and with sufficient rigor, will allow inferences as to the correctness of incorrectness of neo-Darwinian theory as a theory of human social behavior in general, and as a theory of women's reproductive behavior in particular.

Acknowledgments

I wish to thank Samuel K. Wasser for extensive and valuable editorial assistance in revising earlier drafts of this chapter, and Marjorie Rogasner for aid in proofreading. I also wish to thank Richard Alexander, Martin Daly, and two anonymous referees for much help in revising earlier drafts. Further, I owe a debt of gratitude to the National Science Foundation, the Harry Frank Guggenheim Foundation, and the Ford and Rockefeller Foundations' Program on Population Policy for supporting the research that led to, and nurtured, my interest in the evolutionary theory of social behavior.

References

Alcock, J. *Animal behavior.* (2nd ed.). Sunderland, Mass.: Sinauer Associates, 1978.

Alexander, R. D. The evolution of social behavior. *Annual Review of Ecology and Systematics,* 1974, *5,* 324–383.

Alexander, R. D. The search for a general theory of behavior. *Behavioral Science,* 1975, *20,* 77–100.

Alexander, R. D. Natural selection and the analysis of human sociality. In C. E. Goulden (Ed.), *Changing scenes in natural sciences, 1776–1976.* Philadelphia: Philadelphia Academy of Natural Sciences, 1977.

Alexander, R. D. Evolution and culture. In N. A. Chagnon and W. Irons (Eds.), *Evolutionary biology and human social behavior.* North Scituate, Mass.: Duxbury Press, 1979a.

Alexander, R D. *Darwinism and human affairs.* Seattle: Univ. of Washington Press, 1979b.

Alexander, R. D. Hoogland, J. L., Howard, R. D., Noonan, K. M., and Sherman, P. W. Sexual dimorphism and breeding systems in pinnipeds, ungulates, primates and humans. In N. A. Chagnon and W. Irons (Eds.), *Evolutionary biology and human social behavior.* North Scituate, Mass.: Duxbury Press, 1979.

Axelrod, R., and Hamilton, W. D. The evolution of cooperation. *Science,* 1981, *211,* 1390–1396.

Barkow, J. H. Culture and sociobiology. *American Anthropologist*, 1978, *80*, 5–20.
Barth, F. On the study of social change. *American Anthropologist*, 1967, *69*, 661–669.
Bates, D. G. (Ed.). Special issue on elopement and bride capture. *Anthropological Quarterly*, 1974, *47*, 250–392.
Bates, D. G., and Lees, S. H. The myth of population regulation. In N. A. Chagnon and W. Irons (Eds.), *Evolutionary biology and human social behavior.* North Scituate, Mass.: Duxbury Press, 1979.
Becker, G. *Human capital* (2nd ed.). New York: Columbia Univ. Press, 1975.
Betzig, L. L. Despotism and differential reproduction. *Ethology and Sociobiology*, in press.
Blurton Jones, N., and Sibly, R. M. Testing adaptiveness of culturally determined behaviour. In N. Blurton Jones and V. Reynolds (Eds.), *Human behaviour and adaptation.* London: Taylor & Francis, 1978.
Bohannon, L. *Return to laughter.* New York: Doubleday, 1954.
Burley, N. The evolution of concealed ovulation. *American Naturalist*, 1979, *114*, 835–858.
Carniero, R. L. A theory of the origin of the state. *Science*, 1970, *169*, 733–738.
Carr-Saunders, A. M. *The population problem.* London/New York: Oxford Univ. Press (Clarendon), 1922.
Chagnon, N. A. Genealogy, solidarity, and relatedness. *Yearbook of Physical Anthropology*, 1975, *19*, 95–110.
Chagnon, N. A. Mate competition, favoring close kin, and village fissioning among the Yanomamo Indians. In N. A. Chagnon and W. Irons (Eds.), *Evolutionary biology and human social behavior.* North Scituate, Mass.: Duxbury Press, 1979a.
Chagnon, N. A. Is reproductive success equal in egalitarian societies? In N. A. Chagnon and W. Irons (Eds.), *Evolutionary biology and human social behavior.* North Scituate, Mass.: Duxbury Press, 1979b.
Chagnon, N. A. Kin selection theory, kinship, marriage and fitness among the Yanomamo Indians. In G. W. Barlow and J. Silverberg, *Sociobiology: Beyond nature/nurture?* Boulder, Co.: Westview Press, 1980.
Chagnon, N. A. Terminological kinship, genealogical relatedness and village fissioning among the Yanumamo Indians. In R. D. Alexander and D. W. Tinkle (Eds.), *Natural selection and social behavior.* New York: Chiron Press, 1981.
Chagnon, N. A., and Bugos, P. E., Jr. Kin selection and conflict. In N. A. Chagnon and W. Irons (Eds.), *Evolutionary biology and human social behavior.* North Scituate, Mass.: Duxbury Press, 1979.
Chagnon, N. A., Flinn, M. V., and Melancon, T. F. Sex-ratio variation among the Yanomamo Indians. In N. A. Chagnon and W. Irons (Eds.), *Evolutionary biology and human social behavior.* North Scituate, Mass.: Duxbury Press, 1979.
Cohen, R. Marriage instability among the Kanuri of northern Nigeria. *American Anthropologist*, 1961, *63*, 1231–1249.
Colson, E. *Marriage and the family among the Plateau Tonga of northern Nigeria.* Manchester: Univ. of Manchester Press, 1958.
Colson, E. Plateau Tonga. In D. M. Schneider and K. Gough (Eds.), *Matrilineal kinship.* Berkeley: Univ. of California Press, 1961.
Coon, C. S. *Caravan* (Rev. ed.). New York: Holt, Rinehart & Winston, 1958.
Daly, M. Review of *Human sociobiology* by D. G. Freedman. *Nature*, 1980, *284*, 681.
Daly, M., and Wilson, M. *Sex, evolution, and behavior.* North Scituate, Mass.: Duxbury Press, 1978.
Darwin, C. *The descent of man and selection in relation to sex.* New York: Appleton, 1871.
Davis, K., and Blake, J. Social structure and fertility. *Economic Development and Culture Change*, 1956, *4*, 211–235.
Dawkins, R. *The selfish gene.* London/New York: Oxford Univ. Press, 1976.

Divale, W. T., and Harris, M. Population, warfare, and the male supremacist complex. *American Anthropologist*, 1976, *78*, 521–538.

Draper, P. Social and economic constraints on child life among the !Kung. In R. B. Lee and I. DeVore (Eds.), *Kalahari hunter-gatherers*. Cambridge, Mass.: Harvard Univ. Press, 1976.

Flinn, M. V. Uterine vs. agnatic kinship variability and associated cousin marriage preference. In R. D. Alexander and D. W. Tinkle (Eds.), *Natural selection and social behavior*. New York: Chiron Press, 1981.

Fox, R. Alliance and constraint. In B. Campbell (Ed.), *Sexual selection and the descent of man 1871–1971*. Chicago: Aldine, 1972.

Fox, R. The evolution of mind. *Journal of Anthropological Research*, 1979, *35*, 138–156.

Freedman, D. G. *Human sociobiology*. New York: Free Press, 1979.

Gaulin, S. J. C., and Schlegel, A. Paternal confidence and parental investment. *Ethology and Sociobiology*, 1980, *1*, 121–135.

Goodale, J. C. *Tiwi wives*. Seattle: Univ. of Washington Press, 1971.

Greene, P. J. Promiscuity, paternity, and culture. *American Ethnologist*, 1978, *5*, 151–159.

Hames, R. B. Relatedness and interaction among the Ye'kwana. In N. A. Chagnon and W. Irons (Eds.), *Evolutionary biology and human social behavior*. North Scituate, Mass.: Duxbury Press, 1979.

Hamilton, W. D. The evolution of altruistic behavior. *American Naturalist*, 1963, *97*, 354–356.

Hamilton, W. D. The genetical evolution of social behavior, I and II. *Journal of Theoretical Biology*, 1964, *7*, 1–52.

Hart, C. W. M., and Pilling, A. R. *The Tiwi of north Australia*. New York: Holt, Rinehart & Winston, 1960.

Hartung, J. Paternity and inheritance of wealth. *Nature*, 1981, *291*, 652–654.

Howell, N. *Demography of the Dobe !Kung*. New York: Academic Press, 1979.

Irons, W. *The Yomut Turkmen*. Ann Arbor: Univ. of Michigan Museum of Anthropology, 1975.

Irons, W. *Evolutionary biology and human fertility*. Paper presented at the annual meeting of the American Anthropological Association, Houston, December 1977.

Irons, W. Natural selection, adaptation, and human social behavior. In N. A. Chagnon and W. Irons (Eds.), *Evolutionary biology and human social behavior*. North Scituate, Mass.: Duxbury Press, 1979a.

Irons, W. Investment and primary social dyads. In N.A. Chagnon and W. Irons (Eds.), *Evolutionary biology and human social behavior*. North Scituate, Mass.: Duxbury Press, 1979b.

Irons, W. Cultural and biological success. In N. A. Chagnon and W. Irons (Eds.), *Evolutionary biology and human social behavior*. North Scituate, Mass.: Duxbury Press, 1979c.

Irons, W. Political stratification among pastoral nomads. In l'Équipe d'Écologie et Anthropologie des Sociétés Pastorales (Ed.), *Pastoral production and society*. London/New York: Cambridge Univ. Press, 1979d.

Irons, W. Is Yomut social behavior adaptive? In G. W. Barlow and J. Silverberg (Eds.), *Sociobiology: Beyond nature/nurture?* Boulder, Co.: Westview Press, 1980.

Kinsey, A. C., Pomeroy, W. B., Martin, C. E., and Gebhard, P. H. *Sexual behavior in the human female*. Philadelphia: Saunders, 1953.

Kurland, J. A. Paternity, mother's brother, and human sociality. In N. A. Chagnon and W. Irons (Eds.), *Evolutionary biology and human social behavior*. North Scituate, Mass.: Duxbury Press, 1979.

Lack, D. *The natural regulation of animals*. London/New York: Oxford Univ. Press, 1954.

Lack, D. *Population studies of birds*. London/New York: Oxford Univ. Press, 1966.

Lack, D. *Ecological adaptation for breeding in birds*. London: Methuen, 1968.

Lee, R. B. !Kung Bushmen subsistence. In A. P. Vayda (Ed.), *Environment and cultural behavior*. New York: Natural History Press, 1969.

Lee, R. B. Population growth and the beginnings of sedentary life among the !Kung Bushmen. In B. Spooner (Ed.), *Population Growth*. Cambridge, Mass.: MIT Press, 1972.
Lee, R. B. *The !Kung San*. London/New York: Cambridge University Press, 1979.
Loehlin, J. C., Lindzey, G., and Spuhler, J. N. *Race differences in intelligence*. San Francisco: W. H. Freeman, 1975.
Livingstone, F. B. Anthropological implications of sickle cell gene distribution in West Africa. *American Anthropologist*, 1958, *60*, 533–562.
Livingstone, F. B. Cultural causes of genetic change. In G. W. Barlow and J. Silverberg (Eds.), *Sociobiology: Beyond nature/nurture?* Boulder, Co.: Westview Press, 1980.
Lumsden, C. J., and Wilson, E. O. *Genes, mind, and culture*. Cambridge, Mass.: Harvard Univ. Press, 1981.
Malinowski, B. *Argonauts of the western Pacific*. New York: Dutton, 1961. (Originally published, 1922.)
Murdock, G. P. *The ethnographic atlas*. Pittsburgh: Univ. of Pittsburgh Press, 1967.
Orians, G. H. On the evolution of mating systems in birds and mammals. *American Naturalist*, 1969, *103*, 589–603.
Polgar, S. Population history and population policy. *Current Anthropology*, 1972, *13*, 203–211.
Richards, A. Some types of family structure amongst the central Bantu. In A. R. Radcliffe-Brown and D. Forde (Eds.), *African systems of kinship and marriage*. London/New York: Oxford Univ. Press, 1950.
Richerson, P. J., and Boyd, R. Review of Evolutionary biology and human social behavior. *Human Ecology*, 1980, *8*, 221–223.
Robinson, H. B., Woods, S. C., and Williams, A. E. The desire to bear children. In J. S. Lockard (Ed.), *The evolution of human social behavior*. Amsterdam/New York: Elsevier, 1980.
Sahlins, M. *The use and abuse of biology*. Ann Arbor: Univ. of Michigan Press, 1976.
Spooner, B. J. (Ed.), *Population growth*. Cambridge, Mass.: MIT Press, 1972.
Spuhler, J. N. *Genetic diversity and human behavior*. Chicago: Aldine, 1967.
Symons, D. *The evolution of human sexuality*. London/New York: Oxford Univ. Press, 1979.
Trivers, R. L. The evolution of reciprocal altruism. *Quarterly Review of Biology*, 1971, *46*, 35–57.
Trivers, R. L. Parental investment and sexual selection. In B. Campbell (Ed.), *Sexual selection and the descent of man, 1871–1971*. Chicago: Aldine, 1972.
van den Berghe, P. L. *Human family systems*. Amsterdam/New York: Elsevier, 1979.
Verner, J. The evolution of polygamy in the long-billed marsh wren. *Evolution*, 1964, *18*, 252–261.
Wasser, S. K. Reciprocity and the trade-off between associate quality and relatedness. *American Naturalist*, 1982, *119*, 720–731.
Weisner, P. *A study of reciprocity among the Kahalari Bushmen*. Unpublished doctoral dissertation, University of Michigan, 1977.
Weiss, M. L., and Mann, A. E. *Human biology and behavior*. (3rd ed.). Boston: Little, Brown, 1981.
Whiting, J. W. M. Effects of climate on certain cultural practices. In W. H. Goodenough (Ed.), *Explorations in cultural anthropology*. New York: McGraw–Hill, 1964.
Williams, G. C. *Adaptation and natural selection*. Princeton, N. J.: Princeton Univ. Press, 1966.

8

A Life-Stage Analysis of Taiwanese Women: Social and Health-Seeking Behaviors

MARY L. WATERHOUSE

I. Introduction

Much of the past ethnographic and theoretical literature on humans implies that females are subordinate, passive, and dependent on males, and that their strategies have developed in response to male strategies and needs. Although numerous researchers have recently pointed out the

SOCIAL BEHAVIOR OF
FEMALE VERTEBRATES

ISBN 0-12-735950-8

fallacies in such statements (cf. Rosaldo and Lamphere, 1974; Reiter, 1975; Iglitzin and Ross, 1976; Jacobs, 1981), relatively few authors have addressed the potential consequences of these value judgments for people presently experiencing rapid westernization. The intention of this chapter is to identify some of the sources and possible effects of such male-focused value judgments (see Wasser and Waterhouse, this volume) on women's health care in the patrilocal–patrilineal system of the Taiwanese.

A. Stereotypic Images of Women

As is typical in patrilocal–patrilineal cultures, a Taiwanese woman usually lives with her husband and his family after marriage, and inheritance is passed along the father's lineage. This mode of inheritance and residence is also characterized by the existence of different social strategies and sex-roles between women and men that are vulnerable to misinterpretation (Wolf, 1972). Wolf (1974) cites several examples of male-focused stereotypes regarding women in such patrilocal societies:

1. Women and their roles are often seen as secondary to those of men.
2. Female children are said to be less desirable than male children.
3. Women are reported to participate in meaningless social groups.
4. Women are also said to impart little sociopolitical influence.

However, alternative views of women in this culture have been provided by Wolf (1972, 1974, 1975), Ahern (1975), and Harrell (1981a, 1981b). The work of these authors reveals that such stereotypes often reflect the use of value judgments by researchers when evaluating both the sexual division of labor in Taiwan and the ways in which Taiwanese women cope with their particular residence system (see Section II). Unfortunately, many of the male-focused values continue to persist today, and their negative consequences for the Taiwanese health-care system, as it undergoes westernization, are now becoming readity apparent, as explained below.

B. Sex-Role Differences, Traditional Values, and Health Care

Indigenous medical systems evolve according to the ways in which societies are organized (Worth, 1975; Kunstadter, 1975) and in response to the perceived needs of the people (Kleinman and Kunstadter, 1978). As a result, they tend to meet the psychosocial and medical needs of those individuals who adhere to traditional beliefs (Gallin, 1978). Prior to the introduction of Western medicine in the nineteenth century (Gale, 1978), both women and men presumably employed traditional healing methods to similar degrees. However, Taiwanese women are currently twice as likely as men to use traditional methods of health care (Tseng, 1975; Kleinman, 1980). This sex difference in health-seeking behavior appears to result, in part, from sex differences in adherence to traditional sex-roles, and in differences in degrees of exposure to westernization.

Taiwanese men tend to have a relatively broader range of outside enterprises available to them than do women (Wolf, 1975), thereby exposing them to a variety of Western ideals, medical systems, and value judgments. This is especially the case for men who work in wage labor. Women, on the other hand, typically remain more family-oriented, even when employed outside of the home, because Taiwanese philosophies hold women responsible for family cohesion. The maintenance of family cohesion is accomplished, in part, through adherence to traditional concepts and values, including traditional medical treatments. However, the amount of information about and accessibility to health-care methods are commonly dependent on an individual's age, sex, status, and the value placed on his or her work (Odum, 1971). As a result, the relatively greater westernization of men, combined with Western ideologies that promote a higher value on male activities, may be depleting the availability and credibility of the traditional health-care methods. The availability of traditional medicine in Taiwan is further reduced because many social scientists and government officials, who organize and operate medical licensing and health services, tend to promote Western medical programs over traditional programs (Gallin, 1978). This loss could be particularly disruptive for many Taiwanese women and others who are less westernized, since traditional healing is often more culturally appropriate for them.

In this chapter, I first address Wolf's (1974) four points regarding the subordinate images of Taiwanese women (cited in Section I, A) to illustrate why previously applied value judgments and portrayals of women have been misleading. I then show how life-stage and role differences between the sexes, combined with the accessability and appropriateness of medical systems, contribute to sex differences in health-seeking behaviors. Finally, I discuss how the influx of Western values and medicine, in the presence of an erroneously male-focused value system, has contributed to a dualistic health-care system that is becoming increasingly dominated by Western medicine. The discussion will emphasize the unfortunate consequences that could result from the loss of traditional medical systems—systems that could benefit many Taiwanese people when properly used in conjunction with the more technologically advanced medicine of Western cultures.

II. The Environment, Division of Labor, and Sex Roles

A. Ch'i-nan's Complementary Sexual Division of Labor

An ample body of work has been done on the village of Ch'i-nan, Taiwan. This village is discussed here as a representative "model" of how a sexual division of labor that is beneficial to both sexes has developed in that area of Taiwan. Sex roles are quite complementary and thus, female roles are not seen as secondary to those of men.

When the inhabitants of Ch'i-nan immigrated from mainland China to Taiwan approximately 200 years ago (Harrell, 1981a), they were already rice farmers engaging in a distinct division of labor and patrilocal residence. The environmental conditions in Taiwan favored the continuation of those previously established characteristics. However, farmable land was scattered in the Ch'i-nan area, thereby promoting the formation of discrete social units of related individuals for its cultivation and inheritance (Harrell, 1981a). At the same time, other environmental constraints promoted cooperative efforts between these separate social units. For example, invading neighboring settlements often forced the individual agnatic units of Ch'i-nan to unite for defensive measures (Harrell, 1981a). Intergroup cooperation was also facilitated by the establishment and maintenance of communal irrigation systems, which were critical to the development and production of wet rice farming. The patrilocal residence system, which assures a stable community of male members, was particularly advantageous under these conditions.

B. The Allocation of Sex Roles

In addition to the activities above, other needs had to be fulfilled. These included (*a*) health-care maintenance, (*b*) socialization of young, and (*c*) the production and processing of other foods critical to a balanced diet. However, the performance of these latter tasks was not always compatible with wet rice farming, defense, and irrigation activities. For example, working in, and sometimes defending, the water-soaked rice fields would present obvious difficulties for nursing mothers and their infants. Thus, a sexual division of labor, whereby all needs could be met, would seem most efficient and hence beneficial for both women and men. As a result, men worked in the fields, while women were responsible for a variety of tasks that were more compatible with child-rearing and maintaining family cohesion (Wolf, 1975). Moreover, family cohesion appears to be critically important in the Taiwanese culture (see Section III).

III. A Woman's Life-History Stages

Given the preceding sex differences in roles and natal dispersal patterns, one might suspect that child-rearing practices would differ according to the sex of the child and the ways each sex copes with the environment at each life stage. In fact, as will be shown, some of the major reasons for the male-focused stereotypes in this culture cited by Wolf (Section I, A) appear to result from a failure by Westerners describing this culture to take the need for these sex differences into consideration. For example, young girls contribute to many family needs while they are simultaneously being prepared for the patrilocal residential transition. Since the eventual transition results in a complete alteration of social and environmental resources available to women, young girls should be developing a heightened awareness of social and environmental conditions during this first stage (Wolf, 1972). Females then use the learned domestic and social skills to develop appropriate strategies in coping with the new social and environmental conditions. In essence, each life stage prepares them for the next. In their third life stage, women utilize the power, influence, and knowledge that they have accumulated through interpersonal relationships with family and social group members.

Section III details each of the three life stages of women: daughter, mother, and grandmother, to illustrate how women cope with their particular social system. This section also shows how women's health-care strategies are interwoven with their roles—women respond to family illness in accordance with their biological, environmental, and cultural needs, and according to the health-care methods that are available to them.

A. Stage One: The Daughter

Patrilocal residence and adult role expectations in Taiwan require a more demanding socialization process for daughters than for sons. For example, daughters are expected to perform many more domestic activities at an earlier age than sons (Wolf, 1975). Furthermore, the performance of domestic responsibilities is supplemented with the necessary fulfillment of a girl's own developmental needs (e.g., forming peer relationships). The structured and seemingly restrictive norm for daughters could be perceived as an indicator of female subordination. However, the responsibilities of familial and personal activities seem to provide a young girl with valuable skills and abilities (e.g., organizational skills, patience and creativity, and an overall heightened awareness of her physical and social environment) that will eventually enhance her success in the transition to adulthood. The acquisition of such abilities requires that young girls explore diverse behavioral options and social techniques (Wolf, 1972), which in essence provide her with a foundation of diverse and broad-ranging skills to use in response to life changes.

Another potential indicator of female subordination could be extrapolated from the seemingly less emotional aspects of a girl's socialization process. However, in patrilocal societies, daughters are often temporary members of their natal group. The development of strong emotional and dependent ties with a daughter could increase the stress of her inevitable departure. A mother may, therefore, expend differential emotional investments in daughters and sons (see the following discussion). At the same time, a daughter is not solely dependent upon her mother for social and domestic learning or for emotional support. A young girl may spend a considerable amount of time observing or being with her mother's social network (Wolf, 1974), from which she is likely to acquire information on child care, food producing and preparing techniques, husband–wife relationships, health-care, and a wide variety of other sociocultural, reproductive, and subsistence information.

By contrast, sons are permanent residents of their household and parents are often partially dependent on their sons during their elder years. This makes it advantageous for a mother to nurture long-lasting bonds with male offspring, which is accomplished through intensive and emotional socialization processes (Wolf, 1972, 1974). Mothers may also invest differentially in sons because there is relatively little paternal investment in the early childhood socialization process: fathers spend a great deal of time in the rice fields, and there is a traditional tendency for "formal" father–son relationships (Wolf, 1974).

The more demanding physical and emotional socialization process of girls relative to boys, plus patrilineal descent and Western ideals (that place a heightened value on economically based subsistence roles relative to child care and other domestic responsibilities) has contributed to the belief that male children are more valued than female children in Taiwan and many other patrilocal societies. However, this view does not consider all factors that reflect the value of a child. Sex preferences for children should instead be a function of the number and sex ratio of existing offspring and the time lag between parental investments given to offspring and the offspring's reciprocation back to its parents. For example, a newlywed woman in Taiwan should desire a son because his birth will alleviate pressures from her in-laws to produce a male descendant who will assure their ancestral worship (Wolf, 1974; Johnson, 1975). A father, on the other hand, may be eager for a son who will eventually provide farming assistance, as well as inherit his land. These benefits from male offspring, however, will not be totally realized until he marries as an adult. When a grown son marries, his bride will be of great assistance to his mother. Furthermore, adult sons typically provide their elderly parents with political, economic, and emotional security (Wolf, 1972). These factors predict that younger parents will tend to prefer sons over daughters.

Conversely, the anticipated domestic contributions from a daughter suggest that a woman should have a strong desire for a female child after she has produced at least one healthy son. In other words, girls may be as valuable to their families as sons, but in different ways. Sons are more helpful in their later years, whereas daughters are more immediately beneficial to their parents during their earlier developmental years.

Daughters contribute to their families in numerous ways. They perform many domestic tasks at a relatively young age, such as sewing, cooking, sib caregiving, and cleaning (Wolf, 1975). By performing these activities, daughters are contributing to the hygienic and nutritional health and cohesion of the basic Taiwanese social unit, the family. A daughter's contribution is further exemplified when mothers become ill, as the latter are usually unable to escape daily responsibilities even when sick (Kleinman, 1980). Since a woman's social value is often correlated with her ability to fulfill these responsibilities, thereby maintaining a cohesive family unit (Wolf, 1972; Ahern, 1978; Kleinman, 1980), failure to do so can be most stressful. In fact, women most often commit suicide for family-related reasons (Wolf, 1975), whereas men tend to do so for politically-related reasons (King, 1975). Thus, without supplemental help from a competent daughter, a mother's health and well-being could be further jeopardized during illness. Due to the many contributions of a daughter, a woman's reaction to the birth of a girl is predicted to be a function of her age, as well as of the number and sex of her other children. However, these views have yet to be put to empirical test.

The ways in which a daughter learns and uses her skills influences the health, cohesion, and stability of her uterine family. They also influence the quality of the husband to whom she can be matched and her progress into the next life stage. Presumably, similar evaluations are taking place with respect to the qualifications of her potential groom. When marriage takes place, the woman moves in with her husband's family, a transition which embodies the true test of her competencies.

B. Stage Two: The Wife

Moving into her husband's home may be stressful for a young woman. Unlike her husband, she is suddenly isolated from her own family and must somehow adjust to a new social environment. The adjustments will be especially difficult if conflicting interests develop between the bride and her new mother-in-law. These mother-in-law and daughter-in-law conflicts also appear to have contributed to the interpretation of female subordination. Thus, it becomes important to consider what factors might account for such conflicts. Factors to be considered include (*a*) the mother-in-law's defense of her vested interests in her son; (*b*) the bride's vested interest in herself and her future; and (*c*) the relative reproductive potential of the two women. The following discussion will address these issues, as well as the possible ways a new bride might ease her adjustment into a new living environment.

When a young woman moves in with her husband's family, she benefits the mother-in-law with her domestic contributions. Such benefits may not, however, outweigh the threat to the mother-in-law of sharing her son and home with another adult woman (Wolf, 1975). The mother-in-law has invested a great deal in her son for future economic, social, and emotional security. Thus, the presence of a bride may be perceived as a psychological threat to the family structure and its function (Wolf, 1972). This initial phase of the patrilocal transition often creates a potentially stressful atmosphere for both the wife and the mother-in-law. Over the long term, this stress could have an impact on the inclusive fitness (Hamilton, 1964) of both women; continual psychological (or physical) stress can disrupt a person's health, child-rearing abilities, personality, sexuality, and maintenance of interpersonal relationships (Kleinman, 1975). However, these two women need not necessarily succumb to the consequential stress if a conflict does develop between them. For example, the mother-in-law is likely to receive emotional support from other family members. Her husband, married son and any other family members in the home are likely to favor her in conflict situations because of their long-term kinship bonds and desire to protect the family structure and its function. The concern for, and protection of, such family interests may have contributed to the social norms, for example, that forbid open displays of affection between wives and husbands (e.g., Irons, 1979).

Alternatively, Hsu (1965) has suggested that restrictions on public behavior between spouses were established for the maintenance of public female subordination. However, Hsu's view ignores the likely possibility that wives may also have vested interests and ways of protecting themselves from such stresses. For example, a wife can disrupt her husband's perception of family cohesion by trying to divide the loyalty between him and his mother (Wolf, 1972; Ahern, 1975). A woman can also disrupt her husband's sociopolitical authority thorough activities in her developing female social network (see Section III, C).

A compounding consideration regarding the possible conflicts between a wife and her mother-in-law concerns their relative abilities and desires for future offspring. Since the wife will be bearing the mother-in-law's grandchildren, inclusive fitness theory predicts the existence of a reproductive conflict between them as long as the latter's reproductive value is more than one-half that of the daughter-in-law's. The young bride's reproductive potential should be quite high at the time of her marriage. Thus, if the mother-in-law's reproductive value is still high *and* she still desires more young, the bride's presence could facilitate reproductive competition between them. Inclusive fitness theory, therefore, predicts that conflicts between a bride and her mother-in-law should correlate with the mother-in-law's age, and hence her reproductive value.

Another consideration regarding reproductive competition between these two women includes the consideration of other adult sons in the home. If the mother-in-law has only one son and, therefore, only one daughter-in-law, the two women's overlapping interests could provide the impetus for a congenial relationship. On the other hand, if there are other married sons, such conflicts may be predictably high. The overall potential for interpersonal conflicts in the family does seem to increase with the number of other daughters-in-law in the home (Wolf, 1972). At the same time, if all of the married sons and their wives have comparable reproductive potentials, the mother-in-law should have equal interests in all of the daughters-in-law and their children. The daughters-in-law may thus strive for harmonious relationships with one another to ensure minimal domestic conflicts. In fact, for this very reason, mothers-in-law should strive to reduce conflicts between their daughters-in-law.

In spite of the potential for interpersonal conflicts, they do not always develop, and even if they do, they are likely to dissipate when mutually satisfying gains can be realized. For example, a harmonious relationship between the bride and her mother-in-law could influence the relationship between the bride and her husband, thereby affecting their fertility potential. The couple's fertility potential affects the inclusive fitness of both women, making cooperation between them advantageous. A complementary or cooperative relationship between these women is also going to reflect the bride's acceptance into the local women's social community and in her overall credibility in the village. Since both of these women derive much of their emotional support and information base from their networks (Wolf, 1972) (see Section III, C), they both have much to gain by contributing to the network's cohesion.

The Relevance of Female Social Networks

Throughout much of Taiwan, women frequently come together to perform both domestic and economically related chores (Wolf, 1972). During these gatherings, women typically exchange, transmit, and consciously use their individual and collective bodies of information for personal gain (Wolf, 1974). For example, women are most often responsible for family and personal health care. This responsibility is exemplified by the tendency for women, more than men, to represent relatives at ritual healing sessions (Kleinman, 1980). Because of a woman's responsibility to maintain her family's physical and emotional harmony and because of the immediate consequences of her own ill-health, a woman should want access to the most available and personally relevant health-care treatment. Presumably, the importance of the women's health-care role will encourage the exchange of personal or self-directed health-care techniques and information on traditional practitioners between women in social networks. These two forms of healing (i.e., personal or self-care and traditional practitioners and methods) are particularly important for women in their reproductive years (Ahern, 1975; Kleinman, 1980). Indigenous practitioners who emphasize public-style clinical settings are most accessible to women, as well as to children and elderly populations both in rural and urban settings (Dunn, 1976). Traditional healing methods and practitioners also tend to link one's perceived symptoms with more culturally relevant causes (Pockert, 1976). This provides the patients with more meaningful diagnoses and treatments for their illnesses, as well as for continual self-care (Unschuld, 1976) and is therefore particularly relevant to those individuals responsible for their family's health care.

Female social networks also allow women to affect local political, economic, and interpersonal matters, largely through their collective influences on male behaviors. Two of the primary ways of accomplishing this are through (*a*) the advice women bestow upon their sons, thereby influencing the latter's decisions on investments and social affairs (Ahern, 1975) and (*b*) the culturally recognized phenomenon of "loosing face" (Wolf, 1974; Ahern, 1975). The second method is crucial to a woman's social influence because when a man loses face, his social, economic, and political credibility is devalued among other village members. Losing face occurs through "gossip" between women in networks regarding a man's inappropriate behavior. This "news" is eventually shared with the women's family members, which can be shameful to the man in question (Ahern, 1975). For example, if a man does not properly fulfill his role of father and husband, is abusive, or is in the process of making an unpopular political or economic decision, women can use their social networks as a threat to get him to either change or risk social alienation. In this way, women are not as isolated as a patrilocal system may initially project; they are also unlikely to be as vulnerable to physical and emotional abuse or total dependency on men as has been suggested in the past.

In summary, female social groups in Taiwan are not meaningless (Wolf, 1972, 1974, 1975; Ahern, 1975). The information exchanged in such networks can contribute to family cohesion and health care and provides an avenue for female involvement in political, economic, and social matters. In these ways, female networks contribute to the survival and reproductive success of both family and other village members.

C. Stage Three: The Female Elders

The third stage in a woman's life cycle is the "regal" grandmother role (Harrell, 1981b). In a general sense, postreproductive women or grandmothers are honored for their wisdom and are sought-after for advice on child care, health care (Alland, 1970), family problems, matchmaking, and religious activities (Wolf, 1974), as well as for money management and social affairs (Ahern, 1975). But many of these attributions went unnoticed until researchers recognized the value of women beyond their reproductive years (see also Harrell and Amoss, 1981).

Assuming that a woman no longer has the physical capacity or emotional desire for more offspring, the presence of grandchildren will likely be a welcome change for her. The freedom from obligatory or full-time offspring care, plus the domestic help from her daughters-in-law, will enable the older woman to focus more of her attention toward other matters, such as her social network activities and family relations. Such conditions allow a grandmother time to educate grandchildren on matters that are of particular concern to her. She can, for example, transmit her refined knowledge to future generations, thereby contributing to sociocultural transmission and development (cf. Neisser, 1967). An elder woman's social network is especially important at this time and is likely to increase the influence she has on her husband and other family and village members as well (Ahern, 1975).

Relationships that women have nurtured with sons also become fully realized during this life stage. As adults, men more often seek and accept advice from their mothers on political, economic, and social issues (Wolf, 1974; Ahern, 1975). The opinions that men voice in the public arena then, are not necessarily just their own, but also represent those of their mother and her social network.

In many ways then, women can be seen as subtly, but effectively influencing the sociocultural developments of their family and perhaps even their entire village.

IV. Changing Sex Roles, Health-Seeking Behaviors, and a Pluralistic Medical System

This chapter suggests that sex-specific strategies and roles have been quite complementary. However, the influx of Western, male-focused ideals and a changing economy in Taiwan have contributed to the devaluation of women's roles. Such male-focused value judgments will have unfortunate consequences for members of the less-valued sector of society. In Taiwan, this is particularly recognizable in the health-care system. Thus, the following section suggests that the devaluation of traditional health-care practices in Taiwan, which are now predominantly but not exclusively employed by women, is a consequence of a male-focused value system, primarily influenced by Western culture.

A. The Relevance of Traditional Medicine

Traditional Chinese medicine has been evolving for twenty-five centuries (Tseng, 1975). Since medical systems develop according to the ways in which a society is organized (Kunstadter, 1975) and in response to the perceived health needs of the people, traditional practitioners and patients should share similar medical concepts and beliefs. Although some traditional practitioners in Taiwan today provide advice on financial investments and subsistence activities (Tseng, 1975), their relevance is generally recognized in their ability to reinforce moral behaviors and return a patient and her or his family to health and order (Gallin, 1978). Diagnostic sessions are common in which patients discuss problems and their sources in order to alleviate stress (Gallin, 1978) in ways that are culturally meaningful (e.g., public-style group sessions involving the participation of the patients, their relatives, and perhaps even a few neighbors). When an illness or stressful state originates from an interpersonal conflict with a relative or neighbor, these sessions can be revealing and even therapeutic. This is particularly important since culture influences our perception of what is stressful (Kleinman, 1980), and stress can contribute to physical disabilities and even death (Sterling and Eyer, 1981).

Cultural relevance in medical care can also be attained, in part, through doctor–patient agreement as to the causes of, and treatments for, an illness (Faberga, 1975; Zola, 1981). In fact, a patient's decision whether or not to comply with a physician's treatment recommendation is often embedded in the ability of the doctor and patient to negotiate between their respective perceptions of the illness itself and its treatment (Katon and Kleinman, 1981). Such negotiations are best accomplished through insightful patient-oriented interviews, a necessary skill that is accountable for up to 90% of accurate diagnoses (Twaddle, 1981). This requires that doctors use terms and concepts that are meaningful and familiar to the patients, which makes traditional Taiwanese healers particularly valuable to the more traditional members of society (Ahern, 1978). Belief in a cure and the ability to communicate shared beliefs in the cure have a strong effect on the success rate of the medical intervention (Katon and Kleinman, 1981).

B. Sex Differences in Health-Seeking Behaviors

Medical needs of people differ from one population to another (Harwood, 1981), according to both biological and sociocultural factors (Eisenberg and Kleinman, 1981). Individuals within a culture are also expected to differ in their health-seeking behaviors to the degrees that different amounts and types of medical information are available and acceptable to them (Chrisman, 1977; Kleinman, 1980). Westernization in Taiwan has brought with it Western medicine. At the same time, it has also allowed male roles to become westernized more quickly, giving men greater exposure to, and acceptance of this new form of medicine. The following summary describes how health seeking behaviors currently differ between women and men in Taiwan.

1. Women are less able to escape daily responsibilities even when sick (Kleinman, 1980), and are, therefore, dependent on self-care, access-

ible practitioners, and reliable family members when ill.

2. Women in general (Kleinman, 1980), but especially during their reproductive years (Ahern, 1975), are more likely to see traditional medical practitioners than men.

3. Women represent relatives at ritual healing sessions more than men (Kleinman, 1980), reflecting the greater role of the former in Taiwanese family health care.

4. Women represent the majority (70–80%) of temple visitors for aid from diviners (Tseng, 1975; Unschuld, 1976).

5. Female social groups appear to provide sources of traditional "female-specific" health care information. There does not seem to be a parallel network system operating for men.

The preceding health-seeking behaviors of women reflect their vital role in maintaining familial cohesion and health care. However, older people of both sexes and the more traditional individuals, in general, are also inclined to seek traditional-style practitioners rather than Western-style practitioners (Ahern, 1978).

C. Reasons for and Consequences of Western Medical Domination

The use of traditional health care systems, combined with the increasing number of westernized biomedical practitioners in Taiwan, is contributing to its pluralistic medical system. Both systems are necessary as long as they serve the needs of the people. Nevertheless, the domination of the technologically advanced, but not always most appropriate, Western medicine is contributing to the reduction of available indigneous practitioners.

If traditional medicine is still utilized by many individuals, then what accounts for the prevalence of Western medicine? Part of the answer lies in the identification of those most likely to use Western medicine; that is, men and individuals of higher socioeconomic status (Gallin, 1978). Since these individuals are also most likely to be elected as government officials (Gallin, 1978), their promotion of Western medical systems may represent attempts to fulfill their own health, economic, and political interests. Furthermore, Taiwanese government support is almost entirely given to Western medical programs, and the provincial health department, which operates health clinics throughout Taiwan, is primarily staffed with individuals trained in Western medical procedures (Gale, 1978). These Western medical personnel and programs may be inappropriate for patients who adhere to traditional beliefs and values because Western medical concepts regarding illnesses and treatments often contradict those maintained by traditional patients (Ahern, 1978). These Western health-care orientations will inevitably be male-focused, as is the health-care system in the United States (Mendelson, 1981).

There is little doubt that the emphasis on biomedical healing will have unfortunate consequences, especially for traditional women who retain their faith in indigenous health-care methods. For example, some of the values traditionally attributed to grandmothers in Taiwan are jeopardized by the changing health-care system. Women elders have always been honored for their indigenous medical knowledge. Thus, if the promotion of westernized medicine is reducing the credibility of those methods, female elders in Taiwan (much like the elders in many Western cultures) may soon feel the projected inadequacies of the knowledge for which they were historically honored. Such changes may also have an impact on those who benefited from a grandmother's information to the degree that their knowledge also contributed to the prevention of, or treatment for, common illnesses.

A mother's crucial role in family-oriented and self-directed health-care delivery may also become devalued as a result of such westernization. The consequential stress effects from its devaluation may then be further exaggerated by simultaneous loss of her options for traditional, stress-related treatments. The traditionally oriented women who prefer indigenous practitioners during their reproductive years may also face the possibility of reduced pre- and postnatal care for themselves and their infants if traditional practitioners are unavailable.

Of course, we must address the question of just how effective traditional healing methods are for present-day Taiwanese health-care problems before we can freely assess the consequences of its loss. The answer to that question can only be determined by careful comparative health-care studies. One could also argue that individuals who choose traditional medicine only need to be educated or oriented to the concepts of Western medicine for it to become an effective method. However, the rising costs and urban-centered distribution of biomedicine reduce its availability for many potential patients. Furthermore, traditional sociocultural (including medical) beliefs are often interwoven with child-rearing techniques, sex roles, religious concepts, and subsistence activities. To disrupt these belief systems further through a greater Western value orientation, may have consequences that lie far beyond the projected benefits of exclusively implementing biomedical systems.

V. Summary

The "perceived" value of women, their roles, and their strategies have inadvertently declined as societies increasingly measure an individual's worth according to the immediate relative economic gains resulting from their role activities. Such value judgements are readily apparent in Taiwan, where a patrilocal residence pattern and the rapid influx of Western values and ideals have contributed to a subordinate image of women. This chapter takes issue with these value judgments, emphasizing the complementary nature of sex-roles in Taiwan, while unmasking some of the unfortunate consequences these value judgments are having on women's health care. I take a life history approach, which offers alternative explanations for the existence and social value of women's strategies at each life stage, emphasizing those that women use when adjusting to their patrilocal residence system. I then describe how the traditional health-care system in Taiwan provides links between traditional roles and strategies, and an individual's sociocultural perceptions of health and illness. However, current political, idealistic, and economic support for Western medicine is contributing to its domination over the availability of the traditional health-care systems in Taiwan. At the same time, a continued emphasis of traditional role expectations for women and elders, relative to men, combined with the cultural naivete of many Western practioners, is potentially limiting the positive effects of Western medicine for traditionally oriented patients. This is reflected by the tendency for women and elders to utilize various forms of traditional healing at a greater rate than most men. It is argued here that the application of Western, male-focused value judgments is contributing to the domination of Western medicine over traditional health-care systems, even though the former may be presently inappropriate for the more traditionally oriented Taiwanese patients.

Acknowledgments

I would like to thank the following people for commenting on this manuscript: E. Ahern, C. Greenway, S. Harrell, J. Heerwagen, W. Hill, W. Irons, S. E. Jacobs, A. Kleinman, S. Wasser, and P. Williams.

References

Ahern, E. M. the power and pollution of Chinese women. In M. Wolf and R. Witke (Eds.), *Women in chinese society*. Stanford, Calif.: Stanford Univ. Press, 1975.

Ahern, E. M. Chinese style and western style doctors in Northern Taiwan. In A. Kleinman, P. Kunstadter, E. R. Alexander, and J. L. Gale (Eds.), *Culture and healing in asian societies. Anthropological, psychiatric and public health studies*. Cambridge, Mass.: Schenkman, 1978.

Alland, A. *Adaptation in cultural evolution: An approach to medical anthropology*. New York: Columbia Univ. Press, 1970.

Chrisman, N. The health seeking process: An approach to the natural history of illness. *Culture, Medicine and Psychiatry*, 1977, *1*, 351–377.

Dunn, F. Traditional Asian medicine and cosmopolitan medicine as adaptive systems. In C. Leslie (Ed.), *Asian medical systems: A comparative study*. Berkeley: Univ. of California Press, 1976.

Eisenberg, L., and Kleinman, A. (Eds.), *The relevance of social science for medicine*. Dordrecht: Reidel, 1981.

Faberga, H. The need for an ethnomedical science. *Science*, 1975, *189*, 969–975.

Gale, J. L. Patient and practitioner attitudes toward traditional and western medicine in a contemporary Chinese setting. In A. Kleinman, P. Kunstadter, E. R. Alexander, and J. L. Gale (Eds.), *Culture and healing in Asian societies. Anthropological, psychiatric and public health studies*. Cambridge, Mass.: Schenkman, 1978.

Gallin, B. Comments on contemporary sociocultural studies of medicine in Chinese societies. In A. Kleinman, P. Kunstadter, E. R. Alexander, and J. L. Gale (Eds.), *Culture and healing in Asian societies. Anthropological, psychiatric and public health studies*. Cambridge, Mass.: Schenkman, 1978.

Hamilton, W. D. The genetical theory of social behavior: I and II. *Journal of Theoretical Biology*. 1964, *7*, 1–52.

Harrell, S. Social organization in Hai-Shau. In E. M. Ahern and H. Gates (Eds.), *The anthropology of Taiwanese society*. Stanford, Calif.: Stanford Univ. Press, 1981a.

Harrell, S. Growing old in rural Taiwan. In S. Harrell and P. Amoss (Eds.), *Other ways of growing old*. Stanford, Calif.: Stanford Univ. Press, 1981b.

Harrell, S., and Amoss, P. (Eds.), *Other ways of growing old*. Stanford, Calif.: Stanford Univ. Press, 1981.

Harwood, A. Guidelines for culturally appropriate health care. In A. Harwood (Ed.), *Ethnicity and medical care*. Cambridge, Mass: Harvard Univ. Press, 1981.

Hsu, F. The effect of dominant kinship relationships on kin and non-kin behavior: A hypothesis. *American Anthropologist*, 1965, *67*, 638–661.

Iglitzin, L., and Ross, R. (Eds.), *Women in the world: Studies in comparative politics* (No. 6). Santa Barbara: Clio Books, 1976.

Irons, W. Investment and Primary Social Dyads. In N. A. Chagnon and W. Irons (Eds.), *Evolutionary biology and human social behavior: An anthropological perspective*. North Scituate, Mass: Duxbury Press, 1979.

Jacobs, S. E. Memorandum on gender roles in U.S. society. New York: Ward Cannel, Crown and Bridge, 1981.

Johnson, E. Women and childbearing in Kwan Mun Haw village: A study of social change. In M. Wolf and R. Witke (Eds.), *Women in Chinese Society*. Stanford, Calif.: Stanford Univ. Press, 1975.

Katon, W., and Kleinman, A. Doctor–patient negotiation and other social science strategies in patient care. In L. Eisenberg and A. Kleinman (Eds.), *The relevance of social science for medicine*. Dordrecht: Reidel, 1981.

King, H. Selected epidemological aspects of major diseases and causes of death among

Chinese in the U.S. and Asia. In A. Kleinman, P. Kunstadter, E. R. Alexander, and J. L. Gale (Eds.), *Medicine in Chinese cultures: Comparative studies of health care in Chinese and other societies.* Washington, D.C.: DHEW, 1975.

Kleinman, A. Social, cultural and historical themes in the study of medicine in Chinese societies: Problems and perspectives for the comparative study of medicine and psychiatry. In A. Kleinman, P. Kunstadter, E. R. Alexander, and J. L. Gale (Eds.), *Medicine in Chinese cultures: Comparative studies of health care in Chinese and other societies.* Washington, D.C.: DHEW, 1975.

Kleinman, A. *Patients and healers in the context of culture.* Berkeley: Univ. of Caifornia Press, 1980.

Kleinman, A., and Kunstadter, P. Introduction. In A. Kleinman, P. Kunstadter, E. R. Alexander, and J. L. Gale (Eds.), *Culture and healing in Asian societies. Anthropological, psychiatric and public health studies.* Cambridge, Mass: Schenkman, 1978.

Kunstadter, P. The comparative anthropological study of medical systems in society. In A. Kleinman, P. Kunstadter, E. R. Alexander, and J. L. Gale (Eds.), *Medicine in Chinese cultures: Comparative studies of health care in Chinese and other societies.* Washington, D.C.: DHEW, 1975.

Mendelson, R. S. *Mal(e) practice: How doctors manipulate women.* Chicago: Contemporary Books, 1981.

Neisser, E. *Mothers and daughters: A life long relationship.* New York: Harper & Row, 1967.

Odum, H. *Environment, power and society.* New York: Wiley–Interscience, 1971.

Porkert, M. The intellectual and social impulses behind the evolution of traditional Chinese medicine. In C. Leslie (Ed.), *Asian medical systems: A comparative study.* Berkeley: Univ. of California Press, 1976.

Reiter, R. (Ed.), *Toward an anthropology of women.* New York: Monthly Review Press, 1975.

Rosaldo, M., and Lamphere, L. (Eds.), *Woman, culture and society.* Stanford, Calif.: Stanford Univ. Press, 1974.

Sterling, P., and Eyer, J. Biological basis of stress related mortality. *Social Science and Medicine.* 1981, *15E,* 3–42.

Tseng, W.-S. Traditional and modern psychiatric care in Taiwan. In A. Kleinman, P. Kunstadter, E. R. Alexander, and J. L. Gale (Eds.), *Medicine in Chinese cultures: Comparative studies of health care in Chinese and other societies.* Washington, D.C.: DHEW, 1975.

Twaddle, A. Sickness and the sickness career: Some implications. In L. Eisenberg and A. Kleinman (Eds.), *The relevance of social science for medicine.* Dordrecht: Reidel, 1981.

Unschuld, P. The social organization and ecology of medical practice in Taiwan. In C. Leslie (Ed.), *Asian medical systems: A comparative study.* Berkeley: Univ. of California Press, 1976.

Wolf, M. *Women and the family in rural Taiwan.* Stanford, Calif.: Stanford Univ. Press, 1972.

Wolf, M. Chinese women: Old skills in a new context. In M. Z. Rosaldo and L. Lamphere (Eds.), *Woman, culture and society.* Stanford, Calif.: Stanford Univ. Press, 1974.

Wolf, M. Women and suicide in China. In M. Wolf and R. Witke (Eds.), *Women in Chinese society.* Stanford, Calif.: Stanford Univ. Press, 1975.

Worth, R. M. The impact of new health programs in disease control and illness patterns in China. In A. Kleinman, P. Kunstadter, E. R. Alexander, and J. L. Gale (Eds.), *Medicine in Chinese cultures: Comparative studies of health care in Chinese and other societies.* Washington, D.C.: DHEW, 1975.

Zola, I. Structural constraints in the doctor–patient relationship: The case of non-compliance. In L. Eisenberg and A. Kleinman (Eds.), *The relevance of social science for medicine.* Dordrecht: Reidel, 1981.

PART III

Reproductive Competition and Cooperation among Females

9

Female Roles in Cooperatively Breeding Acorn Woodpeckers*

WALTER D. KOENIG
RONALD L. MUMME
FRANK A. PITELKA

I. Introduction

Although the strategies employed by females for individual gain are often unexpectedly complex, even in species that breed as pairs (see e.g., Lumpkin, this volume), it is in group-living species that both cooperation

*Financial assistance was provided by NSF Grants DEB 78-08764 and DEB 81-09128 and by a graduate fellowship to R. L. M.

SOCIAL BEHAVIOR OF
FEMALE VERTEBRATES

ISBN 0-12-735950-8

and subterfuge can be developed in their richest and subtlest forms. In birds, the most highly developed group-living societies are found among cooperative breeders—species in which more than a simple male–female pair assists in the production and/or care of young at a single nest (Wilson, 1975). Although the name of this type of social organization focuses attention on the unusual degree of phenotypic altruism observed in these species, recent studies have shown that the manifestations of competition within groups of cooperative breeders are as dramatic as those of cooperation. In groove-billed anis (*Crotophaga sulcirostris*), for example, up to four females lay their eggs together in a communal nest and share in caring for the young. However, this apparent cooperation is belied by a complex pattern of egg tossing, increase in clutch size, and delay in the interval between successive eggs, behaviors employed by females in order to increase their own reproductive success at the expense of others in their group (Vehrencamp, 1977). Similarly, the seemingly altruistic hen ostrich (*Struthio camelus*) incubates the eggs of one to six additional females along with her own. But, she may in fact be taking advantage of the survival benefits to her own eggs of having additional eggs present in the nest and, by judicously placing her own eggs in the center and pushing others to the periphery, be manipulating the reproductive efforts of the other females for her own benefit (Bertram, 1979).

Despite these dramatic examples, the cooperative and competitive interactions among females in cooperatively breeding species are poorly known. Ideally, we wish not only to describe these interactions, but to analyze their ecological bases and discover how the strategies employed by females differ from and complement those of males in the population.

The acorn woodpecker (*Melanerpes formicivorus*) is a group-living species whose behavioral ecology makes it a promising candidate for such an analysis. Not only is helping at the nest by nonbreeding group offspring—the most common diagnostic feature of cooperative breeding—common in the acorn woodpecker, but mate sharing by two or more individuals of the same sex is frequent as well. Both these phenomena may occur together in the same group.

Groups of acorn woodpeckers consist of a breeding core, composed of one to four males, and one or two females, among which mating is apparently promiscuous (Stacey, 1979b). The core is accompanied by a variable number of nonbreeding nest helpers, both male and female, which are offspring of the core birds from prior years (Koenig and Pitelka, 1979). Only one nest is made at any one time by a group, regardless of the number of reproductively active individuals it contains; thus, true communal nesting (more than one female laying in the same nest) occurs. This mating system, which can be called cooperative polygamy, contrasts with that of many other cooperative breeders (see reviews by Brown, 1978; Emlen, 1978), which breed either as one or more monogamous pairs or as two or more males sharing a single female (i.e., cooperative polyandry; see section V and Faaborg and Patterson, 1981).

The genetic relationships between individuals within groups of acorn woodpeckers are close: core males are usually siblings or, less commonly, a father and son; core females are also usually siblings, but originate from a group other than that of the males. Group offspring of both sexes remain in the natal group for up to several years; thereafter they may either emigrate and become breeders elsewhere (often with siblings of the same sex) or remain and breed on their natal territory. The latter may occur in either sex but only after all possible parents of the opposite sex have died or disappeared; thus inbreeding among close relatives is usually avoided (Koenig and Pitelka, 1979).

Because differences in the strategies of males and females are not absolute, acorn woodpeckers provide an opportunity to compare the performances and behavior of males and females that help at the nest, emigrate in unisexual units, and share mates. Such comparisons permit the direct testing of hypotheses related to the adaptive significance of the sexual roles and sexual asymmetries that exist in this species. Specifically, we will examine the tendency of female acorn woodpeckers to (*a*) aid more at the nest while they are nonbreeding nest helpers, (*b*) disperse farther and share mates less often than do males, and (*c*) avoid nocturnal incubation and nest sanitation. To the extent that these same asymmetries are paralleled elsewhere, such an analysis may help to explain the patterns of cooperation and competition in other group-living species as well.

II. Methods

Since 1971, a population of 25 groups of acorn woodpeckers at Hastings Natural History Reservation, Monterey County, California, has been color-banded and monitored in a series of studies, beginning with the work of MacRoberts and MacRoberts (1976) and followed by our own, which began in 1974 and continues at the present time. Data on sex differences in behavior have been derived from over 350 birds banded as nestlings since 1972.

In 1979, we began intensive work in conjunction with N. E. Joste on the roles of individuals in the communal nesting effort as a function of age, sex, and status within groups (Joste, Koenig, Mumme, and Pitelka, in press). In addition to data from our long-term demographic work, some of our preliminary results from this newer study are discussed and combined with the earlier findings of MacRoberts and MacRoberts (1976) when appropriate. Further details on the study area and methods can be found in MacRoberts and MacRoberts (1976) and Koenig (1978).

III. Results

A. Helping at the Nest

1. *Length of Time Spent Helping*

Because of the lack of clearly defined courtship or pairing behavior in acorn woodpeckers (MacRoberts and MacRoberts, 1976; Stacey, 1979b), it is rarely possible to determine unambiguously the reproductive status of individuals in groups from field observations. However, inferential knowledge of the rules governing reproductive status within groups (Koenig and Pitelka, 1979), combined with complete knowledge of the origins of group members, allows the assignment of individuals to breeding or nonbreeding (i.e., helping) status with reasonable confidence. This is not to say that all individuals can or do take advantage of the status accorded to them. For example, we assume that all males who are members of the core are fertile. However, our more detailed analysis of the behavior of such males during and prior to egg laying suggests that there is often skewness in mating success among such males and that subordinate individuals may often fail to father any of the group offspring (Joste *et al.*, in press). Similarly, "helpers" are defined as individuals who, because their parent of the opposite sex is still a member of the core, are presumed not to be potential breeders (see Koenig and Pitelka, 1979), regardless of whether they in fact help at the nest or not.

Based on these criteria, the fates of group offspring, banded either as nestlings or juveniles on their natal territory, can be determined. These data can then be used to estimate the proportion of male and female offspring that are nonbreeding helpers as a function of age (Fig. 1).

Figure 1 shows that virtually identical proportions of males and females help at the nest in their first and second years: 75 and 41% for first-year and second-year males, respectively, compared to 74 and 43% for first-year and second-year females (the curves in Fig. 1 are obtained by multiplying successively the proportion of offspring in each age class that

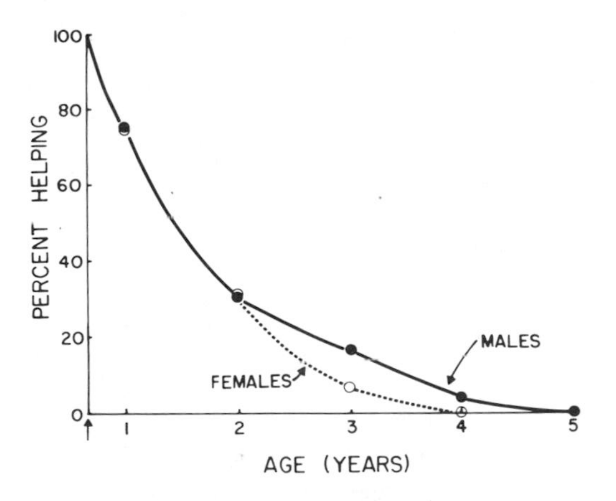

Fig. 1 Estimated proportion of a cohort of male and female acorn woodpeckers who have survived to their first spring that are still nonbreeding helpers at the nest in their natal group as a function of age. Arrow indicates spring of the first year.

were helpers). In the third year, however, a significantly higher proportion of males are still helping, compared to females (11 of 20 males helped in their third year compared to 3 of 18 females, $\chi^2 = 4.5$, $p < .05$). No females helped beyond their third year whereas a small proportion of males (4%) helped up to their fourth year. On average, male group offspring surviving to their first spring serve 1.3 seasons as helpers, whereas females serve only 1.1 seasons as helpers.

2. *Amount of Aid Given by Individual Helpers*

The relative amount of aid provided by individual helpers at different stages of the nesting cycle is graphed in Fig. 2. All but two of the individuals considered were first-year birds, and neither of the two-year-olds deviated significantly from the pattern shown by the younger individuals. In contrast to the finding that females remain helpers for a slightly shorter period of time, female helpers contribute a significantly higher proportion of the total feeding visits at nests than do males (10.5 versus 1.3%, respectively, $p < .001$, two-tailed Mann-Whitney U test; all subsequent statistical tests are also two-tailed when possible). Although neither sex contributes much to incubation or brooding, female helpers still outperform males at both these tasks, significantly so for the proportion of total brooding time ($p < .002$, Mann-Whitney U test).

These results are supported by three measures of helping effort which, unlike the proportion of the total group effort contributed by

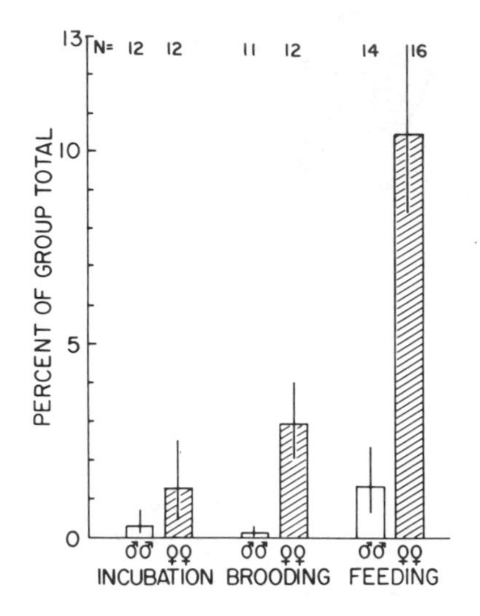

Fig. 2 Percentage of the total group effort during three phases of nest care provided by nonbreeding nest helpers. Bars indicate means; centerlines indicate standard errors, as calculated by arcsine transformation.

individual helpers, are independent of group composition. First, the number of individual male and female helpers still living in the natal territory that actually helped at different stages of the nesting cycle are listed in Table I. The proportion of females helping is greater than that of males in all cases, significantly so during brooding and feeding. Second, females fed nestlings on the average more than three times as frequently as did males. (For females, $\bar{x} \pm SE = 2.8 \pm .8$ visits/hr [$n = 10$ birds]; for males,

TABLE I
Number of Helpers Who Did or Did Not Help at Different Stages of the Nesting Cycle

Stage	Did help	Did not help	Percent helping	Significance*
Incubation				
Males	3	9	25	n.s.
Females	6	6	50	
Brooding				
Males	1	10	9	<.001
Females	11	1	92	
Feeding				
Males	7	7	50	<.01
Females	16	0	100	

*Fisher exact probability test.

$\bar{x} \pm SE = .9 \pm .5$ visits/hr [$n = 8$ birds]; $p < .10$, Mann-Whitney U test.) Third of all pairwise comparisons possible between male and female helpers in the same group, females fed more than males in seven of eight cases (88%) and brooded more than males in four of four cases (100%). Thus, the data presently available clearly support the assertion that female helpers contribute more to raising younger siblings than do male helpers in both absolute and relative terms.

B. Dispersal

1. *Distance of Dispersal*

The converse of the tendency for males to remain as helpers longer than females is for females to disperse earlier than males. Direct data confirming this trend are obscured, however, by a tendency for females to disperse farther from their natal group than males, and thus to be lost to our records even if surviving. For example, of 15 males that survived to March of their first year but did not remain in their natal group through the subsequent breeding season, 12 (80%) are known to have transferred groups locally, and only 3 (20%) disappeared. For first-year females, however, only 4 of 16 individuals (25%) are known to have moved locally whereas 12 (75%) disappeared; the difference is significant ($\chi^2 = 7.3$, $p < .01$). Unless there is severe differential mortality right at this stage, this difference is likely to be attributable to a high proportion of females moving off the study area.

Even among the subset of individuals that move within the study area, females move significantly farther, on the average, than males. (For females, $\bar{x} \pm SD = .51 \pm .31$ km [$n = 22$ birds]; for males, $\bar{x} \pm SD = .33 \pm .21$ km [$n = 41$ birds]; $p < .05$, Mann-Whitney U test.) Put differently, 36 of 41 males (88%) known to have moved dispersed to adjacent territories, whereas only 12 of 22 emigrant females (55%) did so ($\chi^2 = 7.0$, $p < .01$). This disparity would most likely increase considerably if the fates of the individuals that disappeared were known.

2. *Dispersal in Unisexual Units*

Females are less likely to move in unisexual units than are males. Only 18 of 33 (55%) females filling reproductive vacancies during the study transferred with another female, compared to 36 of 42 (86%) males ($\chi^2 = 7.4$, $p < .01$). The size of female emigrating units is also significantly smaller than the size of males units. (For female units, $\overline{x} \pm SD = 1.4 \pm .5$ birds [range 1–2, $n = 24$ units]; for male units, $\overline{x} \pm SD = 2.2 \pm 1.1$ birds [range 1–4, $n = 19$ units]; $p < .01$, Mann-Whitney U test.) Emigrating unisexual units of either sex are usually composed of siblings (Koenig and Pitelka, 1979).

3. *Competition to Disperse*

Both sexes of acorn woodpeckers frequently compete for reproductive vacancies arising in the population via contests called power struggles (Koenig, 1981b). However, the intensity of the contests involving female vacancies is slightly, but not significantly, greater than those involving male vacancies. (On a scale of 0–3 based on length of the contest and number of participants [see Koenig, 1981b], the mean intensity of contests involved with filling female vacancies was 1.3 ± 1.2 [$n = 21$ contests]; for filling male vacancies, the mean was $.9 \pm 1.1$ [$n = 14$]; $p > .10$, Mann-Whitney U test.) Also, the number of females present at power struggles involving a male vacancy (48% of 25 banded intruders identified) is significantly greater than the corresponding proportion of males present during power struggles for female vacancies (13% of 23 banded intruders identified, $\chi^2 = 6.5$, $p < .05$). Although the reason for this bias is unclear, the difference suggests that females may be more willing to take the risks associated with participation in power struggles—even when they stand only a small chance of being allowed into the group where the contest takes place—than are males.

In summary, female acorn woodpeckers disperse sooner, farther, more solitarily, and possibly more aggressively than do males. As a result, females are likely to suffer greater mortality in the process of dispersal than are males (see Woolfenden and Fitzpatrick, 1978). This source of differential mortality, along with greater mortality of females during breeding (Section IV, C), is the cause of a pronounced sex bias (approximately 1.5 males per female) in the breeding population (see Section III, C, 2).

C. Breeding Behavior

1. *Inheriting the Natal Territory*

Individuals assume breeding status either after successfully immigrating into a group other than that in which they were born or after the death and replacement of their parent of the opposite sex by unrelated individual(s) (see Section I). In the latter case, the bird is said to have inherited its natal territory, regardless of whether its parent of the same sex is still alive. The proportion of males that eventually inherit their natal territory is approximately 15% (11 cases recorded) compared to only about 8% (5 cases) for females. This may be partly a result of higher mortality among breeding females (see Section IV, C), as well as the tendency for males to breed with siblings more than females (Section III, C, 2). Thus, not only do females that emigrate move a longer distance than do males, but a lower proportion of females are able to remain and inherit their natal territory.

2. *Mate Sharing*

In conjunction with the tendency for females to emigrate in smaller units than males, the number of females in the breeding core is correspondingly smaller than the number of males ($p < .001$, Mann-Whitney *U* test, Table II). Put differently, of all groups during breeding seasons since 1972 whose composition was known, 59 contained equal numbers of core males and females, 36 contained 1 extra male, 14 contained 2 extra males, and 4 contained 3 extra males; only 6 groups contained 1 extra female and none contained 2 or more extra females ($p < .001$, Wilcoxon matched pairs signed ranks test). Thus, although groups of acorn woodpeckers may consist of a pair, a polygynous trio, a polyandrous trio, or any other combination of breeding individuals up to at least 4 males and 2 females, there are on the average fewer potentially breeding females in the population than breeding males (approximately 1.5 males per female).

TABLE II
Number of Male and Female Breeders in Acorn Woodpecker Groups

Number of male or female breeders in group	Number of groups with that number of breeders	
	Males	Females
1	57	99
2	37	19
3	20	0
4	5	0
Mean ± SD*	1.8 ± .9	1.1 ± .4

*$p < .001$, Mann-Whitney *U* test.

3. *Amount of Aid Given by Breeders at Nests*

The degree to which core males and females attend their nests depends in many cases on the intrasexual dominance hierarchy and each individual's apparent confidence of having contributed genetically to the brood (Joste *et al.*, in press; Mumme and Koenig, unpublished data). In order to avoid these complications in assessing relative parental care by breeding individuals, we observed and quantified the behavior of 6 groups of acorn woodpeckers that consisted of a simple male–female pair (Table III).

In these 6 pairs, females fed nestlings and incubated during the day somewhat less, on the average, than did their mates. The ranges, however, were considerable and the differences were not significant. The time spent brooding offspring was identical for the two sexes. Significant differences appear only in the removal of fecal sacs from the nest, an activity rarely undertaken by females ($\chi^2 = 65.0$, $p < .001$), and in spending the night in the nest. Females did roost in the nest along with their mates on 4 of 15 occasions (27%), but never roosted in the nest by themselves.

IV. Discussion

There is a rich repertoire of asymmetries between the sexes involving all phases of the life history of the acorn woodpecker (Table IV). However, many of these asymmetries are interrelated. Thus, the behaviors discussed

TABLE III
Contributions of Females to Nests in Six Simple Pairs of Acorn Woodpeckers

Percentage of total time or activity contributed by females	Mean	Range	Total time nests watched or number of times activity observed
Percentage of total incubation time during day	39.0[a]	10.2–64.8	82 hours
Percentage of time brooding	50.0[a]	45.6–63.7	100 hours
Percentage of total feeding visits	46.5[a]	33.0–62.0	140 hours, 1212 visits
Percentage of fecal sacs removed	3.6	.0–18.2	94 times
Percentage of times roosting in nest	26.7[b]	.0–50.0	15 times

[a]Calculated using arcsine transformation of raw data (Sokal and Rohlf, 1969).
[b]Females always roosted simultaneously with the male.

TABLE IV
Sexual Asymmetries in Acorn Woodpecker Behavior

	Females	Males
Helping (predispersal)	Help up to third year; $\bar{x} = 1.1$ seasons	Help up to fourth year; $\bar{x} = 1.3$ seasons
	Incubate and brood a small amount	Usually do not incubate or brood
	Feed a significant amount	Feed negligibly or not at all
Dispersal	Leave by end of third year	Leave by end of fourth year
	Move an average of $\gg .5$ km, often out of the study area	Move an average of $> .3$ km, often to a neighboring territory
	Emigrate alone or with at most 1 sibling	Emigrate alone or with 1–3 siblings
	Compete extremely vigorously for breeding vacancies	Compete vigorously for breeding vacancies
Breeding (postdispersal)	About 8% inherit natal territory	About 15% inherit natal territory
	Breed alone or with 1 other sibling	Breed with up to 3 other siblings
	Rarely remove fecal sacs or incubate at night	Remove most fecal sacs and spend all nights in nest

above and listed in Table IV can be subjected to three main questions: (*a*) Why do female helpers help more than male helpers? (*b*) Why is there sex-differential dispersal, with females moving farther and more solitarily than males? (*c*) Why do female breeders only rarely take part in nocturnal incubation and in nest sanitation?

A. Asymmetries in Helping Behavior

At least three hypotheses that could explain a sexual asymmetry in helping behavior have been suggested in the literature. First, an individual of the sex most likely to inherit all or part of the natal territory might be expected to gain more by aiding younger siblings, who could help defend the territory until the older helper gained breeding status. This hypothesis has been proposed to explain, in part, the finding that male Florida scrub jays (*Aphelocoma coerulescens*), who are more likely to acquire eventually a part of their parents' territory for their own, help more at the nest than their sisters (Stallcup and Woolfenden, 1978; Woolfenden and Fitzpatrick, 1978).

Second, the sex more likely to disperse with other siblings should gain more by helping to raise them. This hypothesis is suggested indirectly by Ligon and Ligon (1978a), who proposed that helping in green woodhoopoes (*Phoeniculus purpureus*) is in part a means of interacting with and forming bonds with nestlings. Such nestlings are valuable resources to older siblings in that they may later emigrate with and help older birds in their own reproductive efforts.

Third, a sexual asymmetry in helping behavior might be expected if individuals gain reproductively by breeding in territories adjacent to those of relatives, perhaps through mutual alarm-calling or lessened territorial defense (Greenwood and Harvey, 1976; Greenwood *et al.*, 1979). If this were true, then individuals of the sex more likely to live near relatives after dispersal would be expected to gain more by raising relatives, and correspondingly to invest more as helpers.

All three of these hypotheses predict that male acorn woodpeckers should help more than females because the former are more likely to inherit their natal territory, disperse in unisexual groups with siblings, and live near relatives after dispersal. However, the data discussed previously suggest that the opposite is true. It is still possible that some variant of these hypotheses might be important: for example, perhaps offspring raised in part by an older female sibling may later aid the older bird in inheriting her natal territory. Since breeding vacancies may be more valuable to females than to males (Section IV, B), females might then be expected to help more than males even though they do not inherit their natal territory as frequently. At present, however, we have observed too few cases of females inheriting their natal territory to assess this hypothesis critically.

We will consider three additional hypotheses that might affect the relative investments of the sexes while helping at the nest. First, females might help at the nest while males defend the territory. As a consequence of this or some other kind of division of labor, the total effort expended by male and female helpers during the breeding season might be equal, or even skewed toward males. If this were true, the appropriate problem would be to determine why there is differential allocation of effort expended by the helpers rather than why females help more than males.

Although we cannot reject this hypothesis, preliminary data suggest that both sexes of helpers take an active role in both intra- and interspecific territory defense (Mumme and Koenig, unpublished data). Division of labor may still be important in ways we have not quantified, but at present it appears unlikely to provide the primary explanation for the asymmetry in feeding and brooding behavior observed among helpers.

Second, if the experience gained by helping were more important to the subsequent success of females than of males, females would be expected to invest more in such cooperative behavior than males. A similar idea (the "learning-to-mother" hypothesis) has been proposed to explain the prevalence of allomothering in primates (Lancaster, 1971; Blaffer Hrdy, 1977; but see Wasser and Silk and Boyd, this volume).

We have few direct data from acorn woodpeckers bearing on this hypothesis. However, the relatively even contributions of males and females in breeding pairs does not support the hypothesis that females must be more experienced than males to breed successfully. The successful breeding during the study by at least two first-year females who were without any prior helping experience also suggests that such experience is not essential for reproductive success. On the other hand, the greater mortality suffered by breeding females (see Section IV, C) does suggest that females may be more stressed while breeding and thus that experience could be of greater importance to females than males.

Third, females may be less likely to survive and breed on their own than males. If the variance in reproductive success of females in the population is greater than that of males, it might be relatively more to the benefit of females to help raise siblings since this would be the only opportunity for many individuals to invest in genetic relatives (we thank A. B. Clark [personal communication, 1980] for suggesting this hypothesis to us). In effect, this hypothesis proposes that helping is in part a "bet-hedging" strategy (Stearns, 1976) by females to ensure that they are survived by close relatives (i.e., siblings) even if they die prior to gaining reproductive status of their own. This hypothesis, unlike those previously considered, relies on the concept of inclusive fitness (Hamilton, 1964), and thus requires that kin selection contribute substantially to the evolution of helping behavior (e.g., Brown, 1974; Ricklefs, 1975).

We know of no way to prove this hypothesis directly, but it is at least consistent with the data discussed previously. Because of the sex-differential dispersal, a greater proportion of females probably die without ever breeding than do males. Also, the greater tendency for males to share mates with siblings will further decrease the variance in reproductive success for males relative to females (Wade and Arnold, 1980).

Unfortunately, this hypothesis is little more than an ad hoc explanation for the sexual asymmetry in helping behavior observed in acorn woodpeckers, and it does not appear likely to be supported in other cooperative breeders. For example, variance in reproductive success of females is also likely to be greater, because of female-biased dispersal, in both Florida scrub jays and green woodhoopoes. Yet males help more than females in the former (Stallcup and Woolfenden, 1978), whereas no apparent sexual asymmetry has been detected in the helping behavior of the latter (Ligon and Ligon, 1978b).

In summary, the tendency for female helpers to assist more than male helpers at nests of acorn woodpeckers is not consistent with most previous hypotheses relevant to the evolution of helping behavior. At present, at least three hypotheses must still be considered of potential importance in explaining this asymmetry in behavior: (*a*) there is division of labor, with females helping at the nest while males defend the territory, (*b*) females benefit more from the experience gained by helping than do males, and (*c*) females help more because of the apparently high variance in female reproductive success, thereby resulting in a relatively high proportion of females whose only contribution to raising genetic relatives is likely to be indirect, through helping to raise siblings. Of these three hypotheses, we consider the last to be most promising. This hypothesis is not, however, mutually exclusive of other hypotheses, and indeed, it is not supported by data from the only two other cooperative species for which comparable data are presently available. Thus, it is likely that a complex of ecological and genetic factors are responsible for the phenomenon of helping at the nest in cooperative breeders.

B. Polyandry and Asymmetries in Dispersal

Why is there sex-differential dispersal? And why, given a sexual asymmetry in dispersal, do females move earlier, farther, and more solitarily than males?

We have proposed elsewhere (Koenig and Pitelka, 1979) that sex-differential dispersal in acorn woodpeckers is an adaptation to reduce inbreeding (Zahavi, 1974; Greenwood and Harvey, 1976; Greenwood *et al*., 1978; Dhondt, 1979; Packer, 1979; but cf. Shields, 1983 for a qualified view). Thus, here we will be concerned primarily with the latter question.

The pattern of female-biased dispersal found here is, with only a few known exceptions, general among birds. However, the tendency of groups to be polyandrous rather than polygynous, which we consider to be in part a result of the sexual asymmetries in dispersal, is clearly not a general avian phenomenon. Thus, we consider it appropriate to examine the behavior of acorn woodpeckers in search of clues that may help to explain these patterns in dispersal and mating.

Greenwood (1980) recently reviewed several proposed explanations for female-biased dispersal in birds. Included were Whitney's (1976) hypothesis based on the analogy of the sex determination mechanism to haplodiploidy, which could predispose the homogametic sex (males in birds) to be more cooperative and sedentary, and Gauthreaux's (1978) hypothesis based on behavioral dominance of males over females, which could allow males to outcompete related females for nearby territories and force the females to disperse farther in order to avoid inbreeding. We agree with Greenwood (1980) that neither of these hypotheses provides a satisfactory explanation for sex-biased dispersal. In our view, the pattern of behavioral dominance is likely to shift, depending on ecological factors affecting the sexes (e.g., Smith, 1980). Also, the minor genetic asymmetry resulting from the sex determination mechanism in birds (where the sex chromosomes make up <2% of the genome [Matthey, 1950]) is likely to be easily reversed if ecological conditions select for an alternate dispersal pattern.

An alternative explanation for female-biased dispersal, proposed by Greenwood and Harvey (1976) and elaborated by Greenwood (1980), is that of resource defense by males. If males defend a resource necessary to females in order to breed, philopatry would benefit males relatively more than females by allowing them to acquire territories in nearby, familiar areas. Meanwhile, females, presumably not limited in their ability to acquire mates, would then be forced to disperse greater distances in order to avoid inbreeding with their philopatric male relatives. This hypothesis, combined with the converse situation of mate defense resulting in male-biased dispersal, is able to account for the striking general differences in the dispersal pattern of birds (where dispersal is nearly always female biased) compared to mammals (where dispersal is nearly always male biased), as well as many of the known exceptions to these patterns (Greenwood, 1980).

However, resource defense by males does not appear to us to be sufficient to explain female-biased dispersal in acorn woodpeckers, or possibly in many other cooperative breeders as well. This hypothesis presupposes two conditions: (*a*) that females do not defend resources, and (*b*) that females are the limiting sex. Thus, presumably males are frequently excluded from reproducing, but a female is almost always able to acquire a male willing to mate with her. Neither of these postulates appear to be fulfilled in many cooperative breeders. In the acorn woodpecker, for example, territory establishment and defense are performed by either males or females; in particular, we have observed rare cases of birds of either sex maintaining a territory for periods of up to nearly a year in the absence of group members of the opposite sex. Furthermore, breeding opportunities are limited for individuals of both sexes, as evidenced by the presence of both male and female nonbreeding nest helpers, as well as the striking contests that take place for breeding vacancies of either sex (see

Section III, B, 3 and Koenig, 1981b). Thus, it is not obvious why males should benefit more than females from remaining in their natal area and obtaining a territory on familiar ground. Similarly, we are not convinced that resource defense can explain female-biased dispersal in Florida scrub jays, where again both sexes are limited in their breeding opportunities, or in other cooperative breeders, in which both males and females may be helpers at the nest.

Instead, we propose that female–female competition may be playing an important role in causing the female-biased dispersal pattern and also the tendency toward polyandry seen in many cooperative breeders. This possibility was first proposed by Frame *et al.*, (1979) to explain female-biased dispersal and the rarity of females breeding together in groups of wild dogs (*Lycaon pictus*), a cooperatively breeding mammal with a social organization remarkably similar to that of acorn woodpeckers. These authors proposed that competition among females attempting to reproduce simultaneously in a group was likely to exceed that between males. The primary reason for this is that females reproducing communally produce extra offspring, who then compete directly for the limited parental attention of adults. Thus, female–female competition and interference potentially extend throughout the entire nesting cycle (see also Wasser, this volume). In contrast, male–male competition occurs primarily prior to the production of young, in the form of competition for matings with the female(s). Thus, female–female competition is more likely to be detrimental to the reproductive success of the group than is male–male competition. Because of the considerable competition between females who nest together, females will be forced to search farther and more vigorously for breeding vacancies that they will not have to share with a sibling.

Once eggs are laid, female acorn woodpeckers nesting communally invest almost equally in nesting activities (Mumme and Koenig, unpublished data). However, we have identified several characteristics of such nests that support the competition–interference hypothesis. First, a significantly higher proportion of eggs laid in the nests of groups in which females are nesting communally are abnormally small "runt" eggs (Koenig, 1980). Using all available data from nests at Hastings Reservation, 8 of 97 (8.3%) eggs laid in group nests were runts, compared to only 3 of 300 (1.0%) eggs laid by females nesting by themselves ($p < .01$, Fisher exact probability test). The production of runts in all cases may simply be the result of accidental interference (Koenig, 1980); however, whether interference is accidental or deliberate, a significantly higher proportion of a female's reproductive effort is wasted on inviable runt eggs when she is nesting communally than when she nests singly.

Second, a lower proportion of eggs hatch in nests in which two females nest communally than in nests of single females. Our data on this are not precise, since we do not always know the fate of eggs that disappear within a few days of when the rest of the clutch hatches. However, of 287 eggs laid by females nesting singly, 253 (88%) apparently hatched, whereas of 90 eggs laid by females nesting communally, only 71 (79%) hatched. ($\chi^2 = 4.1$, $p < .05$; Koenig 1982; these figures exclude runt eggs.) This difference is in the direction predicted if there is considerable interference among females nesting together.

Third, a more direct manifestation of female–female competition within groups containing two nesting females is egg tossing (Mumme and Koenig, unpublished data). Although the details surrounding several of the nine instances of egg tossing we have observed are incomplete or confusing, in general the evidence suggests an interpretation similar to that proposed by Vehrencamp (1977) for groove-billed anis. That is, females that lay second may toss some or all of the eggs laid by the first female up to the time that both females have begun laying eggs, after which no more tossing occurs. Egg tossing thus is likely to reduce the reproductive skew between communally nesting females and also helps to synchronize the clutches of such females. Synchronization subsequently makes it less likely that offspring of either female will gain an age–size advantage as nestlings.

Our discovery of this phenomenon in acorn woodpeckers was particularly surprising. Such overt interference might be expected among anis, where communally nesting females are apparently unrelated (Vehrencamp, 1978). However, as discussed earlier, most communally nesting acorn woodpeckers are closely related, and the females involved in all cases of egg tossing observed thus far have been sisters. Thus, close kinship ties do not moderate the intense competition that appears to exist between communally nesting females (Mumme *et al.*, n.d.).

As a result of these manifestations of female–female competition, groups with a single breeding female fledge as many young as those with two communally nesting females (Koenig, 1981a). Of particular interest here, however, is how the reproductive success of groups varies with increasing numbers of males and/or females sharing mates. Figure 3 provides such data for groups containing a variable number of males sharing a single female (none of the groups contains nonbreeding nest helpers). The total heights of the columns are the total number of young fledged by groups and, at the same time, the reproductive success of the

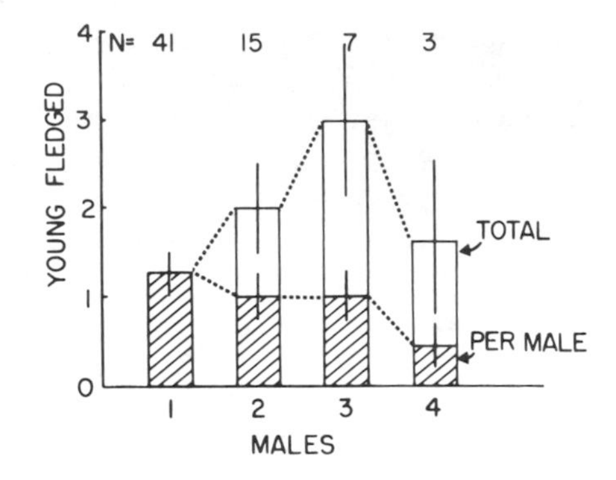

Fig. 3 Reproductive success of groups with 1 breeding female, no nonbreeding nest helpers, and 1–4 core males. Total height of bars is the total number of young fledged per group (number per female); hatched portion is number of young fledged per male, assuming no skew among individual mating success. Bars indicate means; centerlines indicate standard errors.

females in these groups. Total young fledged increases with more males, up to three and then drops off sharply, although groups with four breeding males are still slightly more successful than pairs. Thus, females should attempt to breed with more than one male when possible, and optimally together with three males.

The hatched portions of the histograms in Fig. 3 are the number of young produced by each male, assuming that there is no skew in paternity regardless of how many males are in the group. Young fledged per male is greatest for males nesting singly, but decreases only modestly in groups with two or three males. Each male in these intermediate-sized groups sires, on the average, 72% of the offspring he would have if he had not been sharing a female. Males sharing a female with three other males do considerably worse than this, but still average nearly .5 offspring per year.

To the extent that the assumption of equal paternity is violated, dominant males will presumably fare better and subordinates correspondingly worse than the averages graphed in Fig. 3. However, the position of subordinates will be moderated somewhat by inclusive fitness considerations, since communally nesting males are almost always close relatives. Furthermore, there is likely to be a similar skew in the reproductive success of communally nesting females (e.g., Vehrencamp, 1977). Thus, although the variance in the reproductive success for individual males graphed in Fig. 3 is certainly an underestimate, we do not feel that skewness among the males will affect our conclusions based on the comparison of multimale versus multifemale groups.

Figure 4 graphs complementary data for groups containing one or two communally nesting females, a single male, two breeding males, and three breeding males; again, none of the groups contain nest helpers. In all three cases, total group reproductive success decreases when an additional female is added, and reproduction per female suffers markedly.

For individuals of either sex, the optimal strategy is to breed singly. However, a comparison of Figs. 3 and 4 suggests that two and even three males sharing a mate do not interfere nearly as much with group reproductive success as do two females nesting communally. Females who share their nests with another female lose an average of 68% of their reproductive success whereas males who share a female with one or even two other males lose only 28% of their reproduction.

Similarly, the optimal strategy for a male is to nest as a pair with a single female. But in contrast, the optimal breeding situation for a female is to breed with three males. The resolution of these differing interests of males and females appears to be a compromise: pairs are in fact the most frequent breeding group size (Koenig, 1981a); groups with two and three males are successively less common; and those with more than three breeding males and/or more than one breeding female are relatively rare.

But why do individuals of either sex ever nest communally when they are more successful by themselves? This paradox is considered in detail elsewhere (Koenig, 1981a). In short, we believe that acorn woodpeckers do not gain directly by breeding in cooperative groups (e.g., through increased foraging efficiency or predator detection), but instead are

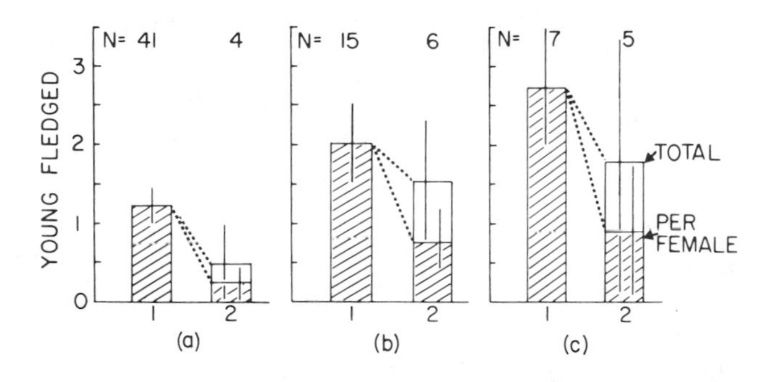

Fig. 4 Reproductive success of groups with (a) 1 breeding male, (b) 2 breeding males, and (c) 3 breeding males; no nonbreeding nest helpers; and 1 or 2 core females. Total height of bars is the total number of young fledged per group; the hatched portion is number of young fledged per female. Bars indicate means; centerlines indicate standard errors.

"forced" to accept the option of group breeding as a result of resource localization (Koenig and Pitelka, 1981; see also Stacey, 1979a). Specifically, we propose that the importance of the specially modified trees used to store acorns to both reproduction and survivorship results in a relative absence of marginal habitats. The lack of such marginal areas, to which "excess" young in most species disperse, means that offspring are ecologically forced to remain in their natal group—and their parents are similarly forced to tolerate them—until they are able to obtain a breeding position elsewhere in the population. Group living and cooperative breeding might thus evolve in the absence of any direct reproductive advantage to breeding in groups as compared to breeding as pairs.

We have shown that reproductive success per capita decreases with increasing group size in this population of acorn woodpeckers (Koenig, 1981a). The results presented here extend those data in an important way: reproductive success also decreases per breeding individual for both males and females sharing a mate. These patterns of decreasing reproductive success per individual with increasing group size are consistent only with the resource localization hypothesis for the evolution of cooperative breeding (Koenig and Pitelka, 1981).

In summary, the available evidence supports the hypothesis that competition and interference between communally nesting females is greater and affects the reproduction of groups significantly more than that among males. This intrasexual competition presumably creates stronger selective pressure for females to search more widely and compete more vigorously for opportunities to breed singly than males. As a result, females avoid cooperating with sisters except, perhaps, when it is necessary in order to obtain a particularly outstanding territory.

C. Asymmetries in Parental Care by Breeding Adults

Increased investment in incubation by males, such as through nocturnal incubation, may characterize breeding situations where the female is particularly stressed (Vehrencamp, 1982). This might occur, for example, if both the cost of producing a clutch and the incidence of nest predation—forcing frequent laying of replacement clutches—were high, or possibly if female–female competition were extreme. In acorn woodpeckers, the incidence of nest predation is low and the production of replacement or even second clutches unusual (Koenig, 1978), but intrasexual competition among females appears to be great, at least when nesting communally (Sections III, C and IV, B). In order to test for undue stress among females, we compared the mortality of known or presumed breeding birds during or in the two weeks immediately following a nesting effort; presumably a higher mortality rate of females would suggest greater stress as a result of breeding.

The results, shown in Table V, show that breeding is clearly associated with higher mortality in females: 11% of breeding females disappear or are found dead while a nesting attempt is in progress; this percentage is over eight times that of males. We are not sure of the causes of this mortality, but we consider it likely that it is at least in part a byproduct of the tendency for females to nest by themselves or with a single sibling rather than to spread the risk with several other individuals, as do males. In any case, these data suggest that female acorn woodpeckers are stressed during breeding, possibly more so than is usual among birds.

On the other hand, nocturnal incubation by the male is the rule among woodpeckers (Kendeigh, 1952); also, males are more active than females in nest sanitation in several other picids (e.g., the sapsuckers *Sphyrapicus varius* [Johnson, 1947] and *S. thyroideus* [Crockett and Hansley, 1977]). Thus, these asymmetries may reflect phylogenic peculiarities and have little relationship to ecological influences unique to acorn woodpeckers or cooperative breeding. Our contined work on the roles of male and female breeders at nests should provide more information on the significance of these asymmetries in the future.

V. Conclusion

Female acorn woodpeckers apparently lead a high-risk, competitive life compared to the relatively secure existence of males (Table IV). Females work hard as helpers, strike out on their own early, compete intensely and over a wide area for breeding vacancies, and usually do not cooperate with siblings when dispersing or breeding. As a result, they

TABLE V
Mortality of Breeders During a Nesting Effort

	Number disappeared	Number surviving	Mortality (%)
Males	2	155	1.3
Females	10	81	11.0
	$\chi^2 = 9.8$, $p < .001$		

incur greater risks and suffer higher mortality than males in both dispersal and breeding. When they do breed in conjunction with other females, interference is sometimes intense, as each tries to outcompete the other and gain the larger proportion of the group's limited resources.

In contrast, males do little as helpers, remain in their natal group relatively longer, frequently move to an adjacent group or inherit their natal territory, compete for vacancies by exploiting the numerical strength of cooperating siblings rather than by individual brute force, and frequently share a female with other siblings. Most males sharing a female do compete for matings during the brief egg-laying period, but such competition has a relatively minor effect on the group's reproductive success.

In addition, some males, particularly subordinates breeding along with two or three other siblings, do not even seem to compete for access to the female during the critical egg-laying stage. Instead, these males more or less ignore the female and subsequently do little at the nest until late in the nesting cycle. Possibly, this behavior may be interpreted as an attempt by these individuals to risk as little as possible while successful parenting is unlikely anyway and instead to try and outlive their male competitors in order to become the dominant breeder in their group (Koenig and Mumme, unpublished data). This strategy of low investment does not appear to be available to females.

We propose that the greater interference among female acorn woodpeckers nesting together forces them to assume the high-risk strategy in dispersal and breeding and possibly the more active role in helping at the nest as well. Except for the few cases discussed earlier, comparable data on sex-roles of helpers in cooperatively breeding species are few and do not yet present a consistent trend. However, numerous other cooperative breeders in which both sexes are active in raising young share the acorn woodpecker's mating pattern of males sharing mates more frequently than do females. Cooperatively polyandrous species in which males frequently, but females rarely, share mates include the Tasmanian native hen (*Gallinula mortierii*) (Ridpath, 1972), the dusky moorhen (*G. tenebrosa*) (Garnett, 1980), the Galapagos hawk (*Buteo galapagoensis*) (de Vries, 1975; Faaborg *et al.*, 1980), the Harris's Hawk (*Parabuteo unicinctus*) (Mader, 1975; Mader, 1979), the noisy miner (*Manorina melanocephala*) (Dow, 1977; Dow, 1979), and probably the yellow-billed shrike (*Corvinella corvina*) (Grimes, 1980).

Although uncommon in other taxa (but see Frame *et al.*, 1979), it is of interest that cooperative polyandry is found in humans (references in Alexander, 1974, p. 371), where it is associated with ecological conditions of resource limitation possibly similar to those suggested to be important to cooperatively breeding birds.

In addition to the above species, there are numerous examples of cooperative polygamy in which, as is the case in the acorn woodpecker, males share a female more frequently than vice versa. These species include white-winged choughs (*Corcorax melanorhamphus*) (Rowley, 1978), Arabian babblers (*Turdoides squamiceps*) (Zahavi, 1974), and pukeko (*Porphyrio porphyrio*) (Craig, 1979). The reverse situation, in which females in a cooperative group regularly share a male but males do not share a female, has been suggested for the magpie goose (*Anseranas semipalmatus*) (Frith and Davies, 1961), but this requires confirmation. Thus, female–female competition may be a dominant factor influencing the mating patterns of many cooperative breeders.

VI. Summary

There is a wide range of sexual asymmetries in the behavior of the cooperatively breeding acorn woodpecker, covering virtually all aspects of an individual's life history. As helpers, females brood and feed young significantly more diligently than do males. However, a lower proportion of these same females eventually breed on their natal territory and those that disperse do so at a slightly younger age than do the males. Females also move farther away from their natal group when emigrating and are less likely to move with other siblings than are males. During breeding, there is no pairing behavior within the breeding core, and females are significantly less likely to share mates with sisters than males are with brothers. Care of eggs and nestlings appears to be evenly divided between the sexes, at least when breeding as a pair, except for a lack of female participation in nocturnal incubation and nest sanitation.

Because of the more aggressive dispersal strategy of females, they presumably suffer higher mortality at this stage than males. Since a higher proportion of females die without reproducing at all, and those that do breed are less likely to share reproduction with other siblings than are males, the variance in reproductive success of females is greater than that of males in the population. Thus, the overall mating pattern is biased toward polyandry, although within any particular group, mating may be monogamous, polygynous, or polyandrous, depending on the composition and size of the core.

We examine several hypotheses that might contribute to the origin and maintenance of these behavioral asymmetries. Nocturnal incubation by males may be a result of physiological stress on breeding females, as evidenced by their greater mortality during a breeding effort. However, this trait, as well as the lack of female involvement in nest sanitation, is widespread in woodpeckers and thus may reflect phylogenetic inertia rather than ecological conditions peculiar to cooperative breeding, at least in the acorn woodpecker. The tendency for helper females to aid more at the nest is opposite to the trend reported for several other cooperative species and is not easily explained by previously suggested hypotheses applicable to the evolution of helping behavior. Possibly the greater investment by helper females may be a "bet-hedging" strategy that enables females to increase the probability of leaving genetic relatives significantly more than would the comparable investment by males. If so, the origin of this asymmetry may lie in the differing payoffs to prereproductives of investing their energy in raising genetic relatives. However, at least two alternatives—that (*a*) the investment of helpers is the same but females help more at the nest while males help in some other way or (*b*) helping behavior provides experience that benefits females in their future reproductive careers more so than males—cannot be eliminated and may also be important to explaining this phenomenon.

The pattern of sex-differential dispersal and male bias in the breeding composition of groups is central to the social organization of acorn woodpeckers. Several kinds of data suggest that females nesting communally compete and interfere with each other significantly more than do males sharing a female. This asymmetry is likely to have been an important ecological factor forcing females to compete more than males for breeding opportunities and to cooperate less with other siblings during breeding attempts.

Numerous other cooperative breeders share the tendency for breeding females to share mates rarely, whereas males do so frequently. Greater interference between females attempting to reproduce in the same group than between males is likely to have been an important factor influencing this bias in mating pattern. Thus, competitive interactions among females have been important determinants of social organization in cooperative breeders.

Acknowledgments

We particularly wish to acknowledge the contributions of M. H. and B. R. MacRoberts to our knowledge of acorn woodpeckers and to the successful initiation of our work on this species. More recently, N. E. Joste has contributed significantly to our research program, and L. M. Hanks, N. E. Joste, G. Moore, and S. E. Nishimura have provided us with competent field assistance. The manuscript has been greatly improved by the comments of S. T. Emlen, J.

Gradwohl, R. Greenberg, S. K. Wasser, P. L. Williams, and D. W. Winkler. W. M. Shields and S. L. Vehrencamp provided access to unpublished material.

References

Alexander, R. D. The evolution of social behavior. *Annual Review of Ecology and Systematics*, 1974, *5*, 325–383.

Bertram, B. C. R. Ostriches recognise their own eggs and discard others. *Nature*, 1979, *279*, 233–234.

Blaffer Hrdy, S. *The Langurs of Abu*. Cambridge, Mass.: Harvard Univ. Press, 1977.

Brown, J. L. Alternate routes to sociality in jays—with a theory for the evolution of altruism and communal breeding. *American Zoologist*, 1974, *14*, 63–80.

Brown, J. L. Avian communal breeding systems. *Annual Review of Ecology and Systematics*, 1978, *9*, 123–155.

Craig, J. L. Habitat variation in the social organization of a communal gallinule, the pukeko, *Porphyrio porphyrio melanotus*. *Behavioral Ecology and Sociobiology*, 1979, *5*, 331–358.

Crockett, A. B., and Hansley, P. L. Coition, nesting, and postfledging behavior of Williamson's sapsucker in Colorado. *Living Bird*, 1977, *16*, 7–19.

de Vries, Tj. The breeding biology of the Galapagos hawk, *Buteo galapagoensis*. *Le Gerfaut*, 1975, *65*, 29–57.

Dhondt, A. A. Summer dispersal and survival of juvenile great tits in southern Sweden. *Oecologia*, 1979, *42*, 139–157.

Dow, D. D. Reproductive behavior of the noisy miner, a communally breeding honeyeater. *Living Bird*, 1977, *16*, 163–185.

Dow, D. D. Agonistic and spacing behavior of the noisy miner *Manorina melanocephala*, a communally breeding honeyeater. *Ibis*, 1979, *121*, 423–436.

Emlen, S. T. The evolution of cooperative breeding in birds. In J. R. Krebs and N. B. Davies (Eds.), *Behavioural ecology: An evolutionary approach*. Sunderland, Mass.: Sinauer Associates, 1978.

Faaborg, J., and Patterson, C. B. The characteristics and occurrence of cooperative polyandry. *Ibis*, 1981, *121*, 477–484.

Faaborg, J., de Vries, Tj., Patterson, C. B., and Griffin, C. R. Preliminary observations on the occurrence and evolution of polyandry in the Galapagos hawk (*Buteo galapagoensis*). *Auk*, 1980, *97*, 581–590.

Frame, L. H., Malcolm, J. R., Frame, G. W., and van Lawick, H. Social organization of African wild dogs (*Lycaon pictus*) on the Serengeti Plains, Tanzania 1967–1978. *Zeitschrift für Tierpsychologie*, 1979, *50*, 225–249.

Frith, J. J., and Davies, S. J. J. F. Ecology of the magpie goose, *Anseranas semipalmatus* Latham (Anatidae). *CSIRO Wildlife Research*, 1961, *6*, 91–141.

Garnett, S. T. The social organization of the dusky moorhen, *Gallinula tenebrosa* Gould (Aves: Rallidae). *Australian Wildlife Research*, 1980, *7*, 103–112.

Gauthreaux, S. A., Jr. The ecological significance of behavioral dominance. In P. P. G. Bateson and P. K. Klopfer (Eds.), *Perspectives in ethology* (Vol. 3). New York: Plenum, 1978.

Greenwood, P. J. Mating systems, philopatry, and dispersal in birds and mammals. *Animal Behaviour*, 1980, *28*, 1140–1162.

Greenwood, P. J., and Harvey, P. H. The adaptive significance of variation in breeding fidelity of the blackbird (*Turdus merula* L.) *Journal of Animal Ecology*, 1976, *45*, 887–898.

Greenwood, P.J., Harvey, P. H., and Perrins, C. M. Inbreeding and dispersal in the great tit. *Nature*, 1978, *271*, 52–54.

Greenwood, P. J., Harvey, P. H., and Perrins, C. M. Kin selection and territoriality in birds? A test. *Animal Behaviour*, 1979, *27*, 645–651.

Grimes, L. G. Observations of group behaviour and breeding biology of the yellow-billed shrike *Corvinella corvina*. *Ibis*, 1980, *122*, 166–192.

Hamilton, W. D. The genetical evolution of social behaviour. I, II. *Journal of Theoretical Biology*, 1964, *7*, 1–52.

Johnson, R. A. Role of male yellow-bellied sapsucker in the care of the young. *Auk*, 1947, *64*, 621–623.

Joste, N. E., Koenig, W. D., Mumme, R. L., and Pitelka, F. A. Intragroup dynamics of a cooperative breeder: An analysis of reproductive roles in the acorn woodpecker. *Behavioral Ecology and Sociobiology*, in press.

Kendeigh, S. C. Parental care and its evolution in birds. *Illinois Biological Monographs*, 1952, *22*, 1–356.

Koenig, W. D. *Ecological and evolutionary aspects of cooperative breeding in acorn woodpeckers of central coastal California*. Unpublished doctoral dissertation, University of California, Berkeley, 1978.

Koenig, W. D. The incidence of runt eggs in woodpeckers. *Wilson Bulletin*, 1980, *92*, 169–176.

Koenig, W. D. Reproductive success, group size, and the evolution of cooperative breeding in the acorn woodpecker. *American Naturalist*, 1981a, *117*, 421–443.

Koenig, W. D. Space competition in the acorn woodpecker: Power struggles in a cooperative breeder. *Animal Behaviour*, 1981b, *29*, 396–409.

Koenig, W. D. Ecological and social factors affecting hatchability of eggs. *Auk*, 1982, *99*, 526–536.

Koenig, W. D., and Pitelka, F. A. Relatedness and inbreeding avoidance: Counterploys in the communally nesting acorn woodpecker. *Science*, 1979, *206*, 1103–1105.

Koenig, W. D., and Pitelka, F. A. Ecological factors and kin selection in the evolution of cooperative breeding in birds. In R. D. Alexander and D. W. Tinkle (Eds.), *Natural selection and social behavior: Recent research and new theory*. Concord, Mass.: Chiron, 1981.

Lancaster, J. Play-mothering: The relations between juvenile females and young infants among free-ranging vervet monkeys (*Cercopithecus aethiops*). *Folia Primatologica*, 1971, *15*, 161–182.

Ligon, J. D., and Ligon, S. H. Communal breeding in green woodhoopoes as a case for reciprocity. *Nature*, 1978a, *276*, 496–498.

Ligon, J. D., and Ligon, S. H. The communal social system of the green woodhoopoe in Kenya. *Living Bird*, 1978b, *17*, 159–197.

MacRoberts, M. H., and MacRoberts, B. R. Social organization and behavior of the acorn woodpecker in central coastal California. *Ornithological Monographs*, 1976, *21*, 1–115.

Mader, W. J. Biology of the Harris' hawk in southern Arizona. *Living Bird*, 1975, *14*, 59–85.

Mader, W. J. Breeding behavior of a polyandrous trio of Harris' hawks in southern Arizona. *Auk*, 1979, *96*, 776–788.

Matthey, R. Les chromosomes des oiseaux. In P.-P. Grassé (Ed.), *Traité de Zoologie* (Vol. 15). Paris: Masson, 1950.

Mumme, R. L., Koenig, W. D., and Pitelka, F. A. Reproductive competition and communal nesting in the acorn woodpecker: Sisters destroy each other's eggs. Manuscript in preparation, n.d.

Packer, C. Inter-group transfer and inbreeding avoidance in *Papio anubis*. *Animal Behaviour*, 1979, *27*, 1–36.

Ricklefs, R. E. The evolution of co-operative breeding in birds. *Ibis*, 1975, *117*, 531–534.

Ridpath, M. G. The Tasmanian native hen, *Tribonyx mortierii*. II. The individual, the group, and the population. *CSIRO Wildlife Research*, 1972, *17*, 53–90.

Rowley, I. Communal activities among white-winged choughs *Corcorax melanorhamphus*. *Ibis*, 1978, *120*, 178–197.

Shields, W. M. Optimal inbreeding and the evolution of philopatry. In I. R. Swingland and P. J. Greenwood (Eds.), *The ecology of animal movement*. London/New York: Oxford Univ. Press, 1983.

Smith, S. M. Henpecked males: The general pattern in monogamy? *Journal of Field Ornithology*, 1980, *51*, 55–64.

Sokal, R. R., and Rohlf, F. J. *Biometry*. San Francisco: Freeman, 1969.

Stacey, P. B. Habitat saturation and communal breeding in the acorn woodpecker. *Animal Behaviour*, 1979a, *27*, 1153–1166.

Stacey, P. B. Kinship, promiscuity, and communal breeding in the acorn woodpecker. *Behavioral Ecology and Sociobiology*, 1979b, *6*, 53–66.

Stallcup, J. A., and Woolfenden, G. E. Family status and contributions to breeding by Florida scrub jays. *Animal Behaviour*, 1978, *26*, 1144–1156.

Stearns, S. C. Life history tactics: A review of the ideas. *Quarterly Review of Biology*, 1976, *51*, 3–47.

Vehrencamp, S. L. Relative fecundity and parental effort in communally nesting anis, *Crotophaga sulcirostris*. *Science*, 1977, *197*, 403–405.

Vehrencamp, S. L. The adaptive significance of communal nesting in groove-billed anis (*Crotophaga sulcirostris*). *Behavioral Ecology and Sociobiology*, 1978, *4*, 1–33.

Vehrencamp, S. L. Body temperatures of incubating versus non-incubating roadrunners. *Condor*, 1982, *84*, 203–207.

Wade, M. J., and Arnold, S. J. The intensity of sexual selection in relation to male sexual behaviour, female choice, and sperm precedence. *Animal Behaviour, 1980, 28*, 446–461.

Whitney, G. Genetic substrates for the initial evolution of human sociality. I. Sex chromosome mechanisms. *American Naturalist*, 1976, *110*, 867–875.

Wilson, E. O. *Sociobiology: The New Synthesis*. Cambridge, Mass.: Harvard Univ. Press, 1975.

Woolfenden, G. E., and Fitzpatrick, J. W. The inheritance of territory in group-breeding birds. *BioScience*, 1978, *28*, 104–108.

Zahavi, A. Communal nesting by the Arabian babbler: A case of individual selection. *Ibis*, 1974, *116*, 84–87.

10

Altruism in Coati Bands: Nepotism or Reciprocity?

JAMES K. RUSSELL

I. Introduction

Coati (*Nasua narica*) bands are associations of several adult females and their juvenile (<1 yr) and subadult (1–2 yr) offspring (Kaufmann, 1962). Female coatis are unique among procyonids (Mammalia: Carnivora) in maintaining persistent social bonds with other adults (Ewer, 1973). Adult male coatis, like other adult male procyonids, are solitary (Russell, 1981). Gregariousness in other procyonids (raccoons in Schneider *et al.*, 1971; Fritzell, 1978; ringtails in Trapp, 1978; others in Ewer, 1973, pp. 269–273) is typically restricted to temporary relationships between mates or between a mother and her most recent offspring. Such relationships also occur in coatis. It is the social relationships among adult females, then, that distinguish the gregarious coatis from their solitary

SOCIAL BEHAVIOR OF
FEMALE VERTEBRATES

ISBN 0-12-735950-8

relatives. These relationships, best referred to as social bonds because of their persistence, are what require explanation for an understanding of the evolution of gregariousness in this species. Inferences about why persistent social bonds occur among adult female coatis require two kinds of information: What benefits are conferred through gregariousness, and how are these benefits distributed? The benefits of gregariousness often depend on altruism within social groups, and this appears to be the case in coatis. In this chapter, I will describe the forms of altruism in coati bands and then analyze their distribution in relation to the histories of social bonds, with a view to discriminating among the three alternative evolutionary routes to altruism: group selection (Wynne-Edwards, 1962), kin selection (Hamilton, 1964), and reciprocity selection (Trivers, 1971). Whichever hypothesis better explains adult bonds will also be implicated in the evolution of gregariousness in coatis, since it is these bonds that set coatis apart from their solitary relatives.

Coatis subsist primarily on invertebrates they find through intensive foraging in forest litter (Russell, 1982). As the principal items in their diet constitute a more or less evenly distributed, sparse resource, they will encounter greater competition by foraging in groups than if they forage solitarily and are unlikely to derive much benefit from sharing information about the locations of food patches (Waser and Wiley, 1979). Unlike the large social carnivores (Mech, 1970; van Lawick and van Lawick-Goodall, 1971; Schaller, 1972), coatis do not share food or cooperate in its capture (Kaufmann, 1962). As for some other small carnivores (Rood, 1974, 1978, 1982), the benefit of group foraging for coatis apparently derives primarily from cooperative vigilance and defense against predators. To be effective, a system of cooperative vigilance and defense requires that individuals help one another when a predator is encountered (Treisman, 1975). In a coati band, adult females take the major responsibilities for vigilance and defense: foraging on the peripheries of bands, devoting a relatively large proportion of their foraging time to vigilance, and attacking potential predators when they encounter them (see Section III,A). These behaviors are directly beneficial to the juvenile members of the band and indirectly beneficial to the juveniles' mothers.

Any social behavior that provides a benefit to its recipient at some cost to its performer is called altruistic (Wilson, 1975). A social system that requires altruism from some of its members to be more advantageous than solitary life poses an evolutionary problem. Unless altruists benefit in some way from performing altruism, genes predisposing individuals to behave altruistically cannot persist in the face of natural selection. Three alternative evolutionary mechanisms have been proposed to explain the occurrence of altruism: group selection, kin selection, and reciprocity selection. Group selection is clearly implausible for coatis because of the high rate of migration between groups, as will be described later. Kin selection and reciprocity selection differ fundamentally in the way that altruists benefit from their own altruism. Under kin selection, an individual directs its altruism to close genetic relatives and actually increases its inclusive fitness (Hamilton, 1964) with each "altruistic" act. Under reciprocity selection, on the other hand, the benefit of performing altruistic behaviors comes through maintaining social bonds with individuals who behave altruistically in return. Reciprocal altruists in effect exchange altruism.

Kin selection and reciprocity selection should be accompanied by altruists' use of different strategies for selecting social partners toward which they will behave altruistically. Kin altruists should restrict their altruism to close genetic relatives. Their propensity to perform altruism should depend solely on degree of genetic relatedness and should not be affected by the behavior of their social partner towards them. The tendency of reciprocal altruists to maintain social bonds must depend on the behavior of their social partners regardless of their genetic relatedness. Particularly, reciprocal altruists must discriminate between social partners who reciprocate their altruism and those that do not, favoring the former and terminating relationships with the latter (Trivers, 1971; Boorman and Levitt, 1973, 1980).

Kin selection and reciprocity selection are not exclusive of one another, although they may conflict in some cases. I do not wish to imply that one or the other is the sole influence on coati social behavior, but rather to determine which has more explanatory power.

II. Methods

I observed the behavior of free-ranging coatis during three periods totaling 26 months, between August 1975 and August 1978, on Barro Colorado Island, Panama. Two habituated bands provided most of the observations. I captured and marked 78 individuals, including all but 2 individuals in the habituated bands, and could identify an additional 10 individuals, including the remaining 2 in the habituated bands, on the basis of unusual coloration, scars from botflies and fights, or other individual peculiarities. Captured individuals were weighed, aged by tooth wear, examined for ectoparasites, and marked with freeze brands and ear tags.

One or two individuals in each band were equipped with radio collars to facilitate location of the bands for observation.

A coati band's normal response to humans is an organized escape (Russell, 1980). Any member of the band can trigger an escape response with an alarm vocalization or rapid running. To prevent such responses, it was necessary to attract all of the members of a band to me by briefly offering them small pieces of banana when I first contacted them on any particular day or after a separation. Ten minutes after all of the members of a band had become aware of my presence and I had ceased feeding them, I began collecting 10-min focal animal samples in alternation with scan samples (J. Altmann, 1974) of spatial organization for as long as I could remain with the band. I recorded the composition of the band when I first contacted it and at the beginning of each focal animal sample. Compositions of unhabituated bands were recorded whenever possible, either by simply observing a passing band quietly or by attracting it to me and feeding it. I recorded aggressive and grooming interactions ad libitum as long as this did not interfere with the collection of focal animal or scan samples. By determining subjects for focal animal samples by rotation through a list of the band members, I selected my vantage point for observation independently of the animal's behavior. Under these conditions, ad libitum samples of infrequent events should be relatively unbiased.

III. Results

A. Forms of Allomaternal Care

A mammalian mother's reproductive success depends in large part on the care she provides her offspring. Such care reduces the time and resources she can devote to her own maintenance and preparation for future reproduction. If other adults also provide care to her offspring, she can afford to devote more time and resources to herself. Such care, often called allomaternal care, can be considered a service provided to the mother. In this section, I will consider two forms of allomaternal care prevalent in coati bands: protection from predators and allogrooming. I will discuss services provided directly to other adults in a later section.

Coati females separate from their bands to nest solitarily a few weeks before parturition. They remain isolated from their bands for several weeks postpartum, returning to the band with their new juveniles when weaning is nearly complete. Allomaternal care is typically restricted to the postnesting season. Occasionally, however, two females will nest together or one female will join another on its nest after having lost her own litter to predators. In such cases, which are relatively rare, both females apparently care for all the young without regard to parentage. Communal nursing occurs in this context and also occasionally after the nesting season. In the postnesting season, the most important forms of care juveniles receive are protection from predators and from ectoparasites.

Juvenile coatis suffer a much higher predation rate than adults (Russell, 1982). Adult females help each other protect juveniles by foraging in a spatial organization highly advantageous to juveniles. A coati band foraging on the ground assumes an elliptical outline (Fig. 1). This spatial organization does not conform to the rank formation predicted as optimal for reduction of foraging competition (S. A. Altmann, 1974), but corresponds closely to the circular formation predicted as optimal for resistance to predation (Treisman, 1975). Adults tend to take positions on the periphery while juveniles tend to remain in the center (Russell, 1980). By foraging on the periphery, adults reduce the chance that a juvenile can be attacked and at the same time increase their own exposure to predators (Hamilton, 1971). Some peripheral positions will provide reduced foraging competition, so I have divided peripheral positions into three categories based on the direction of the band's movement (Fig. 1). The van position would appear to have advantages and disadvantages. Animals in the van would have the least foraging competition, but would be the first to encounter predators lying in wait. Animals in the tail of the band would encounter the greatest foraging competition, as they would always forage in areas already searched by the remainder of the band. They might also be susceptible to pursuit predators. Animals on the flanks would encounter little foraging competition and would not be more exposed to predators than animals in any other peripheral position. Adults also devote a significantly higher proportion of their foraging time to vigilance than do juveniles (Russell, 1980). By foraging gregariously, group members can afford to devote less of their time to vigilance and more of it to searching for food than if they foraged separately (Pulliam, 1973). Adult coatis devote significantly less time to vigilance when foraging in large groups than when foraging in small groups or alone (Russell, 1980). Juveniles devote more time to vigilance when unaccompanied by adults (Russell, 1980). Thus, by taking positions on the periphery of the band and by contributing to group vigilance, adults help each other protect juveniles and at the same time reduce the amount of time each has to devote to vigilance.

The efficacy of group vigilance depends on an effective response to

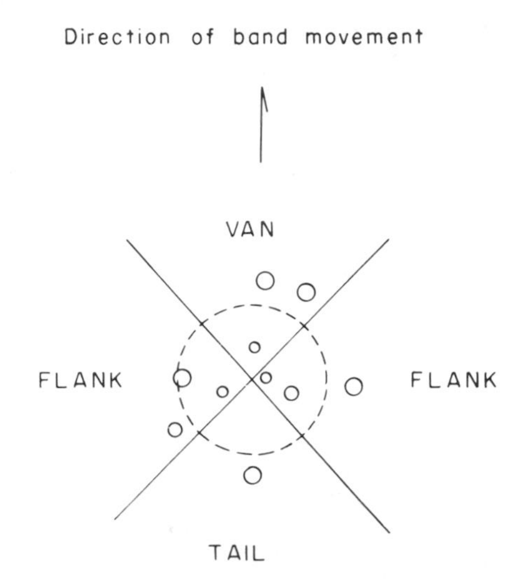

Fig. 1 Spatial organization of coati bands while foraging on the ground. Large circles represent adults; medium-sized circles, subadults; and small circles, juveniles.

predators once they are detected (Treisman, 1975). Coatis respond to the presence of a potential predator in one of two general ways: alarm or confrontation.

In alarm, an individual that detects a source of danger increases the rate and loudness of the foraging contact vocalization—grunting. The other band members cease foraging, become silent, and watch the first individual. The first individual silently watches the source of alarm and then either returns to foraging or suddenly runs away from the source of alarm. If it returns to foraging, the others do also. If it begins running, all the band members start running nearly simultaneously, radiating away from the source of alarm. After they have run a few meters they all freeze. At this point one or two adults are usually still on the ground while the rest of the band members are clinging to small tree trunks 1–2 m off the ground and watching the adults on the ground. Usually, one adult female initiates an orderly retreat, leading the juveniles away from the source of alarm. Other adults take up positions on the flanks of the retreating band. One adult usually remains behind watching the source of alarm until the band is safely away; then it runs to join them.

In confrontation, one or several adults charge the potential predator. I observed this response only to adult male coatis, but Janzen's (1970) observation of a group of coatis attacking a boa constrictor suggests a similar response to at least some other predators. Other adults will support the first attackers if they are not immediately successful. The immature band members usually remain behind or climb up trees. The importance of adult protection can be seen during the nesting season when juveniles are unaccompanied by adults. During this period juveniles are attacked relatively frequently by males and must resort to flight to the terminal branches of trees to save themselves. Attacks by males during the nesting season last much longer than when adult females are present (Russell, 1981).

Coatis forage actively for most of each day, but near midday, bands usually rest together for ½–1 hr. During this time, adults mostly rest or allogroom, while juveniles and subadults rest, allogroom, and engage in play chases and exploration. As during foraging, adults tend to be on the periphery of the band more often than juveniles. All classes of individuals spend more time in vigilance during rest, probably because it costs them little. Still, adult females spend significantly more time in vigilance than do younger age-classes (Russell, 1980).

During rest periods, adults also protect the offspring of other adults from parasites through allomaternal grooming. The principal ectoparasites of coatis are ticks and mites. Although grooming does not reduce the incidence of mites on coatis, it significantly reduces the incidence of ticks compared to their incidence on solitary males (Table I). Although the pathology of ticks on coatis has not been studied, a great variety of endoparasites have been documented as being transmitted by ticks on dogs (Flynn, 1973). These parasites can result in serious debilitation. Presumably, coati ticks also carry endoparasites. Thus, by providing allomaternal grooming to juveniles, adult females provide an important service to the juveniles and indirectly to their mothers. Juvenile coatis on the average receive half of their allogrooming from adults other than their mothers (Russell, 1980). Juveniles also receive some nursing from adults other than their mothers, although they are nearly weaned by the time groups reaggregate after the nesting season. The advantage to juveniles of receiving milk is obvious.

B. Does Allomaternal Care Depend on Kinship?

If kin selection were an important factor in the evolution of allomaternal care in coati bands, adults that are closely related to the juveniles of a band should be more inclined to provide allomaternal care than those that are not. In order to evaluate this possibility, it is necessary to have some information about genealogical relations within bands.

TABLE I
Incidence of Ectoparasites on Coatis Captured in Traps

	Ticks			Mites		
Age–sex class[a]	None	Some small ticks	Some engorged ticks	No infestation noted	Abdominal infestation	Infestation elsewhere only
Juveniles	24	3	1	8	3	13
Subadults	5	0	0	12	0	2
Adult Females	31	6	1	21	5	5
Adult Males	7	5	9	18	0	3

Statistical tests
Presence of ticks (adult males versus band members, $\chi_1^2 = 23.7$, $p < .001$)
Proportion of ticks engorged (adult males versus band members, $\chi_1^2 = 5.3$, $p < .05$)

[a]There are no significant differences by sex for juveniles or subadults.

1. *Genealogical Relationships in Bands 2 and 7*

Although my study was not long enough to document genealogies completely, some qualitative genealogical information is available from the histories of immigration and emigration (Fig. 2). Detailed behavioral observations were taken primarily in Bands 2 and 7 in 1977–1978. In 1975–1976, Band 2 consisted of two breeding adults (F21 and F23), one nulliparous adult (F26, later F76*), and five juveniles, two of whom were females that survived to breeding age (F22 and F27). These two females and the one surviving breeding adult (F23) formed the adult core of Band 2 in 1977–1978. At least one, and perhaps more, of the dyads among adult females in Band 2 in 1977–1978 therefore must have represented a close genetic relationship; that is, at least one sister–sister or one mother–daughter relationship. In 1976 two nulliparous females (F24 and F25, later F74 and F75) immigrated into Band 2. They were apparently unrelated to the remainder of the band (for further details, see Section III,C). In 1977, these two females and the other nulliparous female from Band 2 (F26/F76) left to form Band 7, in which they were joined by yet another dispersing female (F72), who was also apparently unrelated to them or to F76. Thus, Band 7 in 1977–1978 included at most one close genetic relationship between adult females, that between F74 and F75.

*The first digit of an individual's designation represents its band and changes when the individual changes bands.

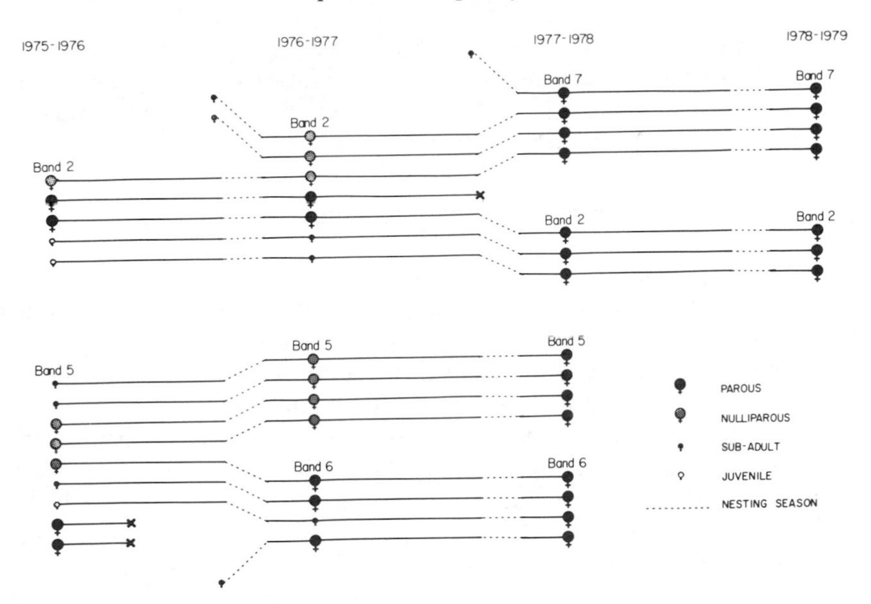

Fig. 2 Genealogical histories of composition of four coati bands. Only the histories of adult females are shown.

2. *Communal Nesting and Communal Nursing*

In 1977, F24 and F25, who immigrated into Band 2 together in the previous year, shared a single nest and essentially pooled their litters. The nest was difficult to observe, but there was no apparent tendency for them to discriminate among the young in any way, and both were observed nursing on the nest with approximately equal frequency. In 1978, these two females (now F74 and F75) nested separately at first, but F74 joined F75 on her nest after she apparently lost her own litter to predation. Both of these forms of communal nesting have been observed in captivity (McToldridge, 1969; Smith, 1980), but have not been previously reported from nature. None of the other females I observed nested communally.

In addition to its occurrence during communal nesting, communal nursing takes place occasionally in the postnesting season. Six of the 17 occasions on which I observed nursing during rest periods in the post-nesting season involved communal nursing. Most of these involved F74 and F75, and thus represented a continuation of their communal nesting relationship. However, on one occasion F74 and F75 both nursed one of F72's or F76's offspring. I also observed communal nursing on one occasion in Band 5 in 1977.

Communal nesting and communal nursing are relatively rare phenomena and are difficult to observe. Because of incomplete genealogical information, I could not determine whether genealogical relatedness was an important factor in their occurrence.

3. *Contributions to Protection from Predation*

Direct observations of the behavior of individual adult females in the contexts of confrontation of and flight from predators were too infrequent to support a quantitative analysis. However, some measure of the contributions of various adults to the protection of juveniles can be derived from records of their position with respect to foraging bands of the amount of time they devoted to vigilance. Vigilance takes away from time that could be spent searching for food. Positions on the peripheries of bands presumably involve more exposure to predators than positions in the center. Some peripheral positions will tend to involve less foraging competition than central positions, but an individual will enjoy this benefit only if it takes one of the more advantageous foraging positions (Flank, Van) for a disproportionate amount of time.

The benefits of adults taking peripheral positions and devoting a relatively large amount of time to vigilance are primarily enjoyed by juveniles. Since these behaviors simultaneously protect all of the juveniles of a band, the contributions of individual adults to individual juveniles cannot easily be assessed. However, from August 1978 to March 1979, three females in Band 7 (F72, F74, and F75) had no living offspring in the band and were unrelated to the surviving juveniles in Band 7 (F76's offspring). In Band 2, during the same period, each adult female had at least one surviving juvenile, and at least two females from Band 2 were closely related to some other juveniles in the band as well. Under kin selection, the females with no juvenile relatives should contribute less to juvenile protection, whereas under reciprocity selection, no difference is expected if by contributing to the protection of unrelated juveniles adults maintain relationships with other adults that will help them protect their own offspring in subsequent years (see Section III,C). There was no significant difference in any measure of the contribution to juvenile protection between adults with juvenile relatives and adults without juvenile relatives (Table II).

4. *Contribution to Protection from Parasites*

The same contrast in genealogical histories can be used to assess the importance of kinship in determining the amount of allomaternal groom-

TABLE II

Contributions of Adults with and without Juvenile Relatives to the Protection of Juveniles

Group	During terrestrial foraging					During rest	
	Probability of being peripheral	Probability, if peripheral, of being in the			Proportion of time devoted to vigilance	Probability of being peripheral	Proportion of time devoted to vigilance
		Van	Flank	Tail			
Adult females with juvenile relatives (F22, F23, F27, and F76)	.83	.08	.46	.46	.10	.55	.18
Adult females without juvenile relatives (F72, F74, and F75)	.82	.15	.32	.53	.13	.63	.23
Statistical test for differences between groups	$G = .022$; $df = 1$; n.s.	$G = 1.87$: $df = 2$; n.s.			Mann-Whitney large sample $z = .62$; n.s.	$G = .40$, $df = 1$; n.s.	Mann-Whitney large sample $z = 1.09$; n.s.

ing adults provide. Setting aside the grooming F74 and F75 provided to each other's offspring, because their genealogical relationship was uncertain, there was no significant difference in the percentage of grooming that juveniles received from adults other than their mothers in the two bands (Fig. 3: Band 2, 32%; Band 7, 36%; Mann-Whitney $U = 42.5$; $n_1 = 6$; $n_2 = 16$; n.s.).

C. Patterns of Adult Bond Formation, Maintenance, and Termination

1. *Immigration, Emigration, Band Fission, and Band Fusion*

Social bonds between adult females arose through two processes. Some maturing females remained with their natal bands, establishing at an adult level bonds that were already present when they were immature. I call this process internal recruitment. Alternatively, some maturing females left their natal bands to join others, into which they immigrated. Some females migrated between bands more than once.

I observed four cases of immigration into two bands during my study

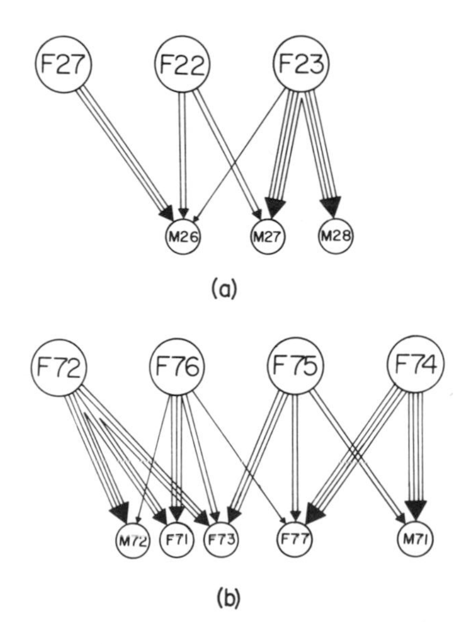

Fig. 3 Percentage of grooming bouts each juvenile (small circles) in (a) Band 2 and (b) Band 7 received from each adult female (large circles) in 1977 and 1978: → <20%, ⇛ 20–39%, ⇛ 40–59%, ⇛ >60%. The adult most directly above a juvenile is its mother.

(Fig. 2). The sources of these immigrating females were unknown, other than that they did not come from known neighboring bands. In 1976, females F24 and F25 immigrated into Band 2, and female F63 immigrated into the newly formed Band 6. In 1977, female F72 immigrated into the newly formed Band 7. All of these females were two-year-olds when they immigrated. Females F72 and F63 began their associations with their respective bands as primiparas with dependent juveniles, while females F24 and F25 were nulliparous when they immigrated. Although none of the females that matured in the habituated bands emigrated, migration from one band to another must be fairly common. Twelve of the 30 dyadic bonds in Bands 2, 5, 6, and 7 arose through immigration. Six bonds arose through internal recruitment. The remaining 12 bonds were established before the beginning of the study, and their ontogeny was unknown.

I did not observe any adult females emigrating singly from bands. Group emigrations or band fissions occurred twice in habituated bands (Fig. 2), and the strong affiliation between Bands 4 and 10 in 1976–1977 suggests that they too were products of a band fission (Figs. 4–6). The fissions I observed resulted in less than six adults remaining in each band. If fissions or other emigrations had not occurred, adult numbers would have been seven or more in each case. Stable bands with more than five adults did not occur, although some bands (2 and 7, 5 and 6) fused temporarily to form unusually large bands for periods of several hours, sometimes several days in succession. These band fusions seemed to occur when litter fauna, the coati's dietary mainstay, were especially abundant. Although bands sometimes intermingled at large fruiting trees, they generally separated when they left the trees, and I did not consider these events instances of band fusion. Band fissions, then, seemed to have the function of maintaining the adult component of band size within its observed narrow range (Russell, 1982). The advantage of this narrow range would appear to be reduction of foraging competition to a minimum compatible with maintenance of groups. The immature members of Bands 2 and 7 fused into a temporary band of from seven to nine members for most of the 1978 nesting season. Although litter fauna are scarce during the nesting season, which coincides with the dry season, in the absence of protective adults the advantages of combined vigilance apparently outweighed the costs of foraging competition for these immature members.

2. *Comparison of Discontinued and Maintained Relationships*

It is evident that adult coatis form and dissolve social bonds relatively frequently. To assess whether these changes represent individuals learning about social relationships and selecting among them on the basis of

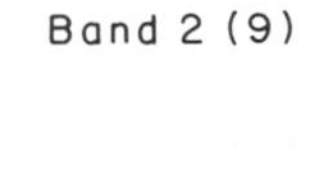

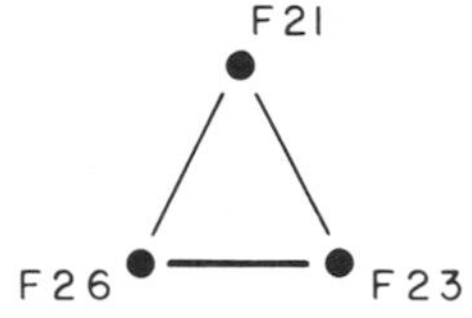

Band 4 (6)

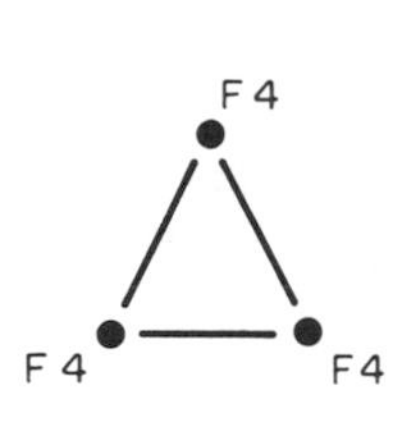

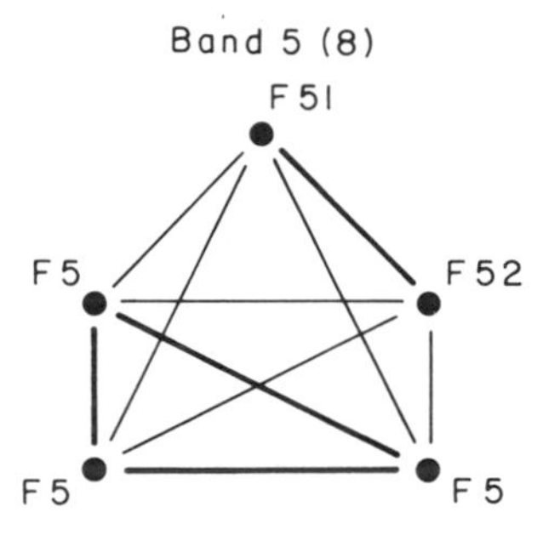

Fig. 4 Strengths of affiliation among adult females in several bands, 1975–1976. Values represent the number of composition records in which both members of a dyad appeared as a proportion of the number of records in which either appeared: — 1.0, — .90–.94. Parenthetical numbers are sample sizes.

their value to them, it is necessary to consider in some detail the relationships that were broken in comparison to those that were maintained. Three kinds of data are pertinent to this analysis: tendencies of individuals to associate with one another, reciprocation in affiliative interactions, and patterns of aggressive interactions. The only broken bonds for which these data are available are those that were broken in the fission of Band 2 in 1977. These relationships can be compared to the relationships these individuals achieved subsequent to fission, as well as to other relationships that were maintained from one year to the next.

a. Tendency to Associate

Although no long-term changes in band membership occurred outside the nesting season, members of a band were not always found

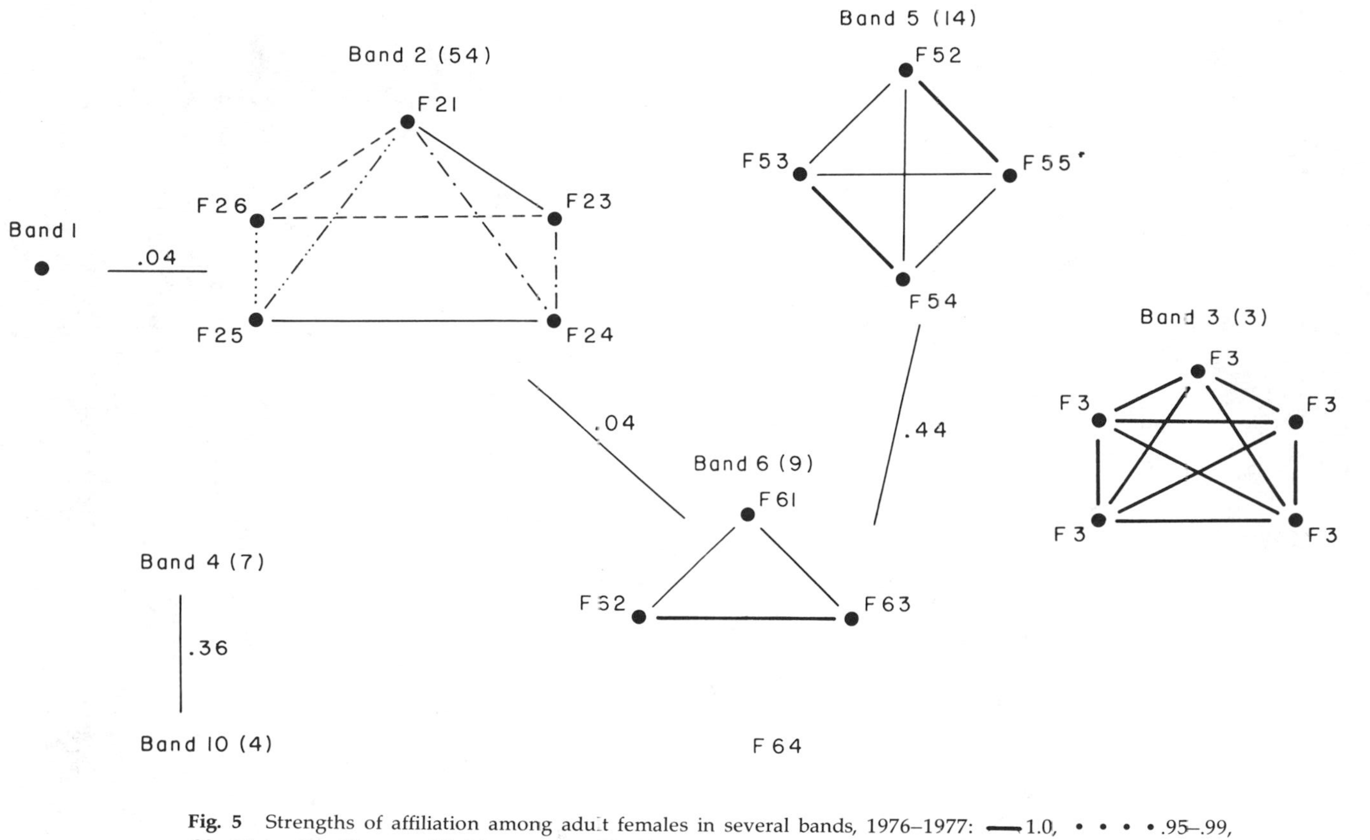

Fig. 5 Strengths of affiliation among adult females in several bands, 1976–1977: ▬ 1.0, • • • • .95–.99, — .90–.94, ----.85–.89, –·–· .80–84, · · ·– · · ·–.75–.79, · · · ·.70–.74.

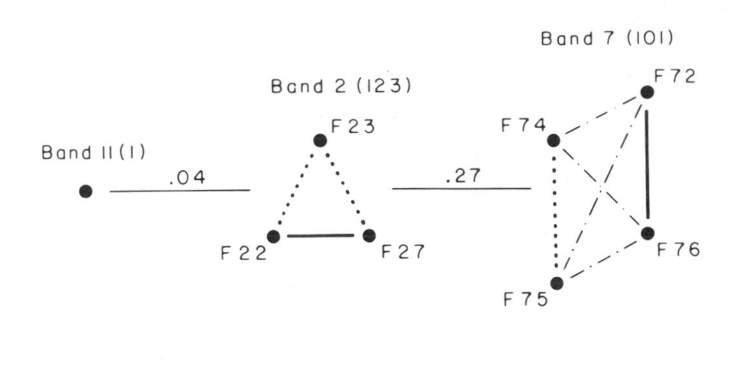

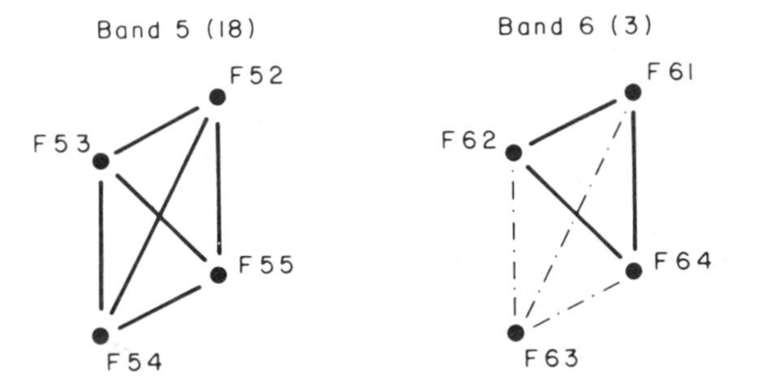

Band 4 (3)

Band 9 (1)

Band 8 (5)

Fig. 6 Strengths of affiliation among adult females in several bands, 1977–1978. Symbols: see Fig. 5.

together. Occasionally, one or several individuals foraged separately from the remainder of the band for periods of from several hours to several days.

The percentages of composition records in which both members of adult dyads were present (Figs. 4–6) show that Band 2 in 1976–1977, prior to its fission, was considerably less stable than other bands or than Band 2 in other years. The immigrants, F24 and F25, maintained a relatively strong association with each other, but their associations with other members of the band were among the weakest recorded. F26's associations with the remaining adults in the band (F21 and F23) were weakened subsequent to

F24 and F25's immigration. Her association with the two immigrants, at this point, was also relatively weak. In fact, F24 and F25 established stronger associations with F21 and F23 than with F26. Thus, F26's associations with the rest of the band weakened in general and did not simply switch from strong association with the original band members to strong association with the immigrants. F21's and F23's association with each other was apparently unaffected by the immigration.

After Band 2 split into Bands 2 and 7 (and females F24, F25, and F26 became F74, F75, and F76 respectively), the two bands maintained an unusually strong association with each other; that is, they tended to fuse into a temporary large band relatively frequently. In Band 2, associations among all adult females, the results of internal recruitment, were very strong. In Band 7, on the other hand, there was a mixture of strong and weak associations. F74 and F75 maintained their strong association with each other. Although their association with F76 increased after the fission, it remained relatively weak. F76 established a very strong association with the new immigrant, F72. F72's association with F74 and F75 was identical with F76's association with them.

b. Reciprocation in Affiliative Interactions

The principal form of affiliative interaction in coati bands is allogrooming. Allogrooming relationships are particularly important among adults. Adults groom other adults more frequently than they groom either juveniles or subadults ($G = 159.0$; $df = 2$; $p < .001$). While grooming in adult–juvenile dyads is polarized in favor of the juveniles (Russell, 1980), grooming in adult dyads is generally highly reciprocal. The mean percentage of the total number of bouts in a dyad that was performed by the more frequent groomer was only 57%.

There were two exceptions to this rule of reciprocity in allogrooming among adults, both of which occurred in Band 2 in 1976–1977, prior to its fission. These were in the F23–F24 and F21–F26 dyads. In both cases, nulliparous females performed much more grooming towards parous females than they received from them (Fig. 7). These two nulliparous females, along with F25, left Band 2 to form Band 7 in the band fission. F25's grooming relations to the parous females were too weak to document adequately with regard to reciprocity. The difference in polarity between the F23–F24 and F21–F26 dyads and all maintained adult grooming dyads is significant (Mann-Whitney $U = 0$; $n_1 = 2$; $n_2 = 14$; $p < .01$).

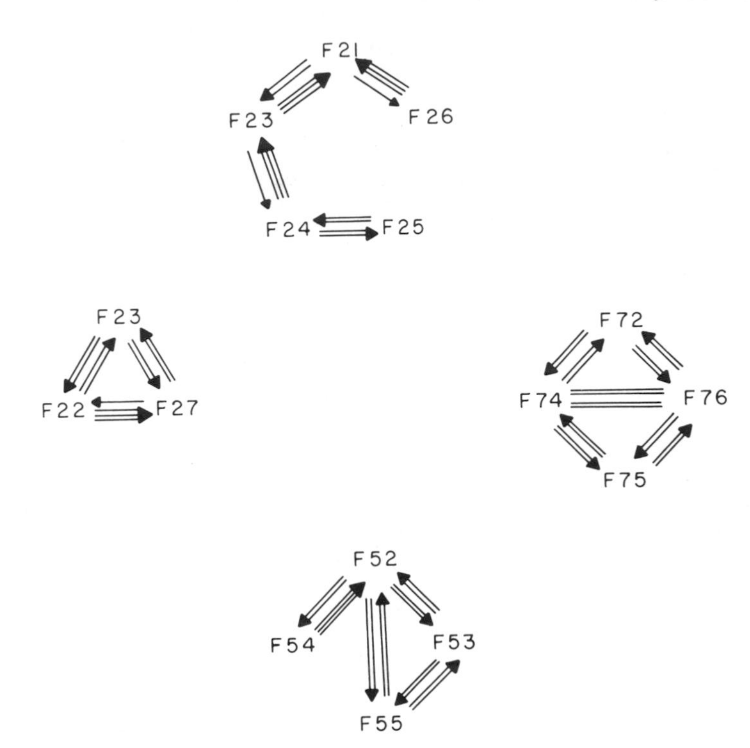

Fig. 7 Grooming relations among adult females. Arrows represent the number of bouts performed by each member of each dyad for which more than five bouts were recorded, as a proportion of the total number of bouts performed by the members of the dyad towards each other. → 20–39%, ⇛ 40–59%, ⭆ 60–79%. Top, 1976–1977; center and bottom, 1977–1978.

c. *Patterns of Aggression*

Intense threats and fights are rare among the adults of a band. During foraging, adults that were excavating or eating food items occasionally threatened other adults that approached, but these interactions always resulted in simple avoidance by the approaching adults. True fights, involving squeaks, squeals, open-mouth threats, lunges, etc., occurred only three times among adult members of the same band under natural circumstances in my observations. Two incidents were apparently accidental by-products of the heightened aggression involved in driving away adult males and were quickly replaced by allogrooming between the aggressive parties. The third involved a vigorous defense by one adult of an unusually large prey item (a spiny rat—*Proechimys semispinosus*) against all but the juvenile members of the band.

Although aggression naturally occurred at a low rate within bands, I provoked aggressive interactions within bands while feeding them for habituation. The distribution of aggressive acts provides further evidence that the relationships between the nulliparous females in Band 2 in 1976–1977 and the remainder of the band were unusually disadvantageous for these nulliparous females. It was much more difficult for me to prevent aggression while feeding Band 2 in 1976–1977 than for either Band 2 or Band 7 in 1977–1978. Most of the aggressive interactions between adults (78%) occurred between the nulliparous females and the parous females ($G = 8.4$; $df = 1$; $p < .01$; Fig. 8). All of the aggressive interactions between parous and nulliparous females for which the initiator could be determined were begun either by a parous female or by a subadult female who elicited the support of a parous female. The subadult females of Band 2 in 1976–1977 were also particularly aggressive to the nulliparous females. They directed 84% of their aggressive acts at the nulliparous females and initiated all aggressive interactions between themselves and the nulliparous females. Band 2 parous females supported subadult females when they initiated aggression against nulliparous females on seven occasions and never supported nulliparous females. Of special importance, the nulliparous females supported each other on seven occasions in aggressive interactions with the remainder of the band. On two of these occasions, F26 supported F25, the only two occasions on which F26 supported any adult in intraband aggression. These observations of F26 in 1976 suggest that she had already formed a stronger bond to F25 than she had with the parous females of Band 2, with whom she had lived for at least the preceding 1.5 yr.

In 1977–1978, most instances of aggression occurred between Band 2 and Band 7 members when I fed the two bands together, either while they were temporarily fused or at a fruiting tree (Fig. 8). There were also some instances of aggression within the newly formed Band 7, most of which consisted of aggression by females F74 and F75 directed towards F76. A few instances of aggression also occurred within Band 2 in 1977–1978, but in general, aggression within bands was much rarer after the fission than before it.

3. *Sources of New Relationships*

To form new social relationships, individuals need to find partners with which they can establish relationships that are better than their current ones. Sometimes this can occur within bands, as in the establishment of relationships among F26, F24, and F25. In other cases, however, this is impossible. Immigrating individuals, by definition, must establish

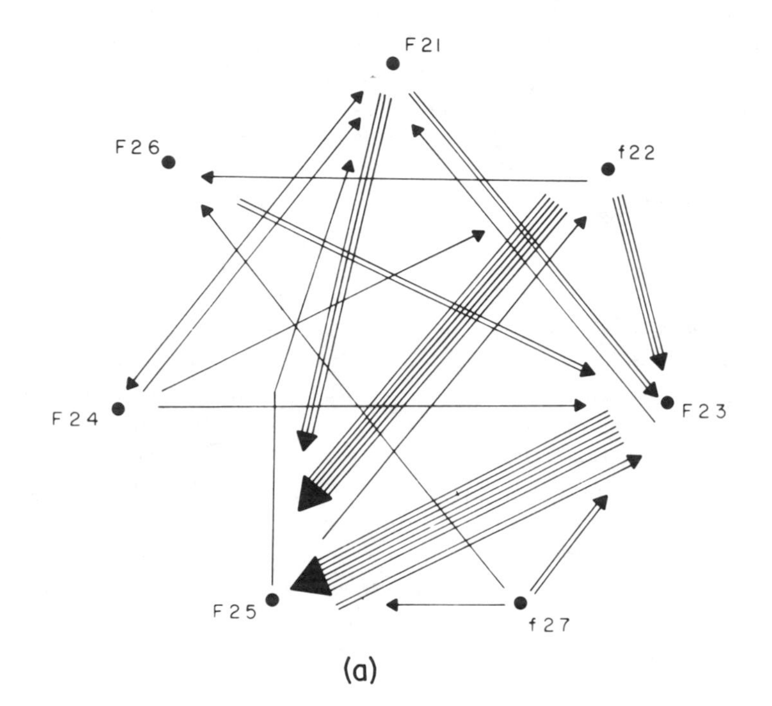

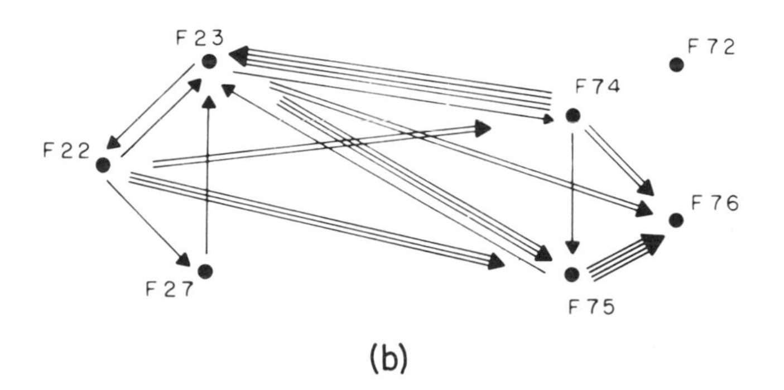

Fig. 8 Incidence of aggression in (a) 1976–1977 and (b) 1977–1978. Incidence was not recorded in 1975–1976 or for Bands 5 and 6. Each line represents one incident of a highly aggressive threat or attack or an uninterrupted series of threats or attacks. F, adult female; f, subadult female.

relationships anew, with individuals with whom they have not previously shared band membership. If an individual's current relationships are bad enough, it may be advantageous to emigrate on the chance that better relationships will be found. Still, it is easy to see that it would be advantageous to know something about potential social partners before making a move. Relationships between bands are interesting in this regard.

When two coati bands approach each other while foraging on the ground or at a fruiting tree, they generally ignore or calmly avoid each other. Occasionally, there is some aggression between bands at concentrated food sources, such as fruiting fig trees. Relations between certain pairs of bands, however, were more friendly. When Band 2 and Band 7 met while foraging in 1977–1978 and 1978–1979, they usually mixed completely and formed a temporary band, with the juveniles tending to gather in the center and adults tending to form a perimeter around them. Bands 2 and 7 regularly foraged together in this way when they used overlapping areas of their home ranges. This mixed band was sometimes observed on several consecutive days, although the bands separated by the end of each day. After a period of several days of foraging in a mixed band, the component bands would separate and forage in other parts of their home ranges independently. I observed Bands 2 and 7 foraging together in eight of the nine months outside the nesting season for which I have more than a week's records.

Group grooming occurred when the two bands met after a long separation. Most or all of the members of both bands would form a single writhing pile, with each individual grooming another and being groomed by one or more other individuals. Grooming partners switched rapidly and repeatedly, so that it was impossible to keep track of who groomed whom. Throughout the group grooming session, one or several individuals uttered the chuckling vocalization also heard during reconciliation after misdirected aggression. Grooming in these sessions appeared to have no effect in removing ectoparasites. Each grooming bout, only a few seconds long, was exaggerated in form, with quick jerky movements, in contrast to the slow and deliberate movements and prolonged bouts of normal allogrooming and mutual grooming. Group grooming occurred between the bands only after long separations. When the bands associated with each other intermittently over a period of several days or weeks, they simply mixed without incident when they encountered each other.

These phenomena were not restricted to the relation between Bands 2 and 7. Other friendly associations between bands are represented in Figs. 4–6. Bands 5 and 6 were observed to group-groom, and on one occasion Band 2 group groomed with part of Band 6. F24 and F25 each traveled with

Band 5 on separate occasions in February 1977. Kaufmann (1962) also observed group grooming between two bands on one occasion.

4. *The Process of Reaggregation*

Observations of bands during the period when changes in adult membership actually occur are difficult to obtain. Juveniles, newly descended from the nest, are particularly shy, and their mothers particularly defensive at this time. Nevertheless, some information from this period is available (Table III). An interesting pattern is apparent in these observations. Social bonds that were strong in the preceding year are reestablished first and most decisively. Despite selective pressure to establish and stabilize social bonds quickly after descent from the nest (Russell, 1982), weaker bonds are reestablished only tentatively at first and are subject to change for several weeks.

IV. Discussion

The social bonds among adult female coatis support a system of allomaternal care primarily beneficial to juvenile members of the band. Through cooperative vigilance, confrontation of predators, and allogrooming, adult females provide care to each other's offspring as well as to their own offspring if they have any. To the exent that individual adults provide care to juveniles other than their own, they behave altruistically.

Altruistic services might be explained by any of three selective mechanisms, or some combination of them: group selection, kin selection, and reciprocity selection. I have not considered group selection explicitly because it requires very low migration rates between groups (Williams, 1966), a condition clearly violated by coatis.

Kin selection is also less likely in the presence of high migration rates between groups because individuals would tend to become separated from their nearest relatives. However, if migrant individuals forego altruistic services until their own reproduction supplies relatives or if related individuals tend to migrate together, kin selection might still explain altruism. I have provided two kinds of evidence that indicate that kin selection is not a sufficient explanation of altruism in coati bands. Adults migrate between bands relatively frequently and establish normal social bonds with apparently unrelated adults. Furthermore, adults unrelated to

TABLE III
The Process of Reaggregation of Bands after Nesting

Year and band	Date	Observation
1977 2, 7	May 12	F24 and F25, without their juveniles, join F23 and her three juveniles. F27 and F22, without their juveniles, join them temporarily. Later, F24 and F25 travel together as one group; F23, her juveniles and her subadult offspring from the previous year travel as another, unaccompanied by F27 and F22.
	May 13	F24 and F25 forage together without their juveniles.
	May 14	F24 and F25 forage together with their juveniles.
	May 17	F24 and F25 forage together with their juveniles.
	May 23	F24, F25, and their juveniles are together with the rest of Band 2 in the early afternoon. By late afternoon they are again separate.
	May 27	F24, F25, and their juveniles are with the rest of Band 2 again.
	June 1	F22, F23, F27, their juveniles, and the subadults from the previous year all forage together. F24, F25, and their juveniles are not with them.
	June 12	F24, F25, and their juveniles are with the rest of Band 2 again.
	June 23	F24, F25, and their surviving juveniles are observed with F26 and a new adult female and some juveniles for the first time. These animals become Band 7. F24, F25, and F26 are called F74, F75, and F76 hereafter. The new female is F72.
5	May 20	F53, F55, and their juveniles forage together. F52 and F54 are not present.
	May 23	F55 forages with her juveniles, without any other adults.
	May 28	F53, F55, and their juveniles forage together. They are joined by F52, F54, and their juveniles.
6	May 12	F62 forages together with her juveniles, without any other females.
	May 18	Band 6 fully reaggregated.
1978 7	May 6	F74 joins F75 on her nest after losing her own litter.
	May 24	F74, F75, and F75's juveniles forage together.
	May 29	F74, F75, and F75's juveniles forage together. F72, F76, and their juveniles forage together in another part of Band 7's range.
	June 2	Band 7 fully reaggregated.

the juvenile members of bands contribute as much to their care and protection as do related adults.

I do not wish to imply that no effects of kin selection can be found in the coati social system. The favoritism F23 displayed toward f22 and f27 in intraband aggression in 1976–1977 may represent nepotism (kin altruism). The unusually strong relationship between F74 and F75 may also have been nepotistic, although the genealogical relation between these females was unknown. What is evident is that kin selection is insufficient to account for many of the social bonds among adult coatis and the mutual aid these bonds entail. For this reason, I have provided information regarding the patterns of adult bond formation, maintenance, and termination that may be used to evaluate the possibility that reciprocity selection has been an important factor in the evolution of the tendency to form these bonds. The history of the fission of Band 2 into Bands 2 and 7 provides several kinds of evidence that individuals that discontinue relationships do so because the relationships are weak or disadvantageous to them. The nulliparous females that left Band 2 to form Band 7 had relatively weak tendencies to associate with the other members of the band prior to the fission. The available data also indicate that these females had disadvantageous grooming relationships with the other adults in Band 2, in contrast to the highly reciprocal grooming relationships that generally obtain among adults. The emigrating adults were also the recipients of aggression before they left. The aggression they performed was entirely defensive in nature. Aggression was most frequently initiated by the subadult females (f22 and f27). If Band 2 had remained together in the following year, when these subadult females became mature, it would have exceeded the observed size range of bands. The aggression by subadult females may be explained as preparation for the impending band fission. These subadult females may have been seeking to create the impending band fission in a manner that assured them of receiving further favoritism from F23. This suggests an influence of kin selection on the ontogeny of social bonds. F23's response to aggression initiated by f22 and f27 was probably quite important. She either ignored the aggressive incidents they initiated or interfered on their behalf if they were counterattacked by the other females. She never took the part of any of the other females against the subadults. Perhaps partly in response to this evident favoritism of F23 for f22 and f27, F26 began forming relationships with F24 and F25 and supported them in aggressive interactions with the rest of the band.

The changes in relationships consequent to the formation of Band 7 indicate that individuals choose to invest in new relationships to a degree dependent on their apparent potential for developing into advantageous equitable relationships. F26's (later F76's) relationships with F24 and F25 (later F74 and F75) did not become as strong and beneficial as typical adult dyadic relationships, but she did establish well-balanced grooming relationships with them. This represented an improvement over her status in Band 2 prior to the fission. Perhaps more important, her prospects for equitable relationships in Band 2 appeared poor at the time of her decision to leave because of the favoritism within the clique composed of F23, f22, and f27.

F26's ability to form advantageous relationships with F24 and F25 was probably limited to some degree by the unusually strong relationship of the latter two females. F24 and F25 immigrated into Band 2 together and also nested together in two of the observed years. Thus, F24 and F25 had a strong mutually supportive relationship already established when F26 required a new one because her current relationships were poor. This put F26 in a weak position while she was seeking to establish new relationships with them. Nevertheless, when the fission occurred, F24 and F25, after some tentative interactions with F22, F23, and F27, chose to emigrate with F26 to form Band 7. They also had been recipients of aggression and participants in disadvantageous grooming relationships while in Band 2.

After the band fission, F76 was still in a position of need because of F74 and F75's mutual favoritism. It is therefore probably not accidental that she formed the strongest relationship of the three with the new immigrant, F72. The relationship between F76 and F72 quickly became as strong as any adult dyadic relationship I observed except for the unusually strong relationship between F74 and F75. In addition to becoming virtually inseparable from F72 and establishing an equitable grooming relationship with her, F76 received substantial allomaternal aid, directed to her offspring, from F72 (see Fig. 3). Thus, when both F76 and F72 were faced with the unusually strong relationship between F74 and F75, they developed a strong new relationship with each other. F72's role in this process can be seen as similar to that of F76. When confronted with three possible interactants, she chose to invest in the relationship that offered the greatest benefit.

Analysis of the histories of adult social bonds in coati bands does indicate that individuals assess the benefits they receive from social relationships and terminate relationships that are not beneficial. This satisfies the principle requirement of reciprocity selection: that individuals must monitor their social relationships and maintain only those that are beneficial to them. In fact, this requirement applies also to simply cooperative relationships (Rapaport, 1960, p. 182). Reciprocal relationships differ from cooperative ones only in that the requirement for equitability in the distribution of the benefits of cooperation to the participants can be relaxed at any point in time, allowing altruism, so long as this requirement is maintained over the course of the relationship. The concordance between the observed patterns of bond maintenance and the pattern expected under reciprocity selection indicates that reciprocity selection has been important in the evolution of social bonds among adult coatis.

Some influence of kin selection on adult social bonds is suggested in the cases of favoritism among F22, F23, and F27 and between F74 and F75. The role of kin selection in these relationships is uncertain because of uncertainties in the genealogical relationships among these females. Further study would be required to document any effect of kin selection in coatis. If kin selection does operate in some social bonds, it would appear to be in conflict with reciprocity selection to some extent. If individuals favor relatives in their social relationships, they may do so at the expense of maintaining reciprocal relationships with nonrelatives. This conflict between kin selection and reciprocity selection has apparently not been considered previously and would merit both empirical and theoretical attention.

Reciprocity and kin altruism are not mutually exclusive, and relationships that combine both effects would bring greater benefits to their participants than either kind of relationship by itself. Thus, even where reciprocity is well developed, one would expect a preference for close relatives as social partners. One might ask, then, why so much migration occurs among coati groups. The answer, mere speculation at this point, would appear to turn on two effects: the unpredictability of survival and the ecological optimality of a small range in group size. Individuals can maintain relationships with their close relatives only if such relatives survive, and to the extent that maintenance of these relationships is compatible with the maintenance of an optimal group size. When either too many or too few close relatives survive, a conflict arises between maintaining relationships with them and only them and maintaining an optimal group size. Emigrating to unrelated groups and accepting nonrelatives as immigrants permit maintenance of an appropriate group size in the face of unpredictable survival of relatives. Individuals who can form reciprocal relationships with nonrelatives can enjoy the benefits of gregariousness, whereas individuals who restrict their relationships to close kin might suffer through living in either too large or too small a group. The importance of reciprocity in the coati social system, thus, probably derives largely from the narrow range of optimal group sizes (3–5 adults). Where a greater range in group size is permissible, one might expect kinship to evolve as a more important determinant of social relationships than is the case among the coatis.

V. Summary

Coati gregariousness, unique among the Procyonidae, derives some of its advantages from altruistic interactions among group members. Altruism requires group selection, kin selection, or reciprocity selection in order to be evolutionarily stable. Each mechanism differs from the others in the constraints it places on social migration. These differences lead to predictions about the amount and form of social migration that should allow discrimination among the mechanisms as sources of altruism in any particular population.

Focal animal samples of behavior, scan samples of spatial organization, and ad libitum samples of infrequent social interactions taken on two habituated coati bands of known individuals provided the data for this study.

Adult female coatis protect and care for each other's young through foraging within a protective spatial organization, participating in cooperative vigilance and defense against predators, and allogrooming and occasionally nursing them during rest periods.

The contributions of adults to the care and protection of juveniles were assessed for effects of kin selection. No preferences for kin were documented, despite strong contrasts between the principal study groups in the degree of genetic relatedness among adults and juveniles.

Cases of social migration were analyzed to detect dependence of the maintenance of relationships on equitability, an important requirement for reciprocity. Several lines of evidence indicate the existence and importance of such dependence among coatis.

The advantageous of gregariousness in coatis depends largely on the benefits of altruistic services that adult females provide each other in caring for and protecting their offspring. Such altruism might theoretically derive from any or all of three alternative evolutionary mechanisms: group selection, kin selection, and reciprocity selection. The frequency of migration among coati groups suggests that group selection is an unlikely source of altruism. Kin selection probably operates to some extent in coati groups, but its effects are too weak to document unambiguously in this study. The dependence of the maintenance of social relations on equitability suggests that reciprocity selection has been the most important of the three mechanisms in shaping relationships among adult female coatis. Conflicts and interactions between kin selection and reciprocity selection are briefly discussed, and a speculation on why reciprocity is more important to coatis than kinship is offered.

References

Altmann, J. Observational study of behavior: Sampling methods. *Behaviour*, 1974, *49*, 227–267.

Altmann, S. A. Baboons, space, time and energy. *American Zoologist*, 1974, *14*, 221–248.

Boorman, S. A., and Levitt, P. R. A frequency-dependent natural selection model for the evolution of social cooperation networks. *Proceedings of the National Academy of Sciences*, 1973, *70*, 187–189.

Boorman, S. A., Levitt, P. R. *The genetics of altruism.* New York: Academic Press, 1980.
Ewer, R. F. *The carnivores.* Ithaca, N.Y.: Cornell Univ. Press, 1973.
Flynn, R. *Parasites of laboratory animals.* Ames: Iowa State Univ. Press, 1973.
Fritzell, E. K. Aspects of raccoon social organization. *Canadian Journal of Zoology,* 1978, *56,* 260–271.
Hamilton, W. D. The genetical evolution of social behavior. *Journal of Theoretical Biology,* 1964, *7,* 1–16.
Hamilton, W. D. Geometry for the selfish herd. *Journal of Theoretical Biology,* 1971, *31,* 295–311.
Janzen, D. H. Altruism by coatis in the face of predation by *Boa constrictor. Journal of Mammology,* 1970, *51,* 387–389.
Kaufmann, J. H. Ecology and social behavior of the coati, *Nasua narica* on Barro Colorado Island, Panama. *Univ. of California Publications in Zoology,* 1962, *60,* 95–222.
McToldridge, E. R. Notes on breeding ring-tailed coatis, *Nasua narica* at Santa Barbara Zoo. *International Zoo Yearbook,* 1969, *9,* 89–90.
Mech, L. D. *The wolf: The ecology and behavior of an endangered species.* New York: Natural History Press, 1970.
Pulliam, H. R. On the advantages of flocking. *Journal of Theoretical Biology,* 1973, *38,* 419–422.
Rapaport, A. *Fights, games and debates.* Ann Arbor: Univ. of Michigan, 1960.
Rood, J. P. Banded mongoose males guard young. *Nature,* 1974, *248,* 176.
Rood, J. P. Dwarf mongoose helpers at the den. *Zeitschrift für Tierpsychologie,* 1978, *48,* 277–287.
Rood, J. P. Social system of the dwarf mongoose. In J. F. Eisenberg and D. G. Kleiman (Eds.), *Readings in mammalian behavior. American Society of Mammologists Special Publication Series* A, 1982.
Russell, J. K. *Reciprocity in the social behavior of coatis (Nasua narica).* Unpublished doctoral dissertation, University of North Carolina, Chapel Hill, 1979.
Russell, J. K. Exclusion of adult male coatis from social groups: Protection from predation. *Journal of Mammology,* 1981, *62,* 206–208.
Russell, J. K. Timing of reproductive effort in coatis in relation to fluctuations in food resource availability. In E. G. Leigh (Ed.), *Seasonal rhythms in a tropical forest ecosystem. Barro Colorado.* Washington, D.C.: Smithsonian Institution Press, 1983.
Schaller, G. B. *The Serengeti lion: A study of predator–prey relations.* Chicago: Univ. of Chicago Press, 1972.
Schneider, D. G., Mech, L. D., and Tester, J. R. Movements of female raccoons and their young as determined by radio-tracking. *Animal Behaviour Monograph,* 1971, *4,* 1–43.
Smith, H. J. Behavior of the coati (*Nasua narica*) in captivity. *Carnivore,* 1980, *3,* 88–136.
Trapp, G. R. Comparative behavioral ecology of the ringtail and gray fox in Southwestern Utah. *Carnivore,* 1978, *1,* 3–32.
Treisman, M. Predation and the evolution of gregariousness. *Animal Behaviour,* 1975, *23,* 779–815.
Trivers, R. L. The evolution of reciprocal altruism. *Quarterly Review of Biology,* 1971, *46,* 35–57.
van Lawick, H., and van Lawick-Goodall, J. *Innocent Killers.* Boston: Houghton Mifflin, 1971.
Waser, P. M., and Wiley, R. H. Mechanisms and evolution of spacing in animals. In P. Marler and J. Vandenberg (Eds.), *Handbook of behavioral neurobiology: Social behavior and communication.* New York: 1979.
Williams, G. C. *Adaptation and natural selection: A critique of some current evolutionary thought.* Princeton, N. J.: Princeton Univ. Press, 1966.
Wilson, E. O. *Sociobiology.* Cambridge, Mass.: Harvard Univ. Press, 1975.
Wynne-Edwards, V. C. *Animal dispersion in relation to social behaviour.* Edinburgh: Oliver & Boyd, 1962.

11

Cooperation and Reproductive Competition among Female African Elephants

HOLLY T. DUBLIN

I. Introduction

The African elephant (*Loxodonta africana*) may have one of the most advanced of all mammalian social systems (Wilson, 1975). For elephants, group living provides individuals with defense from predators and enhances their ability to locate resources. Although group membership offers the advantages of cooperation, it also entails the disadvantages of competition for scarce resources and breeding opportunities. Research on African elephant social organization has emphasized female cooperation but

SOCIAL BEHAVIOR OF
FEMALE VERTEBRATES

ISBN 0-12-735950-8

neglected the obvious potential for intrasexual competition. This chapter reviews a number of reported field observations of cooperative and competitive interactions among herd females and identifies factors that may influence the development of elephant behavioral patterns.

Along with unusually long lifespans and broadly overlapping generations, elephants have developed a complex, matriarchal society (Douglas-Hamilton and Douglas-Hamilton, 1975; Moss, 1976). Elephant herds comprise groups of related females and juveniles and calves of both sexes. The herd's activities are directed by the largest elder female, the matriarch. Relative social status of individuals within the herd is determined by kinship lines, age, size, and individual disposition; a rank order hierarchy results. Female offspring remain in their maternal family units or larger kinship groups throughout their lives. This promotes long-lasting bonds between females. Young males, on the other hand, are forcibly excluded from maternal herds at puberty. Their subsequent associations with female relatives and natal herds may be common but are currently not well understood. Adult bulls spend most of their lives alone or in short-lived, loosely organized bachelor herds, joining female herds only intermittently, usually for breeding.

Elephant societies are complex and embody a dynamic balance of affiliative and agonistic interactions between individual herd members. All individuals share the common goals of survival and reproduction, and their cooperative and competitive behaviors are directed towards this end. The behavioral patterns displayed by a specific individual may be determined by a set of factors including age, reproductive state, and physical condition, all of which contribute to the animal's status in the breeding hierarchy. These variables change over time, altering the relative benefits of cooperation and competition for the individual (Chase, 1980). The behaviors expressed may also be directly influenced by environmental factors, including the quality, dispersion, and abundance of forage; water and mineral availability; and predatory pressures.

The balance of this chapter reviews two basic elements of herd social structure: cooperative and competitive interactions between females holding unequal social status. Section II reviews the long-recognized cooperative behavior of herd females. Group defense (Section II,A) and resource acquisition leadership (Section II,B) are discussed first. Then, allomothering is emphasized, and several types of benefits—and costs—to individuals that might result from this behavior are explored (Section II,C).

Section III discusses competitive interactions in elephant herds that would affect the reproductive success of subdominant females. A discussion of the importance of seasonal timing to reproductive success (Section III,A) precedes an exploration of the relationship between social status and breeding opportunity (Section III,B). Social interactions that may affect differential reproductive success among females of unequal rank are also explored (Section III,C). This section does not attempt to encompass all the types of competitive interactions that may exist in the elephant herd hierarchy. The dearth of research on this topic requires that even this limited discussion be speculative and conceptual.

II. Cooperation

Cooperative behaviors among cow elephants involve group defense (Sikes, 1971; Douglas-Hamilton and Douglas-Hamilton, 1975), locating of and leading to food, water, and other resources (Crook, 1971; Leuthold and Sale, 1973; Douglas-Hamilton, 1973; Hanks, 1979), and allomothering (i.e., one female's caring for the young of another) (Woodford and Trevor, 1970; Sikes, 1971; Douglas-Hamilton and Douglas-Hamilton, 1975; Bryceson, 1980).

A. Group Defense

Herd leaders or matriarchs are well known for their willingness to face danger in defense of their herds. A matriarch, usually the largest herd female, readily assumes the front position during confrontation and the rear position in flight. At other times the leaders arrange their herds into a circular defense posture (Kingdon, 1979). This defensive formation protects younger members of the herd by forcing them into the center, a strategy found in other large herbivores, such as the musk oxen (Tener, 1954) and cape buffalo (Sinclair, 1977). The apparent altruism of herd matriarchs may be explained in several ways. Since the young they guard are likely to be close relatives (Douglas-Hamilton and Douglas-Hamilton, 1975), group defense should enhance their inclusive fitness (Hamilton, 1963, 1964) through kin selection (Maynard Smith, 1964). Matriarchs may also gain from reciprocity (Trivers, 1971; Axelrod and Hamilton, 1981) if, as a result, the offspring of females who guard others receive help in the future.

B. Female Leadership

In societies of long-lived and closely bonded individuals, leaders are often repositories of critical information about the surrounding habitat and its changes over both the short and long term. Douglas-Hamilton and Douglas-Hamilton (1975) and Leuthold (1977) suggest that the matriarchs' many years of experience help them guide daily and seasonal movement patterns in search of food, water, and mineral resources.

Most areas in which African elephants evolved are characterized by highly localized, seasonal rainfall. Prime wet and dry season feeding and watering grounds may be widely dispersed within the home range of a family unit. Consequently, elephant herds show distinct wet and dry season ranges that may vary tremendously in size depending on local habitat conditions and human habitation patterns. Within these ranges, elephant families do not show random utilization patterns. Rather, they display a great preference for some areas over others. Movement patterns may be consistent from year to year, with individual herds returning to specific wet and dry season sites repeatedly (Leuthold and Sale, 1973; Douglas-Hamilton and Douglas-Hamilton, 1975; Dublin, unpublished data, 1974–1976), or may be flexible in response to proximate environmental changes (Gorman, 1981).

Female cooperation and coordination allow efficient herd movement to these resources over sporadically used routes. Nutritious forage and daily water are critically important to cow elephants, who gestate or lactate over half of their adult lives. In fact, any relevant, learned information about the surrounding environment will be beneficial to the individual and the family as a whole, including knowledge of potential threats posed by man and other predators (Leuthold, 1977; Curio *et al.*, 1978a, 1978b).

The importance of this information makes older herd females central to the survival of future generations. Long generation times and the substantial overlap of individual life spans creates ample opportunity for the transmission of learned movement patterns from old to young. In elephants, this may include learned responses to such long-term recurrent events as drought (Leuthold and Sale, 1973; Leuthold, 1977) and fire (Vesey-Fitzgerald, 1972). A herd leader's ability to locate and exploit new or spatially and temporally dispersed resources clearly benefits all group members.

C. Allomothering

Allomothering is common among African elephants (Sikes, 1971; Douglas-Hamilton and Douglas-Hamilton, 1975; Moss, 1976). Though older parous females will act as allomothers, young nulliparous females usually assume this role. Allomothers direct calves' movements and assure that calves are not left behind when the family moves away from resting or feeding sites. Surprisingly, lactating mothers also allow suckling by calves other than their own (Douglas-Hamilton and Douglas-Hamilton, 1975; Leuthold, 1977; Kingdon, 1979; Bryceson, 1980; Dublin, unpublished data, 1974–1976). The behavior of allomothers toward their surrogate calves is often so intensely "motherly" that their true relationship is frequently unclear to observers (Kingdon, 1979). By assuming many mundane care responsibilities, allomothers free true mothers to forage at greater distances from the herd and allow them to suckle their young less often.

The temporal stability of families and the overlapping nature of long generations assure herd females ample opportunity to interact with one another, thereby providing a favorable climate for the evolution of allomothering in elephant societies. This apparent family altruism parallels the associations described among cooperative breeders in many long-lived avian (Emlen, 1978; but see Koenig *et al.*, this volume) and mammalian species (Kühme, 1965; Schaller, 1972; Bertram, 1975, 1976; Rasa, 1977; Frame *et al.*, 1979; MacDonald, 1979; Moehlman, 1979; Rood, 1978, 1980). The demographic portrait drawn by Brown (1974) for cooperative breeders includes traits such as low fecundity, delayed sexual maturation, long life spans, low dispersal, and recruitment of mothers' helpers through retention of young. These characteristic traits are evident in several bird species as well as in elephants. However, cross-species generalizations about the determinants of cooperative breeding can be misleading. For example, in some avian species, fledged birds only remain in their parents' territory and help in the rearing of subsequent broods when environmental conditions make their successful dispersal unlikely. In elephants, however, female young usually remain in their maternal herds and help with the care of younger group members.

1. *Potential Benefits of Allomothering*

Widespread and perpetual allomothering among elephants suggests that benefit is realized by one or all of the individuals involved. But what do allomothers gain relative to others who might not assume such energetic burdens? By tending to young other than their own, allomothers stand to increase any or all of the following; (*a*) inclusive fitness, (*b*) likelihood of reciprocation in the future, (*c*) experience in rearing offspring, and (*d*) access to resources.

a. Inclusive Fitness

A potentially important force behind allomothering is the opportunity to enhance one's inclusive fitness through helping close relatives (Hamilton, 1964). Assessment of fitness gains depends on accurate determination of relatedness between actors and recipients. The actual degree of relatedness between elephant allomothers and the young they care for is especially difficult to determine due to their long life spans, long interbirth intervals, and the dearth of long-term individual case histories. However, tentative assumptions regarding relatedness can be inferred from past studies of elephant population biology (Laws and Parker, 1968; Hendrichs, 1971; Douglas-Hamilton and Douglas-Hamilton, 1975; Laws *et al.*, 1975; Hanks, 1979).

Full sibling relationships are unlikely to exist in allomother–calf dyads unless specific males and females meet and breed several times over extended interbirth intervals. Such long-term associations between individual males and females have not been documented in the African elephant and seem unlikely, given the characteristic sexual segregation of elephant society and the frequently extensive dispersal of males.

Another, more likely, mating pattern would produce a genetically lower form of relatedness that has been overlooked in most considerations of allomothering among elephants. Sibships between allomothers may be based upon shared paternity (see also Altmann, 1979). Dominant bulls may obtain the majority of copulations with estrous females after initial contests between males (Buss and Smith, 1966). Because females breed synchronously, a single dominant bull may inseminate several females in the same herd. Female calves born to these herd mates form a cohort relationship. Such relationships may, in fact, form the tightest ties in elephant kinship groups (Douglas-Hamilton and Douglas-Hamilton, 1975). Over the years, the closeness of these peer associations may promote cooperation and reciprocity (Axelrod and Hamilton, 1981).

In promiscuous species where single-offspring litters are the rule, each female produces a series of half-siblings. A third potential genetic basis for allomothering is that of maternal sibship. Several characteristics of elephant family units suggest that this is the most likely form of relatedness between allomothers and the young for whom they provide care. Intercalf intervals have been estimated from demographic data on several elephant populations (Laws *et al.*, 1975; Smuts, 1975; Hanks, 1969,

1979), and consistent patterns of age, intercalf intervals, and placental scars strongly suggest that calves and allomothers frequently share the same mother. Furthermore, most allomothers are significantly younger than the calf's true mother. If allomother–calf dyads are indeed predominantly half-sibships, then by enhancing the probability of survival for their half-siblings, allomothers could profit by increasing their inclusive fitness.

Just as the likelihood of allomothering may vary with the degree of relatedness, it may also vary according to the degree of need (Hamilton, 1964). By providing help at critical times, allomothers might realize large gains in inclusive fitness with a relatively small investment of energy. Periods of illness, injury, malnutrition, or other stress on mother or calf would present these opportunities, as would difficult periods during calf develoment, such as weaning. Cows and calves experience conflict during weaning, like most mammalian mother–young pairs (Trivers, 1974). Mothers frequently resist nursing attempts and often leave newly weaned calves behind when the herd moves on. If allomothers invest their energy prudently, their efforts should intensify during weaning, thereby aiding the dependent calf and freeing the mother to cycle and conceive again.

Support for the hypothesis of differential care with respect to calf development was obtained from several herds at Tsavo West National Park, Kenya (Dublin, unpublished data, 1974–1976). Table I shows that calves in the final stages of weaning, between the ages of 2 and 4 years, interacted significantly more ($p < .01$) with allomothers than those under the age of 2 or over the age of 4 years ($p < .05$).

Allomother–calf interactions vary as a function of the discrepancy in

TABLE I
Dyadic Interactions between Calves and Allomothers

Age class[a] (months)	No. of focal calves[b]	No. of dyadic interactions per observation day[c]	
0–24	19	5.3	} ** (0–24 vs. 25–48)
25–48	18	36.8	} * (25–48 vs. 49+)
49+	20	19.7	

[a]Calves' ages were estimated using methods developed by Sikes (1966, 1968) and Douglas-Hamilton (1973).

[b]Focal sampling by the author of individual calves in Tsavo West National park, 1974–1976.

[c]One observation day was 12 hours of continuous observation, from 0600 until 1800.

$^{*}p < .05$

$^{**}p < .01$

age between them. Where female helpers are close in age or size to smaller calves in the group, allomothering activities often involve play. Play bouts may help educate youngsters in use of their trunks for dusting, mud-bathing, showering, drinking, and feeding. Appropriate responses to external threats may also be taught. Play may also strengthen intraindividual bonds that facilitate cooperation later in life. When the allomother is substantially older than the calf, her behavior is more maternal. Allomothers nurse, bathe, and direct daily movements of the young in these cases. Age-asymmetric associations have been considered and examined more thoroughly than peer-like relationships in studies of elephant social behavior by Douglas-Hamilton and Douglas-Hamilton (1975) and Laws *et al.* (1975).

It is difficult to determine whether the aid given by allomothers helps to increase their inclusive fitness. This question will become more tractable, however, as data accumulate on survival and its correlation to relatedness.

b. Reciprocation

Long-term associations or bonds between related or unrelated individuals are hypothesized to increase the probability that cooperative acts will be reciprocated (Wrangham, 1980; Axelrod and Hamilton, 1981). In elephants, herd females are in contact for many years. Hence, a female who allomothers may be reciprocated for her efforts, either in the present or the future. Immediate reciprocation might include cooperation in defense of her own offspring or aid in locating and acquiring critical resources. Future reciprocation would involve care offered her own offspring, either from other herd females or perhaps from calves she has allomothered. The remaining question for reciprocity is whether the fitness of allomothers is increased through these cooperative acts beyond that of herd females who do not allomother. Data presented in Section II,C,1,d indicate that allomothers are treated differently by mothers when they are with or without the calf. The information needed to resolve this question would be a long-term tally of cooperative acts between individuals from which the degree of reciprocity might be measured directly.

c. Experience

Young females also may profit through caring for the young of another cow if the experience increases the likelihood of successfully rearing their own offspring (Spencer-Booth, 1970; Lancaster, 1972).

The cost to allomothered infants has been examined in Japanese monkeys (Kurland, 1977), langurs (Hrdy, 1976, 1979), and baboons (Wasser, this volume) and found to be high in some cases. Although primate infants may suffer during allomothering from rough handling, separation from the mother, or decreased access to high quality forage or milk, elephant calves do not. Whereas allomothers could gain experience at the expense of another female's fitness in primates, such exploitative strategies are less likely in elephants because elephant allomothers are not permitted ready access to extremely young calves. Costs of allomothering are discussed in Section II,C,2.

d. Resource Access

Perhaps the most immediate benefit to allomothers is their improved access to resources. Frequently, subdominant females are denied access to certain preferred foods or scarce resources (Douglas-Hamilton and Douglas-Hamilton, 1975; Dublin, unpublished data, 1974–1976). When in the company of a dominant female's young, however, these same subdominants are allowed greater access to water, mineral licks, and favored foods. When not accompanied by a dominant's calf, subdominant allomothers are often forcibly excluded from these resources (Table II).

Elephants live in environments that are periodically resource-limited. If maintaining proximity to dominants' calves provides substantial benefits to allomothers in the manner just described, one would expect a strong inverse correlation between allomothering frequency and habitat quality.

TABLE II
Access to Valuable Resources for Subdominant Females with and without the Accompaniment of a Dominant's Calf

	Dominant female		
Subdominant female	Allows access	Prevents access	Total attempts
With calf	110	18	118
Without calf	9	75	84
Total responses	119	93	212

$\chi^2 = 121.3$, $p < .001$

2. *Potential Costs of Allomothering*

Allomothering may be costly as well as beneficial to the allomother. Allomothering activities consume time and energy. Directing or restraining the movements of young calves and helping them to procure food adds an additional burden to the self-maintenance energy budgets of adolescent cows, as it does for helpers in any species (Hamilton, 1964; Emlen, 1978). Thus, nulliparous or nonlactating females caring for other females' young may have insufficient physical reserves remaining to produce their own offspring in a given year. Consequently, their reproduction may be delayed for reasons other than sexual immaturity alone. Impaired physical condition may result in delayed maturation, failure to ovulate, increased failure of implantation, or heavy *in utero* or postpartum mortality. Poor body condition is known to cause deferred reproduction in mammalian species of many different orders (Sadleir, 1969) including ungulates (Preobrazhenskii, 1961; Wiltbank *et al.*, 1962; Klein, 1970; Lamond, 1970; Geist, 1971; Sinclair, 1977; Belonje and von Niekirk, 1975) and humans (Frisch and McArthur, 1974; Pond, 1978).

Lactating subdominants in the herd will allow the calves of older females to suckle (Douglas-Hamilton and Douglas-Hamilton, 1975; Kingdon, 1979; Bryceson, 1980; Leuthold, 1977). And unconfirmed observations of calves being nursed by nulliparous females have been reported in captive and wild elephants. This suckling may result in lactational anestrus and thus reduce the wet nurse's reproductive capabilities during that time (Kingdon, 1979).

III. Competition

In opposition to the proposed benefits of cooperative interactions among group members, females may also sustain costs from intrasexual competition. Past emphasis on male–male competition in many sociobiological analyses may have diverted attention from the form and function of competition among females, which by nature may be less overt. The most important ramifications of female–female competition (which occurs in various forms) are those that serve to increase the variance in reproductive success of individuals (Kleiman, 1979). Over a female elephant's extensive reproductive lifetime, her competitive interactions will vary as a function of her age, reproductive state, and social status.

Reproductive hierarchies may govern the ways by which variable reproductive success is mediated.

A. The Importance of Timing to Reproductive Success

Successful reproduction in elephants requires adequate food, water, and shade during pregnancy and lactation (Douglas-Hamilton and Douglas-Hamilton, 1975). The timing of breeding is critical for ensuring that females can enhance their own condition prior to the time of conception and bear their young at the most propitious time of year (Williamson, 1976). Females in poor nutritional condition often do not conceive. Young born in the height of the rains are threatened by exposure to severe weather conditions. Those born too late in the dry season face shortages of high quality forage (Laws *et al.*, 1975). As a result, most elephant populations show some degree of breeding seasonality around the rainy period (Laws *et al.*, 1975; Hanks, 1979). Areas with bimodal rainfall distribution experience bimodal peaks of breeding activity (Laws *et al.*, 1975; Kingdon, 1979). Peak breeding activity occurs 1 or 2 months after the onset of the rains. This ensures that 22 months later births will occur either during the short rains or during the onset of the long rains (Hanks, 1969) (Fig. 1). Intrasexual competition for access to breeding opportunities may be expected during these optimal times.

B. Social Status and Breeding Opportunities

Females vary in their relative abilities to acquire the necessary resources for successful reproduction. These resources may include food, water, and mates. Competitive superiority in female elephants appears to be directly related to the individual's status in the breeding hierarchy. The underlying determinants of individual status among elephants are not well understood, but age, size, maternal rank, and the individual's disposition are important (Sikes, 1971; Kingdon, 1979). Large, old cows tend to be dominant individuals, and the largest is usually the matriarch.

In this system, young subdominant females are likely to be the poorer competitors and therefore experience lower reproductive success. Within a given birth season, distinct separation of parturition appears between dominant and subdominant females. In my 1974–1976 Tsavo West study (Dublin, unpublished data), subdominant females gave birth later, on

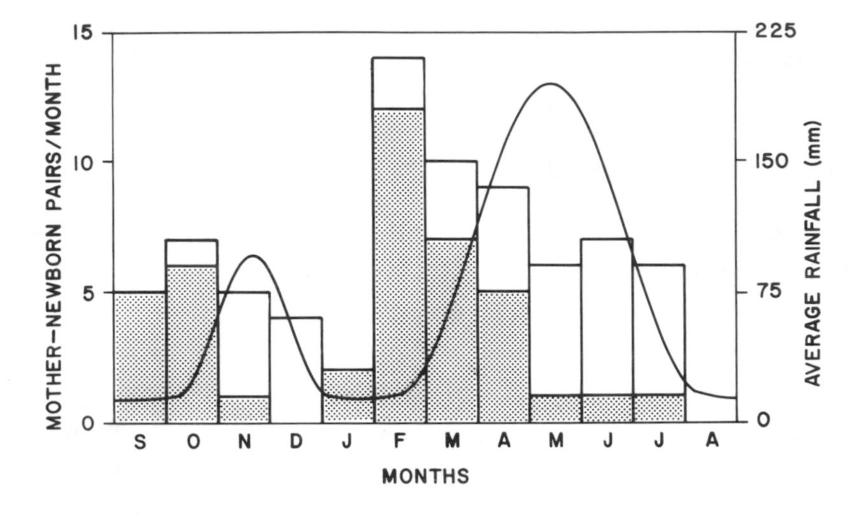

Fig. 1 Distribution of births to dominant (shaded bars) and subdominant (white bars) females in Tsavo West National Park herds in relationship to average monthly rainfall. Social status was determined from dyadic interactions.

average, than older, dominant females (Fig. 1). During the long rains, calves born after the peak in mid-March incurred higher mortality than those born before the peak. Of 15 born after the rainfall peak, 8 (53%) died compared to 5 out of 23 (22%) of those born before the peak ($\chi^2 = 4.03$, $p < .05$). In contrast, during the short rains there was no significant difference in mortality between prepeak and postpeak births. Both groups experienced very low mortality.

Given the unpredictability of resources over the long gestation and preweaning periods of elephants, subdominant females may experience reproductive failure if they cannot breed during peak rainfalls. Where high population densities and/or scarce resources compound this underlying unpredictability, natural selection should favor those females that defer their reproductive efforts until circumstances improve. As resources become more abundant or females gain higher status in the hierarchy, they may acquire greater control over the timing of their own reproduction. This may help explain interpopulational variation in ages at first ovulation (Table III).

In many vertebrate species, young females confronted with high intrasexual competition merely disperse from the maternal home range. However, emigration from the maternal herd is not a viable option for young female elephants. In a social system in which survival depends heavily on cooperative group efforts for protection and location of food,

TABLE III
Average Age at First Ovulation in Free-ranging African Elephant Populations

Population (Country)	Age (years)	Source
Murchison Falls (Uganda)	7	Buss and Smith (1966)
Wankie (Rhodesia)	11	Williamson (1976)
Kruger (South Africa)	12	Smuts (1975)
Tsavo (Kenya)	12	Laws *et al.* (1975)
Mkosmasi (Tanzania)	13	Laws *et al.* (1975)
Luangwa (Zambia)	14	Hanks (1972, 1979)
Murchison Falls North (Uganda)	17	Laws *et al.* (1975)
Murchison Falls South (Uganda)	19	Laws *et al.* (1975)
Budongo Forest (Uganda)	23–25	Laws *et al.* (1975)

young animals are probably unable to survive on their own, or alternatively, to gain access to other herds. Apparently, intergroup segregation is maintained by aggression, which prevents unfamiliar animals from readily joining new groups (C. Moss, personal communication, 1981).

C. Social Mechanisms Affecting Differential Reproductive Success

Female–female competition among individuals takes different forms, expressed overtly at some times and more subtly at others. Kleiman (1980) reviewed the possible methods a dominant female may employ to maintain her reproductive superiority and those that a subdominant may adopt to prevent untimely investment in reproduction. In most cases, the exact physiological mechanism controlling reproductive delay or disruption is not apparent. However, delayed, suppressed, or infertile ovulations; failed implantations; and spontaneous abortions are all possibilities. Evidence from a number of mammalian species suggests that social interactions may mediate the inhibition of subordinate reproduction more often than was formerly realized (see also Wasser; Silk and Boyd, this volume). Inhibition in the presence of their mothers is experienced by juvenile female cactus mice (Skryja, 1978), microtine rodents (Zejda, 1961, 1967; Batzli *et al.*, 1977), common marmosets (Abbott and Hearn, 1978; Kleiman, 1979), and Mongolian gerbils (Payman and Swanson, 1980). Delay of juvenile reproduction in the presence of dominant females living in hierarchical breeding communities has been found in house mice (Drickamer, 1974a, 1979, 1980; Massey and Vandenbergh, 1980), prairie deer mice (Terman, 1969, 1980; Lombardi and Whitsett, 1980; Lloyd, 1980;

Christian, 1970, 1980), rhesus monkeys (Drickamer, 1974b, 1980), dwarf mongoose (Rood, 1980), European badgers (Ahnlund, 1980), wolves (Zimen, 1971; Altmann, 1974; Packard and Mech, 1980), patas monkeys (Rowell, 1978), gelada baboons (Alvarez, 1973; Dunbar, 1979, 1980; Dunbar and Dunbar, 1977), and wild dogs (Frame and Frame, 1976; Frame *et al.*, 1979).

In resource-limited conditions, dominant female elephants may use their competitive superiority to suppress the reproductive efforts of subdominants (Sikes, 1971). Inhibiting the reproduction of subdominant individuals would protect the dominants' own reproductive investment by reducing competition for themselves and their subsequent newborn calves.

Correlations between social rank and reproductive demographics of elephant family units have not been established. However, physiological inhibition of reproduction may be triggered by social stresses imposed on subdominant females by their mothers and other higher-ranking females (Sikes, 1971). In addition to physiological mechanisms, dominants may also limit subdominants' access to fertile mates, thereby further reducing reproductive opportunity. Potential methods of reproductive inhibition are discussed in the following paragraphs.

1. *Delaying or Inhibiting Ovulation*

If stress is a significant factor in delaying or inhibiting estrous cycles, dominants may effectively influence the reproductive efforts of subdominant individuals through directed agonistic behaviors and other forms of social harassment. Estrous cycles may subsequently fall out of synchrony with the optimal breeding periods discussed in Sections III,A and III,B. If estrus in the African elephant resembles that reported for the Asian elephant, *Elephas maximus*, fertile periods are very short and are separated by periods of infertility lasting up to 4 months (Schmidt, personal communication, 1981). Any delay or interruption of the ovulatory cycle may require a cow to cycle one or several more times before successful conception. The interovulation period may be shorter in the African elephant (Short, 1966). Nonetheless, delay could still result in reproductive readiness during suboptimal periods. For example, by the time a cow recycles, cow–calf herds may have already returned to dry season ranges far from breeding males. Also, decreased habitat quality late in the season would lower the physical condition of females and further delay estrus (Laws and Parker, 1968; Sadleir, 1969). Additionally, if conception occurs in the dry season, with parturition following 22 months later, calves would be born under extremely poor conditions, and increased mortality would be expected (see Section III,A and III,B) (Laws *et al.*, 1975; Guinness, Albon, and Clutton-Brock, 1978; Guinness, Clutton-Brock, and Albon, 1978).

2. *Social Aggression*

Seasonal changes observed in female–female interactions support the prediction that dominant individuals influence the reproductive efforts of subdominants through agonistic behavior. In four Tsavo West herds, dominant females significantly increased their agonistic encounters with subdominant females, but significantly decreased their agonism towards adolescent males during the peak breeding period (Table IV). Subdominant females may represent a direct competitive threat for mate access (see Section III,C,4) and optimally timed breeding opportunities, whereas young bulls compete only for food. Herd males, due to their relatively higher energetic demands, may be a greater threat to herd matriarchs and their young outside the breeding season when they vie for access to food, water, and mineral deposits.

If stress can significantly reduce reproductive success, increased aggression could delay or inhibit fertile ovulations. Circumstantial evidence supports this hypothesis (see Section III,C,3).

3. *Inducing Infertile Ovulations*

In 350 B.C., Aristotle noted physiological evidence that first suggested that infertile ovulations occur in elephants. Sikes (1971) noted that adult female elephants and nulliparous or young cows frequently possess

TABLE IV
Differential Aggression of Dominant Females toward Male and Female Subdominants between Prepeak and Peak Breeding Periods

	Subdominant females			Subdominant males		
Herd	n	Trend in aggression	p-value[a]	n	Trend in aggression	p-value[a]
A	5	Increase	<.01	5	Decrease	<.05
B	4	Increase	<.05	3	Decrease	<.01
C	7	Increase	<.001	6	No change	<.10
D	3	No change	<.10	6	Decrease	<.05

[a]Data were analyzed using a paired t-test on observations of known individuals before and during the peak breeding period.

multiple accumulated corpora lutea (Sikes, 1971). These corpora lutea vary in size, shape, and number (Buss and Smith, 1966). Up to 26 have been observed in a single female (Hanks, 1979). Short (1966, 1972); Smith *et al.* (1969); and Hanks (1979) speculated that a "critical mass" of luteal tissue may be required before conception is possible. Fully formed, these corpora lutea have a degenerate appearance and show little or no hormonal activity. Progesterone, the hormone produced by corpora lutea for maintaining pregnancy, is present in only minute amounts relative to that of other mammalian species studied (Short, 1966, 1972; Smith *et al.*, 1969; Sadleir, 1969; Hanks and Short, 1972; Hanks, 1979).

If the abundance or size of corpora lutea masses in individuals is negatively correlated with status in the dominance hierarchy, corpora lutea accumulations may represent an infertile state resulting from social inhibition, as hypothesized by Sikes (1971). These data, however, are presently unavailable. Herds in marginal habitats or dense populations would be most likely to show the correlation if intrasexual competition is heightened under resource limitations.

4. *Limiting Access to Mates*

Opportunities for subdominant females to mate successfully in a given breeding season may also be controlled by dominant females. These females lead their herds from dry season home ranges to prime rainy season sites, where the majority of matings take place (Laws *et al.*, 1975; Kingdon, 1979). Due to their size and social status, dominant females have access to the best forage on the breeding grounds. Consequently, these females are in the best physical condition. They also are the most experienced breeders. Reproductively dominant females thus enter estrus earlier (Smuts, 1975; Hanks, 1979) and breed first. The early breeding efforts of bulls may be monopolized by dominant females. The reduced breeding opportunity of subdominant females may be exacerbated by a resulting temporary loss of libido in breeding males (Schmidt, personal communication, 1981) and perhaps reduced fertility following multiple copulations.

Females may also compete amongst themselves through prolonged sexual receptivity (see Wasser, this volume). In captive populations female elephants will allow mounting and copulation even when unable to conceive. This includes copulations during pregnancy, lactational anestrus, and anovulatory heats (Schmidt, personal communication, 1981). In this way, reproductively dominant females can compete, at little cost to themselves, by monopolizing the reproductive activities of eligible males. The occurrence of copulations outside of fertile periods has been reported in a spectrum of vertebrates including herons (Gladstone, 1979), primates (Kleiman and Mack, 1977; Hrdy, 1977; Rowell, 1978, Wasser, this volume), and lions (Bertram, 1975).

IV. Conclusions

Recent interest in the evolution of female social strategies has drawn attention to the matriarchal system of the African elephant. The relatively long and overlapping lifespans of individual elephants have provided an evolutionary basis for transmission of knowledge between generations and mediation of elaborate social controls between group members. Such advanced forms of social evolution have been most widely reported among primate species and have been attributed primarily to advanced brain development. However, the long lives of elephants, relative to those of other mammals, may have contributed as much to elephants' social development as intelligence alone.

Elephant societies have traditionally been viewed as the epitome of female cooperation. Emphasis has been placed on the overt cooperative acts for which female elephants are well known—group defense and allomothering. Less attention has been directed toward competition among these same family members. Yet the importance of considering both cooperative and competitive interactions has become apparent. As Kingdon (1979, p. 65) succinctly stated with regard to elephants, "Interdependence and mutual support under some conditions are not necessarily incompatible with members being competitive in other situations."

For elephants, selective pressures for transmission of knowledge about spatio-temporal resource distribution, coupled with predator defense, lend weight to the adaptive value of group cooperation. On the other hand, competitive interactions are also likely in a species living under periodic resource constraints. In a dominance-ranked social system in which sociality is essential, competition will be felt most severely by young or subdominant females. Evidence suggests that this competition may have its greatest cost to low-ranking females through social inhibition of their reproductive efforts. The social interactions that lead to suppressed, delayed, or disrupted reproductive attempts can be obvious or subtle but may be the most potent form of intrasexual competition among females.

Among elephants, where individual females may be associated for decades, herd integrity clearly requires a delicate balance of cooperation and competition. Understanding the mechanisms that govern coordination, cooperation, and competition and their associated costs and benefits to individual herd members is thus an important area for further research.

V. Summary

The African elephant (*Loxodonta africana*) has a complex matriarchal social system from which male elephants are excluded at puberty. Female elephants need to live in groups, and their longevity has created conditions favorable for the evolution of complex social interactions among herd members. The nature of these interactions and the factors molding their development are discussed.

Cooperative behaviors include group defense from predators, leadership, and allomothering. It is not now possible to assess the relative importance of kin selection and reciprocity in elephant social evolution because these behaviors have not been analyzed with respect to genetic relatedness. Considering the long-term associations among herd females, both may prove important.

Perhaps because of their subtlety, competitive behaviors among females have received less attention from biologists than have cooperative behaviors. Dominant females may reduce the reproductive success of lower-ranking females either by outcompeting them for resources necessary for reproduction or by direct behavioral intervention. Superior access to limited resources enables dominant females to reach breeding condition first and to breed at optimal times. Direct behavioral intervention may delay, inhibit, or interrupt fertile ovulation, as well as limit subdominants' access to mates. Understanding the negative consequences of matriarchal membership for subdominant females and the evolutionary factors promoting cooperative behaviors requires data that span individual lifetimes.

Acknowledgments

I would like to thank J. B. Sale, formerly of the University of Nairobi, and E. C. Goss, of the Kenya National Parks, for helping me pursue my early interest in the behavioral ecology of African elephants. Since then, many researchers have openly and enthusiastically shared their thoughts and observations on elephant social dynamics. I would like to thank them all for helping me to synthesize and articulate the ideas presented in this chapter. Finally, I am grateful to S. K. Wasser, S. D. West, J. F. Wittenberger, J. S. MacCready, and an outside reviewer for patiently and critically reading and commenting on various drafts of this chapter.

References

Abbott, D. H., and Hearn, J. P. Physical, hormonal, and behavioral aspects of sexual development in the marmoset monkey, *Callithrix jaccus. Journal of Reproduction and Fertility*, 1978, *53*, 155–166.

Ahnlund, H. Sexual maturity and breeding season of the badger (*Meles meles*) in Sweden. *Journal of Zoology, London*, 1980, *190*, 77–95.

Altmann, D. Beziehungen zwischen sozialer Rangordnung und Jungenaufzucht bei *Canis lupus* L. *Zoological Gartens*, 1974, *44*, 235–236.

Altmann, J. Age cohorts as paternal sibships. *Behavioral Ecology and Sociobiology*, 1979, *6*, 161–164.

Alvarez, F. Periodic changes in bare skin areas of *Theropithecus gelada. Primates*, 1973, *14*, 195–199.

Axelrod, R., and Hamilton, W. D. The evolution of cooperation. *Science*, 1981, *211*, 1390–1396.

Batzli, G. O., Getz, L. L., and Hurley, S. S. Suppression of growth and reproduction of microtine rodents by social factors. *Journal of Mammalogy*, 1977, *58*, 583–591.

Belonje, P. C., and von Niekirk, C. H. A review of the influence of nutrition upon the oestrous cycle and early pregnancy of the mare. *Journal of Reproduction and Fertility. Supplement*, 1975, *23*, 167–169.

Bertram, B. C. R. Social factors influencing reproduction in wild lions. *Journal of Zoology, London*, 1975, *177*, 463–482.

Bertram, B. C. R. Kin selection in lions and in evolution. In P. P. G. Bateson and R. A. Hinde (Eds.), *Growing points in ethology*. London/New York: Cambridge Univ. Press, 1976.

Brown, J. L. Alternate routes to sociality in jays with a theory for the evolution of altruism and communal breeding. *American Zoologist*, 1974, *14*, 63–80.

Bryceson, D. N. Visions of paradise. *Animal Kingdom*, 1980, *83*, 24–34.

Buss, I. O., and Smith, N. S. Observations on reproduction and breeding behavior of the African elephant. *Journal of Wildlife Management*, 1966, *30*, 375–388.

Chase, I. D. Cooperative and noncooperative behavior in animals. *American Naturalist*, 1980, *115*, 827–857.

Christian, J. J. Social subordination, population density and mammalian evolution. *Science*, 1970, *168*, 84–90.

Christian, J. J. Endocrine factors in population regulation. In M. N. Cohen, R. S. Malpass, and H. G. Klein (Eds.), *Biosocial mechanisms of population regulation*. New Haven, Conn.: Yale Univ. Press, 1980.

Crook, J. H. Sources of cooperation in animals and man. In J. F. Eisenberg and W. S. Ripley (Eds.), *Man and Beast, Smithsonian Annual III*. New York: Random House (Smithsonian Inst. Press), 1971.

Curio, E., Ernst, U., and Vieth, W. Cultural transmission of enemy recognition: one function of mobbing. *Science*, 1978a, *202*, 889–901.

Curio, E., Ernst, U., and Vieth, W. The adaptive significance of avian mobbing. II. Cultural transmission of enemy recognition in blackbirds: effectiveness and some constraints. *Zeitschrift für Tierpsychologie*, 1978b, *48*(2), 184–202.

Douglas-Hamilton, I. On the ecology and behavior of the Lake Manyara elephants. *East African Wildlife Journal*, 1973, *11*, 401–403.

Douglas-Hamilton, I. and Douglas-Hamilton, O. *Among the elephants*. Glasgow: Collins, 1975.

Drickamer, L. C. Sexual maturation of female house mice: social inhibition. *Developmental Psychobiology*, 1974a, *7*, 257–265.

Drickamer, L. C. A ten-year summary of reproductive data for free-ranging *Macaca mulatta. Folia Primatolica,* 1974b, *21,* 61–80.

Drickamer, L. C. Acceleration and delay of first estrus in wild *Mus musculus. Journal of Mammalogy,* 1979, *60,* 215–216.

Drickamer, L. C. Social cues and reproduction: rodents and primates. In M. N. Cohen, R. S. Malpass, and H. G. Klein (Eds.), *Biosocial mechanisms of population regulation.* New Haven, Conn.: Yale Univ. Press, 1980.

Dunbar, R. I. M. Structure of gelada baboon reproductive units. I. Stability of social relationships. *Behaviour,* 1979, *74,* 72–87.

Dunbar, R. I. M. Demographic and life history variables of a population of gelada baboons. *Journal of Animal Ecology,* 1980, *49,* 485–506.

Dunbar, R. I. M., and Dunbar, E. P. Dominance and reproductive success among female gelada baboons. *Nature,* 1977, *266,* 351–352.

Emlen, S. T. The evolution of cooperative breeding in birds. In J. R. Krebs and N. B. Davies (Eds.), *Behavioural ecology: an evolutionary approach.* Sunderland, Mass.: Sinauer Associates, 1978.

Frame, L. H., and Frame, G. W. Female African wild dogs emigrate. *Nature,* 1976, *263,* 227–229.

Frame, L. H., Frame, G. W., Malcolm, J. R., and van Lawick, H. Park dynamics of African wild dogs (*Lycaon pictus*) on the Serengeti plains, 1967–1978. *Zeitschrift für Tierpsychologie,* 1979, 50(3), 225–249.

Frisch, R. E., and McArthur, J. W. Menstrual cycles: fatness as a determinant of minimum weight for height necessary for their maintenance or onset. *Science,* 1974, *185,* 949–951.

Geist, V. *Mountain sheep. A study in behavior and evolution.* Chicago: Univ. of Chicago Press, 1971.

Gladstone, D. E. Promiscuity in monogamous colonial birds. *American Naturalist,* 1979, *114,* 545–557.

Gorman, J. Elephant watching. *Discover,* April 1981, 72–75.

Guinness, F. E., Albon, S. D., and Clutton-Brock, T. H. Factors affecting reproduction in red deer (*Cervus elaphus*) hinds on Rhum. *Journal of Reproduction and Fertility,* 1978, *54,* 325–334.

Guinness, F. E., Clutton-Brock, T. H., and Albon, S. D. Factors affecting calf mortality in red deer (*Cervus elaphus*). *Journal of Animal Ecology,* 1978, *47,* 817–832.

Hamilton, W. D. The evolution of altruistic behavior. *American Naturalist,* 1963, *97,* 354–356.

Hamilton, W. D. The genetical evolution of social behavior. I and II. *Journal of Theoretical Biology,* 1964, *7,* 1–52.

Hanks, J. Seasonal breeding of the African elephant in Zambia. *East African Wildlife Journal,* 1969, *7,* 167.

Hanks, J. Reproduction of elephants *Loxodonta africana* in the Luangwa Valley, Zambia. *Journal of Reproduction and Fertility,* 1972, *30,* 13–26.

Hanks, J. *The struggle for survival: the elephant problem.* New York: Mayflower Books, 1979.

Hanks, J., and Short, R. H. V. The formation and function of the corpus luteum in the African elephant, *Loxodonta africana. Journal of Reproduction and Fertility,* 1972, *29,* 79–89.

Hendrichs, H. Frielandbeobachtungen zum sozial-system des Afrikanischen elephanten, *Loxodonta africana* (Blumenbach 1797). In H. Hendrichs and U. Hendrichs (Eds.), *Dik-dik und Elefanten.* Munich: Piper, 1971.

Hrdy, S. B. The care and exploitation of nonhuman primates by conspecifics other than the mother. *Advances in the Study of Behavior,* 1976, *6,* 101–158.

Hrdy, S. B. *The langurs of Abu: female and male reproductive strategies.* Cambridge, Mass.: Harvard Univ. Press, 1977.

Hrdy, S. B. Infanticide among animals: a review, classification, and examination of the implications for the reproductive strategies of females. *Ethology and Sociobiology,* 1979, *1,* 13–40.

Kingdon, J. *East African mammals: an atlas of evolution in Africa.* New York: Academic Press, 1979.

Kleiman, D. G. Parent–offspring conflict and sibling competition in a monogamous primate. *American Naturalist,* 1979, *114,* 753–760.

Kleiman, D. G. The sociobiology of captive propogation. In M. E. Soulé and B. A. Wilcox (Eds.), *Conservation Biology.* Sunderland, Mass.: Sinauer Associates, 1980.

Kleiman, D. G., and Mack, D. S. A peak in sexual activity during midpregnancy in the golden lion tamarin, *Leontopithecus rosalia* (Primates: Callithricidae). *Journal of Mammalogy,* 1977, *58,* 657–660.

Klein, D. R. Food selection by North American deer and their response to overutilization of preferred plant species. In A. Watson (Ed.), *Animal populations in relation to their food supply.* Oxford: Blackwell, 1970.

Kühme, W. Communal food distribution and division of labour in African hunting dogs. *Nature,* 1965, *205,* 443–444.

Kurland, J. A. Kin selection in the Japanese monkey. *Contributions to Primatology,* 1977, *12,* 1–150.

Lamond, D. R. The influence of nutrition on reproduction in the cow. *Animal Breeding Abstracts,* 1970, *38,* 359–372.

Lancaster, J. Play-mothering: the relations between juvenile females and young infants among free-ranging vervet monkeys. In F. Poirier (Ed.), *Primate Socialization.* New York: Random House, 1972.

Laws, R. M., and Parker, I. S. C. Recent studies on elephant populations in East Africa. *Symposium of the Zoological Society of London,* 1968, *21,* 319–359.

Laws, R. M., Parker, I. S. C., and Johnstone, R. C. B. *Elephants and their habitats.* London/New York: Oxford Univ. Press (Clarendon), 1975.

Leuthold, W. *African ungulates: a comparative review of their ethology and behavioral ecology.* Berlin/New York: Springer-Verlag, 1977.

Leuthold, W., and Sale, J. B. Movements and patterns of habitat utilization of elephants in Tsavo National Park, Kenya. *East African Wildlife Journal,* 1973, *11,* 369–384.

Lloyd, J. A. Interaction of social structure and reproduction in populations of mice. In M. N. Cohen, R. S. Malpass, and H. G. Klein (Eds.), *Biosocial mechanisms of population regulation.* New Haven, Conn.: Yale Univ. Press, 1980.

Lombardi, J. R., and Whitsett, J. M. Effects of urine from conspecifics on sexual maturation in female prairie mice, *Peromyscus maniculatus bairdii. Journal of Mammalogy,* 1980, *61,* 766–768.

MacDonald, D. W. "Helpers" in fox society. *Nature,* 1979, *282,* 69–71.

Massey, A., and Vandenburgh, J. G. Puberty delay by a urinary cue from female house mice in feral populations. *Science,* 1980, *209,* 821–822.

Maynard Smith, J. Kin selection and group selection. *Nature,* 1964, *201,* 1145–1147.

Moehlman, P. Jackal helpers and pup survival. *Nature,* 1979, *277,* 382–383.

Moss, C. *Portraits in the wild: behavior studies of East African mammals.* Boston: Houghton Mifflin, 1976.

Packard, J. M., and Mech, L. D. Population regulation in wolves. In M. N. Cohen, R. S. Malpass, and H. G. Klein (Eds.), *Biosocial mechanisms of population regulation.* New Haven, Conn.: Yale Univ. Press, 1980.

Payman, B. C., and Swanson, H. H. Social influence on sexual maturation and breeding in the Mongolian gerbil (*Meriones unguiculatus*). *Animal Behaviour,* 1980, *28,* 528–535.

Pond, C. M. Morphological aspects and the ecological and mechanical consequences of fat deposition in wild vertebrates. *Annual Review of Ecology and Systematics,* 1978, *4,* 519–570.

Preobrazhenskii, B. V. Management and breeding of reindeer. In P. S. Zhigunov (Ed.), *Reindeer husbandry.* Springfield, Va.: U.S. Department of Commerce, 1961.

Rasa, O. A. E. The ethology and sociology of the dwarf mongoose (*Helogale undulata rufala*). *Zeitschrift für Tierpsychologie,* 1977, *43,* 337–405.

Rood, J. P. Dwarf mongoose helpers at the den. *Zietschrift für Tierpsychologie,* 1978, *48,* 227–287.

Rood, J. P. Mating relationships and breeding suppression in the dwarf mongoose. *Animal Behaviour,* 1980, *28,* 143–150.

Rowell, T. E. How female reproductive cycles affect interaction patterns in groups of Patas monkeys. In D. J. Chivers and J. Herbert (Eds.), *Recent advances in primatology, Vol. I. Behavior.* New York/London: Academic Press, 1978.

Sadleir, R. M. S. *The ecology of reproduction in wild and domestic mammals.* London: Methuen, 1969.

Schaller, G. B. *The Serengeti lion: a study in predator–prey relations.* Chicago: Univ. of Chicago Press, 1972.

Short, R. H. V. Oestrous behavior, ovulation, and the formation of the corpus luteum in the African elephant, *Loxodonta africana. East African Wildlife Journal,* 1966, *4,* 56–68.

Short, R. H. V. Species differences. In C. R. Austin and R. H. V. Short (Eds.), *Reproduction in mammals. 4. Reproductive patterns.* London/New York: Cambridge Univ. Press, 1972.

Sikes, S. K. The African elephant (*Loxodonta africana*): a field method for estimation of age. I. *Journal of Zoology, London,* 1966, *150,* 279–295.

Sikes, S. K. The African elephant (*Loxodonta africana*): a field method for estimation of age. II. *Journal of Zoology, London,* 1968, *154,* 235–248.

Sikes, S. K. *The natural history of the African elephant.* London: Weidenfeld and Nicholson, 1971.

Sinclair, A. R. E. *The African buffalo: a study of resource limitation of populations.* Chicago: Univ. of Chicago Press, 1977.

Skryja, D. D. Reproductive inhibition in female cactus mice (*Peromyscus eremicus*). *Journal of Mammalogy,* 1978, *59,* 543–550.

Smith, J. G., Hanks, J., and Short, R. V. Biochemical observations on the corpora lutea of the African elephant, *Loxodonta africana. Journal of Reproduction and Fertility,* 1969, *20,* 111–117.

Smuts, G. L. Reproduction and population characteristics of elephants in Kruger National Park. *Journal of the South African Wildlife Management Association,* 1975, *5,* 1–10.

Spencer-Booth, Y. The relationship between mammalian young and conspecifics other than the mother and peers: A review. In D. S. Lehrman, R. A. Hinde, and E. Shaw (Eds.), *Advances in the study of behavior.* New York: Academic Press, 1970.

Tener, J. S. A preliminary study of musk-oxen of Fosheim Peninsula, Ellesmere Island, N.W.T. *Canadian Wildlife Service, Wildlife Management Bulletin,* 1954, 1st. series, No. 9.

Terman, R. C. Pregnancy failure in female deermice related to parity and social environment. *Animal Behaviour,* 1969, *17,* 104–108.

Terman, R. C. Behavior and regulation of growth in laboratory populations of prairie deermice. In M. N. Cohen, R. S. Malpass, and H. G. Klein (Eds.), *Biosocial mechanisms of population regulation.* New Haven, Conn.: Yale Univ. Press, 1980.

Trivers, R. L. The evolution of reciprocal altruism. *Quarterly Review of Biology*, 1971, *46*, 35–57.

Trivers, R. L. Parent–offspring conflict. *American Zoologist*, 1974, *14*, 249–264.

Vesey-Fitzgerald, D. Fire and animal impact on vegetation in Tanzanian National Parks. *Proceedings of the Tall Timbers Fire Ecology Conference*, 1972, *11*, 297–319.

Williamson, B. R. Reproduction in female African elephants in Wankie National Park, Rhodesia. *South African Journal of Wildlife Research*, 1976, *6*, 89–93.

Wilson, E. O. *Sociobiology: the new synthesis.* Cambridge, Mass.: Belknap/Harvard Univ. Press, 1975.

Wiltbank, J. N., Rowden, W. W., Ingalls, J. E., Gregory, K. E., and Koch, R. M. Effect of energy level on reproductive phenomena of mature Hereford cows. *Journal of Animal Science*, 1962, *21*, 219–225.

Woodford, M. H., and Trevor, S. Fostering a baby elephant. *East African Wildlife Journal*, 1970, *8*, 204–205.

Wrangham, R. W. An ecological model of female-bonded primate groups. *Behaviour*, 1980, *75*, 262–300.

Zejda, J. Age structure in populations of the bank vole, *Clethrionomys glareolus. Zoologicke Listy*, 1961, *10*, 249–264.

Zejda, J. Mortality of a population of *Clethrionomys glareolus* Shreb. in a bottomland forest in 1964. *Zoologicke Listy*, 1967, *16*, 221–238.

Zimen, E. *Vergleichende Verhaltensbeobachtungen an Wölfen und Königspudeln.* Diss. zum Erlangen des doktorgrades, Christian-Albrechts Univ., Kiel, Germany, 1971.

12

Cooperation, Competition, and Mate Choice in Matrilineal Macaque Groups

JOAN B. SILK

ROBERT BOYD

SOCIAL BEHAVIOR OF
FEMALE VERTEBRATES

ISBN 0-12-735950-8

I. Introduction

Growing theoretical appreciation of the role that female mammals play in mate choice and the evolution of mating systems (Orians, 1969; S. A. Altmann, Wagner, and Lenington, 1977; Emlen and Oring, 1977; Wittenberger, 1980) has stimulated empirical research on the life histories and reproductive strategies of females of many species. Theoretical predictions about the life histories and reproductive strategies of females are particularly difficult to test among primates, whose long and complex lives span the course of most field projects. However, with data from longitudinal studies of a number of species of the genus *Macaca*, some theories about the life histories and reproductive strategies of female macaques can now be evaluated. In addition, the data stimulate a number of new questions about the evolution of behavior patterns observed among members of the genus. In this chapter, we will present a model to account for one ubiquitous and unusual feature of macaque social organization, corporate matrilineal hierarchies among females, and discuss how this form of hierarchical organization may influence the patterns of cooperation, competition, and mate choice among macaque females.

Female macaques typically form linear dominance hierarchies according to three simple rules:

1. Females outrank all unrelated females outranked by their maternal relatives.
2. Adult females rank immediately above their daughters and immature sons.
3. Adult sisters rank in inverse order of their ages.

These patterns were first described by Kawai (1965) among Japanese macaques (*Macaca fuscata*) and have since proven to be characteristic of all species of macaques for which information about maternal kinship and female dominance hierarchies are presently available (*M. mulatta* in Sade, 1972a; Missakian, 1972; Loy and Loy, 1974; *M. radiata* in Silk *et al.*, 1981a; *M. fasicularis* in Angst, 1975; *M. nemestrina* in Massey, 1977; *M. arctoides* in Estrada, 1978; *M. sinica* in Dittus, 1979). The phenomenon codified by Rule 1 causes the female members of a particular matriline to occupy adjacent ranks and creates a hierarchy of matrilineal units. The relationship between matrilineal kinship and female rank is particularly noteworthy because female rank is positively correlated with reproductive success and lineage size (Drickamer, 1974; Sade *et al.*, 1976; Kurland, 1977; Dittus, 1979; Silk *et al.*, 1980; Silk *et al.*, 1981b) and appears to influence secondary sex ratios (Silk *et al.*, in press b).

This chapter is primarily concerned with the evolution of matrilineal hierarchies and not with the evolution of rank relations within lineages. (For a cogent treatment of the latter see papers by Schulman and Chapais [1980] and Chapais and Schulman [1980]). Our explanation of the evolution of matrilineal hierarchies is based upon the assumption that the pattern of hierarchical organization among macaque females results from cooperation among matrilineal relatives. The essence of our argument can be summarized as follows: All other things being equal, large coalitions are likely to be able to defeat smaller ones in competitive encounters. Since rank (a measure of success in competitive encounters) is positively correlated with reproductive success, selection is expected to favor behaviors that allow an individual to recruit support from a large number of individuals. Any behavior that increases the coefficient of relationship between matrilineal relatives will have this effect. Females who exercise mate choice in such a way as to increase the probability that they and other members of their lineages mate with the same males will substantially raise the average coefficient of relationship within their lineages and increase the number of females that they can recruit in their defense. In the remainder of this chapter, we will consider how this mating pattern, which we will refer to hereafter as "lineage-specific female mate choice" (after McMillan, 1979), or LSFMC, might affect the nature of female dominance interactions, the criteria of female mate choice, and the intensity of reproductive competition among females.

II. Kin Selection, Local Population Regulation, and Female Behavior

We will begin by briefly outlining the theoretical foundation upon which many of the ideas considered in this chapter are based. Existing kin selection models share the assumption that mortality is independent of the size or density of the local group. (See Uyenoyama and Feldman [1980] for an excellent review of theoretical studies of kin selection.) However, there is considerable empirical evidence that the size of macaque groups is regulated at least in part by the availability of local resources. Under stable ecological conditions, free-ranging troops of toque macaques (*M. sinica*) maintained a net population growth rate of 0. During a prolonged drought, little food was available, and the population declined by 15.3%. When the drought ended and food became abundant, the population began to grow (Dittus, 1979, 1980). Other evidence of the limiting role of food availability on population growth comes from sites where macaque groups have been artificially provisioned. At Takasakiyama, for example, the resident troop of Japanese macaques was composed of approximately 200 individuals before provisioning began in 1948. By 1972, this group had grown to 1200 individuals (Kurland, 1977). Provisioning of food to rhesus macques (*M. mulatta*) that had been released on Cayo Santiago Island many years earlier contributed to a 16% increase in the size of the population (Koford, 1965). Similar changes in group size have occurred after provisioning was initiated at several other sites (Fujii, 1975; Koyama *et al.*, 1975; Masui *et al.*, 1975; Sugiyama and Ohsawa, 1975). It appears that provisioning enhances female fertility by reducing the length of the interval between successive births (Takahata, 1980).

If natural groups are limited by the availability of food, as these data suggest, then some aspects of mortality in macaque groups must depend on the size of the local group. Boyd (in press) has shown that when mortality depends upon the size of the local group, the appropriate analog of Hamilton's (1964) $k > 1/r$ rule is

$$\sum_i (r_i - h\bar{r})\, b_i > (1 - h\bar{r})\, c_i \qquad (1)$$

where

r_i = the coefficient of relationship between the initiator and the *i*th recipient of the act;
$\bar{r}$ = the average coefficient of relationship of the initiator to all other members of the local group;
h = the percentage change in mortality with a percentage change in local group size;
b_i = the direct incremental effect of the act on the fitness of the *i*th individual affected; and
c_i = the direct incremental effect of the act on the fitness of the initiator.

This rule was derived using a "family-structured" (Abugov and Michod, 1981) population genetics model in which individuals undergo two episodes of mortality; one in which mortality depends upon the behavioral composition of the group, followed by a second in which mortality depends on the size of the group. The parameters b and c are "direct" effects in the sense that they describe only the change in the probability of surviving the episode of social interaction. Relation (1) states the condition under which alleles that cause small changes in behavioral traits can increase their frequency in the population ("invade" hereafter).

Inequality (1) has a very clear intuitive interpretation in terms of inclusive fitness. The left-hand side of the inequality describes the change in the inclusive fitness of the donor due to both direct and indirect effects of the behavior on the ith recipient. The direct effect is $r_i b_i$—the increase in the fitness of the ith recipient weighted by its relationship to the donor. The indirect effect, which is due to density-dependent mortality, represents the product of the change in fitness per individual $(-hb_i/N)$, the number of individuals affected (N), and the average relatedness of each individual $(\bar{r})$, or $(-hb_i/N)N\bar{r}$. This expression is written more simply as $-hb_i\bar{r}$ in Inequality (1). The interpretation of the right side of the inequality is very similar. The direct effect on the fitness of the donor is c, and by an argument similar to the one above, the indirect effect on the inclusive fitness of the donor is $-h\bar{r}$. Thus, Relation (1) requires that for a behavior to be immune to modification by invading alleles there must be no modification of the behavior that increases the inclusive fitness of ego, taking into account both the direct effects of the behavior and the indirect effects due to changes in density-dependent mortality.

Relation (1) predicts that the kinds of behaviors favored by selection will depend on the parameter h, which specifies the effect of group size on mortality. If resource competition does regulate the number (or density) of females in macaque groups, then $h > 0$. The exact value of h will depend on the details of the processes that cause density-dependent mortality. In the appendix, we argue that values of h within the range of .5–2.0 seem plausible for a variety of models. The results obtained below are relatively insensitive to changes in the value of h within this range.

A. Kin Selection and the Formation of Coalitions

When the same formulation is extended to tripartite interactions, it can be used to generate predictions about the circumstances under which a female is expected to form a coalition on behalf of another female. Consider a particular conflict between two individuals labeled 1 and 2. Suppose that the expected effects of this conflict on the fitnesses of 1 and 2 are w_1 and w_2, respectively. Now suppose that another individual, 3, intervenes in favor of 1. In this case, the expected fitnesses of these three individuals will be designated w'_1, w'_2, and w'_3. Under these conditions, the incremental effects of 3's intervention on the fitnesses of individuals 1, 2, and 3 are

$$b_1 = w'_1 - w_1$$

$$b_2 = w'_2 - w_2$$

$$c = -w'_3$$

Then, Relation (1) can be simplified as

$$b_1 r_{13} + b_2 r_{12} - h\bar{r}(b_1 + b_2) > (1 - h\bar{r})c. \qquad (2)$$

It should be noted that this analysis ignores the possibility that repeated interactions can lead to coalition formation via reciprocal altruism (Trivers, 1971). Axelrod and Hamilton (1981) have recently provided a rigorous analysis of dyadic reciprocity, but to our knowledge no similar analysis exists for tripartite interactions. In any case, selection for kin altruism and reciprocal altruism are not competitive processes. If selection favors reciprocal altruism toward nonkin, then it is likely that it will favor even more extensive reciprocal altruism toward kin.

The nature of dominance hierarchies among macaques suggests that two special cases of Inequality (2) will be particularly relevant:

1. Intervention in conflicts over resources. When the major benefit to the winner of the contest is access to some resource, one would expect that $b \approx b_1 \approx -b_2$. That is, individual 3's intervention increases the probability that individual 1 will win the conflict and obtain access to the resource. In this case, Eq. (2) simplifies to

$$b(r_{13} - r_{12}) > (1 - h\bar{r})c. \qquad (3)$$

This inequality has the same qualitative properties as the usual $k > 1/r$ rule.

2. Intervention in "spiteful" contests. We consider dyadic interactions spiteful when the major benefit to one individual is the harm caused to the other individual.* If individual 3 intervenes in defense of the victim of a spiteful act, one might expect that the effects of intervention would be $b_1 \approx b > 0$ and $b_2 \approx 0$. Thus, there is a direct benefit to the victim of the spiteful act and no effect upon the aggressor. In this case, Eq. (2) simplifies to

$$(r_{13} - h\bar{r})b > (1 - h\bar{r})c. \qquad (4)$$

In spiteful contests there is a threshold value of $r_{13} = h\bar{r}$ below which individual 3 is not expected to intervene in favor of a relative.

In both these cases, the coefficient of relationship between individuals 1 and 3 must exceed a threshold value before individual 3 will intervene in her favor. The magnitude of this threshold value depends upon two factors. First, in both cases increasing c/b (the net cost of intervention) increases the threshold value. Second, when the interactions are purely

*This definition of spiteful behavior differs somewhat from the original definition proposed by Hamilton (1971). The relationship between the two definitions is discussed in more detail by Boyd (in press).

spiteful, there is a minimum threshold value of $h\bar{r}$. Even if the intervention is costless to 3, she should not intervene if r_{13} falls below this value. We will show that for $1/2 < h$, even this minimum value can be important for particular matrilines. Given that *c, b, h, and* $\bar{r}$ are fixed for a given interaction, an individual can potentially recruit all those individuals whose coefficient of relationship exceeds the threshold value.*

B. Coalition Formation, Lineage Size, and Reproductive Success

Whereas recruitment into coalitions is expected to be a function of relatedness, the outcome of competitive encounters among groups of females is likely to be a function of the size of the competing groups. We assume (all other things being equal) that a larger coalition can defeat a smaller one. We recognize that often all other things will not be equal. Differences in the rank, health, or age of the members of coalitions may occasionally allow small groups to prevail over larger ones. Nonetheless, we would expect that behaviors that somehow caused individuals to form coalitions of greater than average size would spread in a population because, on average, such individuals would win a higher than average proportion of contests. This, in turn, is likely to lead to higher rank and greater reproductive success among the members of larger lineages.

There is empirical evidence that lineage size does tend to be correlated with lineage rank (Quiatt, 1966; Kurland, 1977; Silk *et al.*, 1981a), although there are several exceptions to this pattern (e.g., Kawamura, 1965; Sade, 1967). Among female bonnet macaques (*M. radiata*) at the California Primate Research Center, the average size of lineages over a 9-year period was more strongly correlated with present rank than present lineage size (Silk *et al.*, 1981a). This suggests that exceptions to the general rule may possibly represent recent or transient changes in lineage size.

Demographic data now available from several sites indicate that rank is positively correlated with reproductive success. Drickamer (1974) found that high-ranking female rhesus macaques on La Cueva Island begin to reproduce at earlier ages, have greater survivorship among their offspring, and produce successive infants at shorter intervals than do low-ranking females. Over a 10-year period, Sade and his colleagues (1976) documented the fact that individuals of high-ranking rhesus macaque lineages

*The ecological basis of the relationship between competition over limited resources, coalition formation, and matrilineal kinship among females has been discussed by Wrangham (1980) in a paper published after the completion of this chapter.

on Cayo Santiago Island grew at significantly faster rates than those of low-ranking lineages. A similar relationship between rank and reproductive success is also evident among toque macaques in Sri Lanka (Dittus, 1979). Finally, among bonnet macaques at the California Primate Research Center, a female's rank significantly influences the probability that her offspring will survive to reproductive age (Silk *et al.*, 1980, 1981b).

III. Lineage-Specific Female Mate Choice

Mate choice by females provides one obvious means of increasing relatedness among females (Bertram, 1978; Seger, 1977). If competition among matrilines favors more closely related and cooperative lineages, selection may favor members of matrilines who choose to mate exclusively with a small number of males. Given the usual pattern of male dispersal among macaques (Drickamer and Vessey, 1973; Sugiyama, 1976), this preferences will quickly raise the average relatedness within a matrilineage over that which would result from random mating. This strategy will be effective only if members of different matrilines tend to mate with different males. (This last assumption will be defended in Section V, A.)

The effect of a system of lineage-specific female mate choice (LSFMC) is illustrated in the right-hand columns of Tables I and II. In these tables, we used genealogies of the Japanese macaques at Kaminyu, drawn from Kurland (1977). We have computed the relationships of two individuals, Nira and Grite, to all other members of their group under two different assumptions about the mating system: (*a*) females mate randomly with males drawn from a large population, and (*b*) all females of a single generation of a lineage mate with the same male. The second pattern, LSFMC, produces the greatest possible increase in relatedness among matrilineal relatives without entailing any inbreeding in groups in which only nonnatal males are reproductively active.

Each table gives the value of r between the focal individual and each other member of the group under the assumption of random mating. The value of $h\bar{r}$ is given at the bottom of the same column, under the assumption that $h = 1$. In the small Gold matriline, the value of r between Grite and even her most distant matrilineal relative is somewhat greater than $h\bar{r}$. The pattern of relatedness for other members of the Gold matriline is essentially the same. Hence, one would expect that in small lineages, all members would cooperate with one another in competition with members of other matrilines. In the larger Point matriline, Nira has a

TABLE I
Relatedness of Nira to Female Members of Her Lineage and to All Females Outside Her Lineage

Females	Random mating[a] r	LSFMC[b] r
Nira	1.00	1.00
Point	.50	.50
Nale	.13	.25
Perv	.13	.25
Ponko	.13	.25
Na-f	.25	.50
Pv-f	.06	.38
Pv-f	.06	.38
Pk-f	.06	.38
Nr-f	.50	.50
Unrelated females ($n = 36$)	.00	.00
Σr	3.06	4.63
$\bar{r} = \Sigma r/47$	.07	.10

[a]Assumes that females mate randomly with available males.
[b]Assumes that all females of a single generation of a lineage mate with the same male.

TABLE II
Relatedness to Grite to Female Members of Her Lineage and to All Females Outside Her Lineage

Females	Random mating[a] r	LSFMC[b] r
Grite	1.00	1.00
Gold	.50	.50
Grey	.25	.50
Gt-f	.50	.50
Gr-f	.13	.25
Unrelated females ($n = 42$)	.00	.00
Σr	2.38	2.75
$\bar{r} = \Sigma r/47$	.05	.06

[a]Assumes that females mate randomly with available mates.
[b]Assumes that all females of a single generation of a lineage mate with the same mate.

number of matrilineal relatives for whom the value of r is substantially greater than $h\bar{r}$, and a smaller number for whom r is approximately equal to $h\bar{r}$. Thus, given the assumed pattern of random mating, one would predict that members of large matrilines would form coalitions with distant relatives only when the costs of intervention were quite low and there was competition over some resource. When the costs of intervention were high or direct access to resources was not at stake, they would be expected to cooperate only with close relatives.

To see the effect of LSFMC, compare the results just described with those shown in Column 2 of each table. LSFMC has the effect of substantially increasing relatedness within both large and small lineages. For example, in the Point lineage, the value of r for Nira and her cousins is now .38, much larger than $h\bar{r}$ and greater than the value achieved under random mating. The minimum relatedness between Nira and all other members of her matriline is .25, well above that normally expected for full cousins. Thus, LSFMC will substantially increase the number of relatives that a female can recruit (except in the case of "resource-type" interactions with very low c:b ratios).

A. LSFMC and the Evolution of Corporate Matrilines

If we are correct in assuming that selection has favored LSFMC, we can provide an explanation of the fact that females of a single matriline occupy adjacent ranks. Under unselective or random mating, there is a smooth decline in relatedness as one increases the genealogical distance from a particular individual. Thus, there should be little difference in the behaviors directed toward distant matrilineal kin and nonkin. On the other

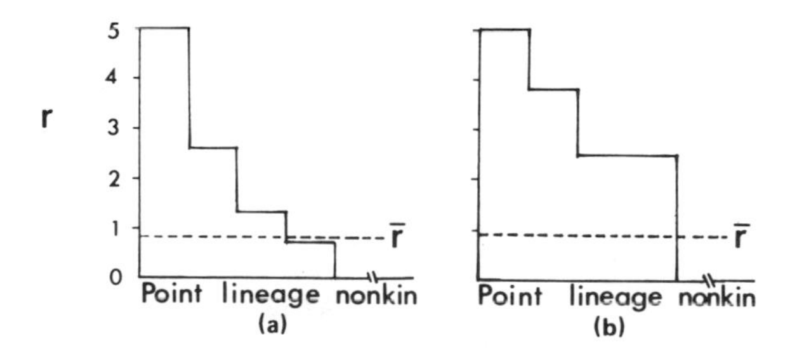

Fig. 1 The values of Nira's relatedness (r, solid line) and average relatedness ($\bar{r}$, dashed line) to the other members of the Point matriline and to all unrelated females under the assumptions of random mating (a) and lineage-specific female mate choice (b) are shown.

hand, the increase in minimum relatedness within lineages due to LSFMC sharply increases the magnitude of the difference in relatedness between matrilineal and nonmatrilineal females (Fig. 1). One would predict that this would lead to a sharp behavioral distinction between matrilineal and nonmatrilineal females, and would produce matrilines that act as corporate units in conflicts that are effectively decided by the relative strengths of the matrilines involved. In this view, members of matrilines are ranked together because the corporate strength of the matriline provides the asymmetry on which the hierarchy is based.

B. Direct Evidence of LSFMC

Support of the LSFMC hypothesis comes from an investigation of the extent of genetic differentiation among lineages of rhesus macaques on Cayo Santiago Island (McMillan, 1979, 1980). Using data on six polymorphic loci published by Olivier, McMillan (1979) computed the "genetic distance" between the matrilineages within five separate groups on the island. Genetic distance is used by population geneticists to measure (among other things) the extent to which subgroups within a population are isolated from each other. If one supposes that females within a group mate only with the males available in the group, one would expect that matrilineages in different groups would be more isolated from each other than matrilineages in the same group. Instead, McMillan (1979) found that the mean genetic distance between the matrilineages within separate groups was greater than the mean distance between the groups. In contrast, the within-lineage distance was comparatively small. This pattern of genetic differentiation within groups suggests either that there are different selective pressures acting on each lineage or that the members of different lineages tend to mate with different males.

McMillan (1979) initially argued that it was unlikely that genetic differences between lineages could be maintained by selection, since all lineages were subject to the same environmental pressures and constraints and were likely to be similarly affected by selection. However, since rank is correlated with access to resources, success in agonistic encounters, and a variety of other factors of considerable adaptive importance, it is conceivable that members of high-ranking lineages do face selection pressures different from low-ranking lineages. McMillan (1980) subsequently reanalyzed her data to determine whether this was the case. If selection were primarily responsible for genetic differences among lineages, high-ranking lineages in different groups would be expected to be more similar genetically than high- and low-ranking lineages within the same group.

Instead, McMillan (1980) found that dominance rank explained virtually none of the variance in genetic distance between lineages, and she concluded that LSFMC was a plausible source of genetic variation between lineages.

C. Behavioral Evidence of LSFMC

Limited behavioral evidence also suggests that LSFMC occurs. Kurland (1977) has described the following mating pattern among Japanese macaques at Kaminyu: "Females in the Faza, Red, and Side matrilines mated preferentially with transient and peripheral males, whereas females in Calm, Knob, and Point matrilines mated more frequently with Dark and Earless" (p. 128). It is particularly noteworthy that Dark and Earless are the highest-ranking resident males and Calm, Knob, and Point are the matriarchs of the three highest-ranking lineages in the group. Fedigan (1976) has also reported that in a large group of captive Japanese macaques, each of the high-ranking males associated regularly with a group of females who often represented an entire lineage.

IV. Effects of LSFMC on Female Behavior

LSFMC will radically alter the patterns of relatedness within macaque groups and may, therefore, affect the evolution of many types of social interactions. In this section, we will outline a number of predictions about the distribution of altruistic and spiteful interactions among females, patterns of paternal investment, and factors influencing the relationships of adult males and females. Several of these predictions are evaluated with data currently available in the literature.

A. Altruistic Interactions

If LSFMC occurs, distant matrilineal relatives will have a higher coefficient of relationship than we would otherwise expect. This is likely to increase the extent of altruism among them. However, S. A. Altmann (1979) has pointed out that knowledge of relatedness is not sufficient to predict the pattern of altruism between two individuals. Detailed quantitative information about the costs and benefits of alternate patterns of behavior is also required. Given that such information is currently unavailable for macaques, it is impossible to use empirical evidence about presumably altruistic behavior to evaluate predictions that would distinguish between hypotheses based upon LSFMC and random mating.

B. Spiteful Interactions

We consider behavior spiteful if the only direct effect of the behavior is to reduce the fitness of both the initiator and the recipient. In terms of Eq. (1), an act is spiteful if $b < c$ and $c > 0$. Since the existence of density-dependent mortality will cause any change in an individual's fitness to have an indirect effect on all other individuals, some spiteful acts can indirectly benefit the initiator. In dyadic interactions, the initiator benefits if

$$(r - h\bar{r})b > (1 - h\bar{r})\ c, \qquad (5)$$

where r is the coefficient of relationship between the two individuals.

There are two reasons to expect that selection should favor spiteful behavior among female macaques. First, given the pattern of relatedness that will result from either LSFMC or random mating and $h > 1/2$, all individuals outside an individual's matrilineage will have $r < h\bar{r}$. These females should, therefore, be the targets of spiteful behaviors according to Eq. (5). This effect will be somewhat stronger with LSFMC. Second, if our argument regarding LSFMC is correct, there is another force favoring spiteful behavior. Recall that we assume (all other things being equal) that large lineages should outrank smaller ones. One method of increasing the relative size of a lineage is to reduce the size of other lineages in the group through spiteful behaviors. We expect that this will be an especially strong force, since a female may thereby increase not only her own reproductive success, but that of her daughters and other matrilineal relatives whose rank (and expected reproductive success) will rise together. In both cases, we expect that spiteful behaviors will be preferentially directed toward females for several reasons: (*a*) adult males outrank most adult females (Sugiyama, 1971; Angst, 1975; Kurland, 1977; Dittus, 1979); (*b*) immature males that will eventually disperse from their natal groups do not provide a permanent source of competition; and (*c*) it is the number of mature females in the lineage that determines its strength.

There is evidence of apparently spiteful aggression toward female macaques at all stages of their reproductive cycles. In one study (Sackett, *et al.*, 1975), it has been discovered that captive pigtail macaque females that require medical treatment (primarily for bite wounds) prior to conceiving subsequently produce signficantly fewer live offspring than females that do not require treatment. In addition, the rate of aggression received by female bonnet macaques prior to conceiving is negatively related to the probability that infants subsequently conceived in the same year by the same females will survive their first 10 days of life (Silk, 1981). These data suggest that aggression prior to conception is related to subsequent reproductive success, but do not prove that aggression by females is directly responsible for reducing the viability of future offspring. Females that are likely to be the targets of aggression may also, for entirely different reasons, be unlikely to produce healthy offspring.

There is evidence of aggression toward females between conception and parturition. Sackett and his collegues (1975) also found that during the second half of gestation, females that are carrying female fetuses require medical treatment three times as often as females carrying male fetuses. Aggression toward pregnant females apparently affects the viability of the fetus, as females who require treatment during pregnancy suffer significantly more spontaneous abortions and deliver more stillborn infants than females that do not require treatment. The female infants of adult females that receive treatment during pregnancy also weigh less and have shorter gestation periods than infant females born to adult females that do not require treatment.

Females are also subject to harassment from other adult females after they give birth. Persistent attempts to take neonates away from their mothers ("kidnapping attempts") have been reported by many workers (e.g., Rowell, *et al.*, 1964; Hinde and Spencer-Booth, 1967; Wolfheim, *et al.*, 1970; Bernstein, 1972; Kurland, 1977; Wolfe, cited in McKenna, 1979; Silk, 1980). The pattern of those interactions suggests that kidnapping attempts by adult females are a form of maternal harassment. Kidnapping attempts are nearly always directed toward unrelated and lower-ranking females (Hinde and Spencer-Booth, 1967; Kurland, 1977; Silk, 1980) and are often resisted by mothers (Hinde and Spencer-Booth, 1967; Bernstein, 1972; Silk, 1980). Wolfheim and her collegues (1970) have observed that females are more protective of their infants in the presence of other females than are females housed alone. They attribute this protectiveness to the attentions of other females. Hinde and Spencer-Booth (1967) noted that "mothers usually resent such attention and may protect their young either by threatening the aunt (kidnapper), or by restricting their movements so the aunt has little opportunity to get near them" (p. 191). Bernstein (1972) describes what occurs after a successful kidnapping attempt: "When a baby is 'kidnapped' by a higher-ranking female, the natural mother usually follows quietly and repeatedly attempts to retrieve her baby" (p. 413). Subsequent analysis of the data reported by Silk (1980) indicates that bonnet macaque females attempt to kidnap female infants at considerably higher rates than they attempt to kidnap male infants of similar maternal rank.

Finally, there is evidence of direct aggression toward infants and juveniles by adult females. Kurland (1977) describes several severe attacks upon infant Japanese macaques as "infanticide in its nascent form." Dittus (1977) has shown that infant female toque macaques are threatened more than twice as often as infant males and that infant and juvenile females are threatened by adult females significantly more often than would be expected on the basis of their distribution in the population. In a subsequent paper, Dittus (1979) reports that when food became scarce during a drought, immature females suffered the greatest mortality. Dittus attributes this to the aggressive behavior of adults and immature males who selectively displaced immature females from food and effectively excluded them from feeding areas. At the California Primate Research Center, kinship and maternal rank influence the severity of aggression by adult females toward infant and juvenile bonnet macaques (Silk *et al.*, 1981c). Females direct a lower proportion of chases and attacks toward related immatures than toward the offspring of unrelated females. Immature females are significantly more likely to receive aggression from unrelated adult females than are immature males of similar maternal rank, but there is no evidence of a similar sex bias in aggression toward related immatures. In the same group, young females are injured, sometimes fatally, at higher rates than males of similar maternal rank. Kurland (1977) has noted a similar difference in the intensity of aggression toward relatives and nonrelatives, and Berman (1980a, 1980b) has noted that infant female rhesus macaques tend to be threatened more often than infant males.

C. Secondary Sex Ratios

Given the conservative nature of female hierarchies, the daughters of low-ranking females have little chance of becoming high-ranking or reproductively successful when they mature. On the other hand, the sons of low-ranking females have some chance of rising in rank and becoming reproductively successful when they transfer to new groups. Low-ranking females might, therefore, increase their relative fitness by altering the sex ratio of their progeny in favor of males or by investing more heavily in their sons than in their daughters. At the California Primate Research Center, low-ranking female bonnet macaques produce significantly more male infants than female infants, while high-ranking females produce males and females in approximately equal numbers (Silk *et al.*, 1980, 1981b). J. Altmann (1980) noted a similar, but nonsignificant, rank-related bias in the sex ratio of yellow baboons (*Papio cynocephalus*) and proposes a similar explanation. Simpson and his colleagues (personal communication) also found that high-ranking rhesus macaque females produce more daughters than sons, and low- and middle-ranking females produce more sons than daughters. There is presently little quantitative information available regarding the extent of maternal investment in sons and daughters. However, among the rhesus macaques in the Madingley colony the interval following the birth of male infants is significantly shorter than the interval following the birth of female infants. Moreoever, interbirth intervals are longest following the birth of female infants to high-ranking females (Simpson *et al.*, 1981). There is limited evidence of a similar pattern among the bonnet macaques at the California Primate Research Center. There, the birth of a surviving female infant to a primiparous female is followed by a significantly longer interval than is the birth of a surviving male infant to a primiparous female or the birth of a surviving infant of either sex to a multiparous female (Silk *et al.*, 1981b). Although these data suggest that females may invest more heavily in their daughters than their sons under some circumstances, the same phenomenon apparently does not occur among rhesus macaques at some other sites (Simpson *et al.*, 1981).

D. Coalition Formation

Coalition formation provides a complex example of the influence of kinship and rank upon the patterns of cooperation and competition among macaque females. Detailed analyses of the form and consequence of coalition formation among Japanese (Kurland, 1977; Watanabe, 1979), rhesus (Kaplan, 1977, 1978; Berman, 1980a, 1980b), pigtail (Massey, 1977, 1979), crab-eating (DeWaal, 1978), and bonnet (Silk, 1981) macaques are now available. Several general patterns emerge from careful consideration of these data.

Coalitions in support of the victims of aggression differ markedly from alliances in support of the original aggressors (Massey, 1977; Kaplan, 1977; DeWaal, 1978; Watanabe, 1979; Silk, 1981). Defense of the victims of aggression is performed almost exclusively by close kin (Massey, 1977; Kaplan, 1977; Kurland, 1977; Watanabe, 1979; Silk, 1981). In coalitions in support of aggressors, the bias in favor of kin is much less pronounced. Given that individuals are more likely to be harassed by higher-ranking individuals than lower-ranking individuals, and that matrilineal relatives share a common rank, defense of victims must more frequently involve aggression toward higher-ranking individuals than does support of aggressors. Coalitions against higher-ranking individuals do appear to involve an appreciable risk to the ally. Kaplan (1978) has observed that individuals who form coalitions against animals higher-ranking than themselves are significantly more likely to be the target of retaliatory attacks than are individuals who form coalitions against lower-ranking animals. Nevertheless, Kurland (1977) found that 73% of all coalitions formed in defense of relatives were directed toward higher-ranking monkeys, whereas only 25% of coalitions formed in defense of nonrelatives were directed toward higher-ranking individuals. Thus, the alliances that involve the greatest risk to the allies are also those that are most likely to involve kin.

Females sometimes take risks when they are unrelated to either the aggressor or the victim. Some of these interactions involve aggression against adult males (Kurland, 1977; Kaplan, 1978; Silk, 1981) (see Section V,C.). Risks are also taken by females who support unrelated females against higher-ranking females. "Females aiding females against other females tended to help those animals ranked closer to them in the dominance hierarchy against animals who were more distantly ranked" (Kaplan, 1977, p. 284). This may be explained if reciprocity in coalition formation is more likely to occur among females that occupy adjacent ranks than among females who occupy widely disparate ranks (Cheney, 1977). However, among the bonet macaques at the California Primate Research Center, this does not appear to be the case (Silk, 1981). There may be at least one other explanation of this observation. While it is generally assumed that individuals form coalitions in order to defend or provide support for their allies, it is also possible that the benefits to the ally are only an incidental outcome of their intervention. Individuals may form alliances in support of nonrelatives in order to reduce the fitness of the aggressor. Insofar as aggression toward higher-ranking individuals involves some risk to the allies and to their opponents, formation of coalitions may represent a form of spiteful behavior.

E. Paternal Investment

Paternal care is generally considered to be of limited importance in polygynous mammals because males make only a small initial investment in their offspring (Trivers, 1972) and may not be able to recognize the offspring they sire. These assumptions have recently been questioned in view of empirical evidence that male baboons do appear to differentiate between their own and other males' offspring and behave altruistically primarily toward the infants they are likely to have fathered (Popp, 1978; J. Altmann, 1980; Packer, 1980; Busse and Hamilton, 1981).

In macaques, LSFMC may not only raise the average degree of relatedness among matrilineal kin, but may increase paternity certainty and alter paternal investment strategies. Paternal investment might take the form of protection from predators (Goss Custard *et al.*, 1972; Bernstein, 1976; Crook *et al.*, 1976), care of infants or juveniles, or protection against aggression by other members of the group.

There have been few investigations of the hypothesis that male macaques invest selectively in their own offspring, in part because observers are usually ignorant of paternity. Although a number of observers have described adoptive relationships between macaque males and infants (Hasegawa and Hiraiwa, 1980; Teas *et al.*, 1980) and extensive male care of infants (Itani, 1959; Kurland, 1977; Taub, 1978), the genetic relationship betwen adult males and the infants they care for is not known. At the California Primate Research Center, a study of the interactions of adult male rhesus macaques and juveniles of known paternity revealed little evidence of selective paternal investment (Berenstain *et al.*, 1981). Thus, there is little evidence supporting the hypothesis that LSFMC will lead to the evolution of more extensive paternal care.

V. Lineage-Specific Mating and Female Choice

We have argued that competition among females for resources and organization into matrilineal units should lead all members of a lineage to mate with a small number of males. The criteria of female choice and evidence of ability to exercise preferences become relevant at this point. Some authors have suggested that the existence of a dominance hierarchy among males in multimale groups eliminates the necessity and possibility of female choice. The outcome of competition among males may demonstrate the genetic superiority of dominant males, though the dimorphism between males and females may make it impossible for females to exercise their preferences (Trivers, 1972).

In macaques, however, female preferences for individual males are evident and apparently effective. "Females can control which males will successfully mate with them by refusing to lift their hips when a male attempts to mount" (Stephenson, 1975, p. 85). Similarly, Eaton (1976) noted that in Japanese macaques, the "ultimate choice of partner appears to be made by the female" (p. 103). Lindburg (in press) documents persistent solicitation of high-ranking males by female rhesus macaques and concludes that mate choice is an important and common female strategy. Evidence of macaque males forcing reluctant females to copulate is uncommon (LeBouef, 1978).

A. Criteria of Female Choice

Although the fact and potential effect of selective mating by females was recognized by Darwin, neither his nor subsequent work has led to general agreement about the basis of female discrimination among potential mates (Mayr, 1972; Selander, 1972). The quality of a male's genes or territory and ability to provide access to resources or paternal care have all been suggested to be important criteria of female choice (Orians, 1969; Trivers, 1972, 1976; Halliday, 1978). There is considerable controversy over the theoretical significance of each of these factors (Williams, 1975; Maynard Smith, 1978; Borgia, 1979), and little empirical information about their actual importance in primates.

It seems plausible that other factors might also influence female mate choice in macaques. We suggest that if females of a single lineage benefit by regularly mating with a small number of males, then female age and male tenure in the group may influence their choice of mates. Young females beginning their reproductive careers will be attracted to males with good prospects for remaining in the group for an extended period of time, whereas old females nearing the end of their reproductive careers will be attracted to males that have already sired a number of their offspring. Middle-aged females that have already produced several offspring, but are likely to produce several more in the future, are expected to consider both previous experience and prospective tenure when choosing a mate.

B. Male Rank and Female Choice

High-ranking males that can compete effectively with other males may have the best chance of maintaining their place in the group. Although the average tenure of adult males is estimated to be less than three years in groups of Japanese macaques (Sugiyama and Ohsawa, 1975; Kurland, 1977), some males are known to have remained in nonnatal troops much longer (Fujii, 1975; Sugiyama, 1976; Kurland, 1977). There is also considerable variance in male tenure among the rhesus macaques of the La Parguera colony (Drickamer and Vessey, 1973). Positive correlations between male rank and tenure exist in both Japanese and rhesus macaques (Drickamer and Vessey, 1973; Norikoshi and Koyama, 1975; Sugiyama, 1976; Kurland, 1977). Thus, high-ranking males are likely to remain in nonnatal groups considerably longer than other males do.

Evidence that high-ranking males can and do prevent subordinate males from mating (Carpenter, 1942; Hanby *et al.*, 1971; Lindburg, 1971; Gouzales, 1974; Stephenson, 1975; Symons, 1978) suggests that competition among males may also influence female choice. Females are expected to prefer males who are likely to be able to maintain access to receptive females. If high-ranking males are most likely to be able to defend exclusive access to their mates (as the data on sexual harassment among males suggest), then females are expected to choose high-ranking males over low-ranking males, all other things being equal.

While there is ample evidence of the positive relationship between male rank and putative reproductive success in macaques (Koford, 1963; Conaway and Koford, 1965; Kaufman, 1965, 1967; Loy, 1969; Lindburg, 1971; Symons, 1978), there are also several exceptions to the pattern that are particularly instructive (Eaton, 1974, 1976; Duvall *et al.*, 1976). Pfeiffer (1978) has reanalyzed data presented by Duvall and her colleagues (1976) and shown that in a newly established group of captive rhesus macques, males that subsequently elevated or maintained their rank fathered 23 of 25 infants born in the group, while males that subsequently lost rank were responsible for the remainder. Similarly, at the California Primate Research Center, increases in the reproductive success of male bonnet and rhesus macaques that subsequently rose in rank, and decreases in the reproductive success of males that subsequently fell in rank, have now been documented (Samuels *et al.*, 1980; Smith, 1981). Assuming that females were able to exercise mate choice, these results suggest that prospective status, as well as present rank, may influence female choice. In another case, older high-ranking females continued to mate with a male even after his rank slipped from first to third (Kaufmann, 1967). Finally, there is evidence that young females prefer young, low-ranking nonnatal males over higher-ranking resident males (Enomoto, 1975).

C. Female Support of Males

Since the effectiveness of female mate choice may be influenced by male rank insofar as rank influences both male tenure and access to females, both males and females are expected to have an interest in the outcome of competitive interactions among adult males. Females may amplify the inherent differences in competitive abilities among males by providing support when males are challenged. They are most likely to do so when a male is established in the group and is already responsible for a number of their offspring. Coalitions involving high-ranking males and females have been observed among Japanese macaques (Imanishi, 1965; Itoigawa, 1973; Eaton, 1976; Fedigan, 1976; Kurland, 1977; Packer and Pusey, 1979), rhesus macaques (Lindburg, 1971), and bonnet macaques (Silk, 1982). In some cases, female support appears to be necessary for males to maintain their dominance rank (Bernstein and Sharpe, 1966; Gouzales, 1980).

Females are also expected to provide support when their preferred mates' status in the group is challenged. Bernstein (1969) provides a colorful description of one such event in a captive group of pigtail macaques:

> In February of 1968, while still submissive to Da (the second ranking male), male Q engaged in a series of fights with the alpha male, Ea. Female Ga (the alpha female) vigourously supported Ea, but male Da avoided the conflicts. Many of the younger animals associated with Q, but were not seen to join in attacks against Ea. The old alpha male was badly injured in the first few fights and thereafter avoided approaching Q and looked away when approached by him. Female Ga, however, persisted in inciting further fights by threatening Q and enlisting the aid of Ea against Q. After several days during which Ea sustained multiple severe injuries, the old alpha male became unresponsive to Q, but Q responded to the threats of Ga by attacking the old alpha male. After three weeks, the alpha male, Ea, was lying comatose on his back while female Ga persisted in attempting to enlist his aid against Q. Q continued to respond by attacking Ea [p. 49].

Females may influence male tenure as well as the acquisition and maintenance of male rank. Macaque females are known to direct aggression toward adult males (Eaton, 1976; Kaplan, 1977, 1978; Packer and Pusey, 1979), and attacks by females upon adult males have preceded the latter's emigration in a number of cases (Imanishi, 1965; Kawai, 1965; Kawamura, 1965; Lindburg, 1971; Kurland, 1977). Packer and Pusey (1979) suggest that the existence of all-male groups of Japanese macaques may be due to the agonistic behavior of females toward peripheral and transient males. Moreover, several authors have noted that the successful integration of males into an established group depends upon their ability to establish affiliative relations with resident females (Yamada, 1963, 1971; Imanishi, 1965; Vessey, 1971; Sade, 1972b; Stephenson, 1975).

Adult females appear to participate in aggressive interactions with emigrant males at considerable risk to themselves. The alpha female, Ga, whose exploits were described previously, subsequently died of wounds she received from the new alpha male and subordinate monkeys after she harassed the new alpha male's consort partners (Bernstein, 1969). Vessey (1971) also reported that several adult females sustained serious wounds when they were attacked by the prospective alpha male. Injuries to resident males and females are sometimes observed when males enter new groups (Lindburg, 1971). The fact that females do participate in aggressive interactions with adult males, despite evident risk of injury, suggests that those interactions may provide important compensating benefits to females.

D. Constraints upon Female Choice

Female choice may be constrained by the choices of higher-ranking females (Seyfarth, 1976; Lindburg, in press). If males have to maintain consortships for several days in order to be assured of inseminating females, high-ranking males may not be able to maintain exclusive access to all simultaneously receptive females. For females of a single generation of a lineage to be able to mate with the same male, they must either be receptive at different times or consort simultaneously with the same male. Given the seasonal constraints upon breeding activities characteristic of macaques (Lancaster and Lee, 1965; Vandenbergh, 1973), females of different lineages are likely to be simultaneously receptive. Faced with the choice, males should preferentially consort with females from high-ranking, reproductively successful lineages. Thus, young or low-ranking females may not be able to exercise their preferences for high-ranking males. Note that this argument fulfills the critical condition of the LSFMC model that females of different lineages mate with different males.

E. Male Choice of Mates

The prediction that males will preferentially consort with higher-ranking females is consistent with observations among rhesus and Japanese macaques. In several groups, high-ranking males consort preferentially with high-ranking and older females, while low-ranking males associate primarily with young and low-ranking females (S. A. Altmann, 1962; Kaufmann, 1965, 1967; Hanby *et al.*, 1971; Stephenson, 1975; Kurland, 1977; Symons, 1978; Samuels *et al.*, 1980). Lindburg (in press) observed that high-ranking males did discriminate in favor of older and reproductively successful females. There is also evidence that high-ranking males maintain exclusive, constant associations with high-ranking females during the period of probable conception, but associate sporadically and nonexclusively with low-ranking females during their conception periods (Fedigan and Gouzales, 1978; Samuels *et al.*, 1980; Wilson *et al.*, 1982). These observations suggest either that low-ranking females avoid high-ranking males, or that high-ranking males devote more time to associations with high-ranking females than low-ranking females.

F. Association between Lineage Rank and Relatedness

Since high-ranking males typically have longer tenures than low-ranking males, associations between members of high-ranking matrilines and high-ranking males should cause these matrilines to be characterized by a higher average relatedness than among members of low-ranking matrilines that mate with low-ranking and often transitory males. This prediction is consistent with observations cited previously that members of high-ranking lineages generally show more extensive cooperation than members of low-ranking lineages, and with the evidence that a positive correlation exists betwen lineage rank and the average degree of relatedness within the lineage (Chepko-Sade and Olivier, 1979).

VI. Conclusions

This model provides a detailed explanation of the evolution of several aspects of the hierarchical organization and social behavior of female macaques and generates a number of empirical predictions. The model can be evaluated by gathering more behavioral observations of the kind cited previously. In particular, it would be useful to obtain further documentation of female–female and female–male cooperation in agonistic encounters, more examples of "spiteful" interactions among females, and evidence that females of a single lineage can and do choose the same mates. Direct measures of relatedness among females and paternity estimates for several groups (based upon immunological or allozyme techniques) could provide a better means for confirming or disproving the model.

It is possible that this model may also apply to other primate species or other mammalian taxa. In general, the model requires that: (*a*) resources be locally limiting, (*b*) groups consist of more than one adult male and more than one female lineage, (*c*) males disperse from their natal groups, (*d*) male tenure is sufficient to allow males to sometimes sire successive offspring of a single female, and (*e*) females exercise effective mate choice. Whenever these conditions are met, the model may logically be applied. We suspect that the data needed to evaluate many of these predictions may already have been collected, but have not been considered with these questions in mind. We hope that this chapter stimulates critical reevaluation of existing data relevant to the predictions of this model and further research on the impact of matrilineality on social behavior.

VII. Summary

Macaque females typically form corporate matrilineal hierarchies in which the members of each lineage collectively rank above or below the members of other lineages and high-ranking members reproduce more successfully than low-ranking members. In this chapter, we developed an evolutionary explanation of the formation of corporate hierarchies and evaluated predictions generated from the evolutionary model, using information available in the literature.

We began with the assumptions that large coalitions are likely to be able to defeat smaller ones in competitive encounters and that macaque groups are limited by the availability of resources. If such contests over resources influence fitness, as the correlation between female rank and reproductive success suggests, selection is expected to favor behaviors that allow an individual to recruit support from the largest number of conspecifics. Behaviors that increase the coefficient of relationship between matrilineal relatives are apt to have this effect. Thus, females that exercise mate choice so as to increase the probability that they and their maternal kin of the same generation mate with the same males will increase the number of females they can recruit in their support. Limited genetic and behavioral evidence suggests that a lineage-specific female mating system may occur. The potential effects of this mating system upon the structure of relatedness were considered, and a number of predictions about the resulting patterns of female cooperation, competition, and mate choice were developed. If, for example, there is a positive relationship between the numerical size and collective strength of a lineage, females may attempt to reduce the size and strength of other lineages by reducing the rate at which they recruit new members. Harassment of cycling, pregnant, and lactating females and aggression toward their offspring may limit recruitment. Moreover, since only females will mature and reproduce in their natal groups, competition among females is likely to be more intense than intersexual competition. Thus, females are expected to focus aggression upon the daughters of unrelated females, rather than on their sons. If selection favors a lineage-specific mating system, females are expected to mate preferentially with males that they have mated with before or are likely to be able to mate with in the future. Insofar as male tenure is positively related to rank and acces to females, most females are likely to choose high-ranking males. However, old females are expected to choose males who have already sired a number of their offspring, regardless of their prospective tenure or current rank, and young females are likely to choose young males with the best prospects for attaining high rank and access to females. Furthermore, females are expected to actively support preferred males in competitive interactions with other males and under some circumstances to compete with other females for access to preferred males.

Appendix

In this appendix, we show that $1/2 < h < 2$ for a variety of reasonable models. Each group begins with n individuals. After social interaction, there are nw_0 individuals, and after density-dependent mortality, there are $nw_0\, d(nw_0)$ individuals. Each of these individuals has m offspring. We seek the value of n when population size is in equilibrium. Clearly, this requires

$$n = mnw_0\, d(nw_0), \tag{A.1}$$

or

$$1 = mw_0\, d(nw_0). \tag{A.2}$$

We consider the following three plausible forms for $d(nw_0)$:

$$d(nw_0 = [1 - (nw_0/k)], \tag{A.3}$$

$$d(nw_0) = \exp[-(nw_0/k)], \tag{A.4}$$

and

$$d(nw_0) = [1 + (nw_0/k)]^{-1}. \tag{A.5}$$

These three forms will be labeled logistic, exponential, and hyperbolic, respectively. The parameter h is defined as

$$h = \frac{-d[\ln d(nw_0)]}{d[\ln (nw_0)]} \tag{A.6}$$

For each of the three forms of $d(\ )$, we calculated h as a function of nw and also calculated n, the equilibrium size of the local group. These can then be combined to yield the following expressions for h:

$$\text{logistic} \quad h = [1 - d(nw_0)]/d(nw_0), \tag{A.7}$$

$$\text{exponential} \quad h = -\ln[d(nw_0)], \tag{A.8}$$

and

$$\text{hyperbolic} \quad h = 1 - d(nw_0). \tag{A.9}$$

From Eq. (A.1) we have

$$d(nw_0)\ w_0 = 1/m, \tag{A.10}$$

where m is the total number of offspring per surviving individual. This suggests that we can write

$$d(nw_0) = (m)^{-\gamma}, \qquad O\ (<)\ \gamma\ (<), \tag{A.11}$$

where γ gives the relative importance of density-independent mortality. For a k-selected animal like a macaque, it seems reasonable to assume that a reasonable proportion of mortality is due to density-dependent factors when the population is in equilibrium.

To calculate a very approximate lower bound on h, note that the hyperbolic model yields the lowest values of h for a given value of $d(nw_0)$. If we suppose that $m = 9$ is a reasonable value for the average complete fertility of female macaques, and that density-independent and density-dependent factors are of approximately equal importance, then $h = 1 - 9^{-1/2} = .67$. To calculate an upper bound, we use the logistic model and suppose that density-independent mortality is somewhat less important than density-dependent mortality ($\gamma = .4$). Then, $h = 1.5$.

Clearly, these calculations provide only very rough approximations of h. It should be kept in mind, however, that the underlying model is itself an approximation. It neglects both age-dependent mortality and fecundity and partial migration. Until a model including these factors is developed, we believe that this calculation of h (and the model that underlies it) provides a better approximation of density-dependent mortality on the evolution of social behavior.

Acknowledgments

We wish to thank a number of people for their comments upon several drafts of this chapter: A. B. Clark, D. Chepko-Sade, J. A. Kurland, C. Packer, P. J. Richardson, P. S. Rodman, M. Rowe, A. Samuels, R. M. Seyfarth, S. K. Wasser, and D. S. Wilson. Insightful criticism by an anonymous reviewer also provided a valuable contribution. D. Dumont and V. Hugo patiently typed the drafts of the chapter.

References

Abugov, R., and Michod, R. E. On the relations of family structured models and inclusive fitness models for kin selection. *Journal of Theoretical Biology,* 1981, *88,* 743–755.

Altmann, J. *Baboon mothers and infants.* Cambridge, Mass.: Harvard Univ. Press, 1980.

Altmann, J., Altmann, S. A., Hausfater, G., and McCuskey, S. A. Life cycle of yellow baboons. *Primates,* 1977, *18,* 315–330.

Altmann, J., Altmann, S. A., and Hausfater, G. Primate infant's effect on mother's future reproduction. *Science,* 1978, *201,* 1028–1029.

Altmann, S. A. A field study of the sociobiology of rhesus monkeys. *Annals of the New York Academy,* 1962, *102,* 338–435.

Altmann, S. A., and Altmann, J. *Baboon ecology.* Chicago: Univ. of Chicago Press, 1970.

Altmann, S. A., Wagner, S. W., and Lenington, S. Two models for the evolution of polygyny. *Behavioral Ecology and Sociobiology,* 1977, *2,* 397–410.

Altmann, S. A. Altruistic behavior: the fallacy of kin deployment. *Animal Behaviour,* 1979, *27,* 958–959.

Angst, W. Basic data and concepts on the social organization of *Macaca fasicularis.* In L. A. Rosenblum (Ed.), *Primate behavior, developments in field and laboratory research* (Vol. 4). New York: Academic Press, 1975.

Axelrod, R., and Hamilton, W. D. The evolution of cooperation. *Science,* 1981, *211,* 1390–1396.

Berenstain, L., Rodman, P. S., and Smith, D. G. Social relations between fathers and offspring in a captive group of rhesus monkeys (*Macaca mulatta*). *Animal Behaviour,* 1981, *29,* 1057–1063.

Berman, C. M. Early agonistic experience and peer acquisition among free ranging infant rhesus monkeys on Cayo Santiago. *American Journal of Physical Anthropology,* 1980a, *52,* 204.

Berman, C. M. Early agonistic experience and rank acquisition among free ranging infant rhesus monkeys. *International Journal of Primatology,* 1980b, *1,* 153–170.

Bernstein, I. S. Spontaneous reorganization of a pigtail monkey group. In C. R. Carpenter (Ed.), *Proceedings of the Second International Congress of Primatology.* Basel: Karger, 1969.

Bernstein, I. S. Daily activity cycles and weather influences on a pigtail monkey group. *Folia Primatologica,* 1972, *18,* 390–415.

Bernstein, I. S. Dominance, aggression, and reproduction in primate societies. *Journal of Theoretical Biology,* 1976, *60,* 459–472.

Bernstein, I. S., and Sharpe, L. Social roles in a rhesus monkey group. *Behaviour,* 1966, *26,* 91–104.

Bertram, B. C. Living in groups: predators and prey. In J. R. Krebs and N. B. Davies (Eds.), *Behavioral ecology, an evolutionary approach.* Sunderland, Mass.: Sinauer Associates, 1978.

Borgia, G. Sexual reproduction and the evolution of mating systems. In M. S. Blum and N. A. Blum (Eds.), *Sexual selection and reproductive competition in insects.* New York: Academic Press, 1979.

Boyd, R. Density, dependent mortality, and the evolution of social interactions. *Animal Behaviour,* in press.

Busse, C., and Hamilton, W. J. Infant-carrying and parental care by male chacma baboons. *Science,* 1981, *212,* 1281–1283.

Carpenter, C. R. Sexual behavior of free ranging rhesus monkeys (*Macaca mulatta*). *Journal of Comparative Psychology,* 1942, *33,* 113–142.

Chapais, B. E., and Schulman, S. R. An evolutionary model of female dominance relations in primates. *Journal of Theoretical Biology,* 1980, *82,* 47–89.

Cheney, D. The acquisition of rank and the development of reciprocal alliances among free ranging immature baboons. *Behavioral Ecology and Sociobiology,* 1977, *2,* 203–218.

Chepko-Sade, B. D., and Olivier, T. J. Coefficient of genetic relationship and the probability of intragenealogical fission in *Macaca mulatta. Behavioral Ecology and Sociobiology,* 1979, *5,* 263–278.

Conaway, C. H., and Koford, C. B. Estrous cycles and mating behavior in a free ranging band of rhesus monkeys. *Journal of Mammalogy,* 1965, *45,* 577–588.

Crook, J. H., Ellis, J. E., and Goss Custard, J. D. Mammalian social systems: Structure and function. *Animal Behaviour,* 1976. *24,* 261–274.

Darwin, C. *Sexual selection and the descent of man.* New York: Appleton, 1871.

DeWaal, F. B. M. Join-aggression and protective-aggression among captive *Macaca fasicularis.* In D. J. Chivers and J. Herbert (Eds.), *Recent advances in primatology. Behaviour* (Vol. 1). New York/London: Academic Press, 1978.

Dittus, W. P. J. The social regulation of population density and age-sex distribution in the toque monkey. *Behaviour,* 1977, *63,* 281–322.

Dittus, W. P. J. The evolution of behaviors regulating density and age-specific sex ratios in a primate population. *Behaviour,* 1979, *69,* 265–302.

Dittus, W. P. J. The social regulation of primate populations: A synthesis. In D. G. Lindburg (Ed.), *The macaques: Studies in ecology, behavior, and evolution.* New York: Von Nostrand Reinhold, 1980.

Drickamer, L. C. A ten-year summary of reproductive data for free ranging *Macaca mulatta. Folia Primatologica,* 1974, *21,* 61–80.

Drickamer, L. C., and Vessey, S. H. Group changing in free ranging male rhesus monkeys. *Primates,* 1973, *14,* 359–368.

Duvall, S. W., Bernstein, I. S., and Gordon, T. P. J. Paternity and status in a rhesus monkey group. *Journal of Reproduction and Fertility,* 1976, *47,* 25–31.

Eaton, G. G. Male dominance and aggression in Japanese macaque reproduction. In W. A. Montagna and W. A. Sadler (Eds.), *Reproductive behavior.* New York: Plenum, 1974.

Eaton, G. G. The social order of Japanese macaques. *Scientific American,* 1976, *235,* 97–106.

Emlen, S. T., and Oring, W. Ecology, sexual selection, and the evolution of mating systems. *Science,* 1977, *197,* 215–223.

Enomoto, T. The sexual behavior of wild Japanese monkeys. The sexual interaction pattern and preference. In S. Kondo, M. Kawai, and A. Ehara (Eds.), *Contemporary primatology, Proceedings of the Fifth International Congress of Primatology, Nagoya.* Basel: Karger, 1975.

Estrada, A. Social relations in a free ranging group troop of *Macaca arctoides.* In D. J. Chivers and J. Herbert (Eds.), *Recent advances in primatology. Behaviour.* (Vol 1). New York: Academic Press, 1978.

Fedigan, L. M. A study of roles in the Arashiyama West troop of Japanese monkeys (*Macaca fuscata*). *Contributions to Primatology*, 1976, *9*. (Monograph)

Fedigan, L., and Gouzales, H. The consort relationship in a troop of Japanese monkeys. In D. J. Chivers and J. Herbert (Eds.), *Recent advances in primatology. Behaviour* (Vol 1). New York: Academic Press, 1978.

Fujii, H. A psychological study of the social structure of a free ranging group of Japanese monkeys in Katsuyama. In S. Kondo, M. Kawai, and A. Ehara (Eds.), *Contemporary primatology, Proceedings of the Fifth International Congress of Primatology, Nagoya.* Basel: Karger, 1975.

Goss Custard, J. D., Dunbar, R. J., and Aldrich-Blake, F. P. G. Survival, mating, and rearing strategies in the evolution of primate social structure. *Folia Primatologica*, 1972, *17*, 1–19.

Gouzales, H. Harassment of sexual behavior in the stump-tail macaque (*Macaca arctoides*). *Folia Primatologica*, 1974, *22*, 208–217.

Gouzales, H. The alpha female: Observations on captive pigtail monkeys. *Folia Primatologica*, 1980, *33*, 46–56.

Halliday, T. R. Sexual selection and mate choice. In J. R. Krebs and N. B. Davies (Eds.), *Behavioral ecology, an evolutionary approach.* Sunderland, Mass.: Sinauer Associates, 1978.

Hamilton, W. D. The genetical evolution of social behavior. I. *Journal of Theoretical Biology*, 1964, *7*, 1–52.

Hamilton, W. D. Selection of selfish and altruistic behavior in some extreme models. In J. F. Eisenberg (Ed.), *Man and beast: Comparative social behavior.* Washington, D. C.: Smithsonian Institution Press, 1971.

Hanby, J. P., Robertson, L. T., and Phoenix, C. H. The sexual behavior of a confined group of Japanese macaques (*Macaca fuscata*). *Folia Primatologica*, 1971, *16*, 123–143

Hasegawa, T., and Hiraiwa, M. Social interactions of orphans observed in free ranging troop of Japanese monkeys. *Folia Primatologica*, 1980, *33*, 129–158.

Hinde, R. A., and Spencer-Booth, Y. The effect of social companions on mother–infant relations in rhesus monkeys. In D. Morris (Ed.), *Primate ethology.* Chicago: Aldine, 1967, pp. 343–364.

Imanishi, K. Identification: A process of socialization in the subhuman society of *Macaca fuscata*. In S. A. Altmann and K. Imanishi (Eds.), *Japanese monkeys.* Atlanta, Ga.: S. A. Altmann, 1965.

Itani, J. Paternal care in the wild Japanese monkey, *Macaca fuscata fuscata*. *Primates*, 1959, *2*, 61–93.

Itoigawa, N. Group organization of a natural group of Japanese monkeys and mother–infant interactions. In C. R. Carpenter (Ed.), *Behavioral regulators of behavior in primates.* Lewisburg, Pa.: Bucknell Univ. Press, 1973.

Kaplan, J. R. Patterns of fight interference in free ranging rhesus monkeys. *American Journal of Physical Anthropology*, 1977, *47*, 279–288.

Kaplan, J. R. Fight interference and altruism in rhesus monkeys. *American Journal of Physical Anthropology*, 1978, *49*, 241–250.

Kaufmann, J. H. A three year study of mating behavior in a free ranging band of rhesus monkeys. *Ecology*, 1965, *46*, 500–512.

Kaufmann, J. H. Social relations of adult males in a free ranging band of rhesus monkeys. In S. A. Altmann (Ed.), *Social communications among primates.* Chicago: Univ. of Chicago Press,

1967.

Kawai, M. On the system of social ranks in a natural troop of Japanese monkeys, basic rank and dependent rank. In S. A. Altmann and K. Imanishi (Eds.), *Japanese monkeys.* Atlanta, Ga.: S. A. Altmann, 1965.

Kawamura, S. Matriarchal social ranks in the Minoo-B troop: A study of the rank system of Japanese monkeys. In S. A. Altmann and K. Imanishi (Eds.), *Japanese monkeys.* Atlanta, Ga.: S. A. Altmann, 1965.

Koford, C. Ranks of mothers and sons in bands of rhesus monkeys. *Science,* 1963, *141,* 356–357.

Koford, C. Population dynamics of rhesus monkeys on Cayo Santiago. In I. DeVore (Ed.), *Primate behavior.* New York: Holt, Rinehart & Winston, 1965.

Koyama, N., Norikoshi, K., and Mano, T. Population dynamics of Japanese monkeys at Arashiyama. In S. Kondo, M. Kawai, and A. Ehara (Eds.), *Contemporary primatology, Proceedings of the Fifth International Congress of Primatology, Nagoya.* Basel: Karger, 1975.

Kurland, J. A. Kin selection in the Japanese monkey. *Contributions to Primatology,* 1977, *12.* (Monograph)

Lancaster, J. B., and Lee, R. B. The annual reproductive cycle in monkeys and apes. In I. DeVore (Ed.), *Primate Behavior.* New York: Holt, Rinehart, & Winston, 1965.

LeBouef, B. J. Sex and evolution. In T. E. McGill, D. A. Dewsbury, and B. D. Sachs (Eds.), *Sex and behavior.* New York: Plenum, 1978.

Lindburg, D. G. The rhesus monkey in N. India: An ecological and behavioral study. In L. A. Rosenblum (Ed.), *Primate behavior, developments in field and laboratory research* (Vol. 2). New York: Academic Press, 1971.

Lindburg, D. G. Mating behavior and estrus in the Indian rhesus monkey. In P. K. Seth (Ed.), *Perspectives in primate biology,* in press.

Loy, J. *Estrous behavior of free ranging rhesus monkeys: A study of continuity and variability.* Unpublished doctoral dissertation, Northwestern University, 1969.

Loy, J., and Loy, K. Behavior of an all-juvenile group of rhesus monkeys. *American Journal of Physical Anthropology,* 1974, *40,* 83–96.

Massey, A. Agonistic aids and kinship in a group of pigtail macaques. *Behavioral Ecology and Sociobiology,* 1977, *2,* 31–40.

Massey, A. The author replies (to Kurland and Gaulin, 1979). *Behavioral Ecology and Sociobiology,* 1979, *6,* 83.

Masui, K., Sugiyama, Y., Nishimura, A., and Ohsawa, H. The life table of Japanese macaques at Takasakiyama. In S. Kondo, M. Kawai, and A. Ehara (Eds.), *Contemporary primatology, Proceedings of the Fifth International Congress of Primatology, Nagoya.* Basel: Karger, 1975.

Maynard Smith, J. *The evolution of sex.* London/New York: Cambridge Univ. Press, 1978.

Mayr, E. Sexual selection and natural selection. In B. Campbell (Ed.), *Sexual selection and the descent of man, 1871–1971.* Chicago: Aldine, 1972.

McKenna, J. J. The evolution of allo-mothering behavior among colobine monkeys: Function and opportunism in evolution. *American Anthropologist,* 1979, *81,* 818–840.

McMillan, C. A. Genetic differentiation of female lineages in rhesus macaques on Cayo Santiago. *American Journal of Physical Anthropology,* 1979, *50,* 461–462.

McMillan, C. A. The possibility of selection as a factor in the patterning of genetic distances between matrilineages of rhesus monkeys on Cayo Santiago. *American Journal of Physical Anthropology,* 1980, *52,* 251–252.

Missakian, E. A. Genealogical and cross-genealogical dominance relationships in a group of free ranging rhesus monkeys on Cayo Santiago. *Primates,* 1972, *13,* 169–180.

Norikoshi, K., and Koyama, N. Group shifting and social organization among Japanese monkeys. In S. Kondo, M. Kawai, A. Ehara, and J. Kawamura (Eds.), *Contemporary primatology, Proceedings from the Fifth Congress of the International Primatological Society.*

Tokyo: Japan Science Press, 1975.

Orians, G. H. On the evolution of mating systems in birds and mammals. *American Naturalist*, 1969, *103*, 589–603.

Packer, C. Male care and exploitation of infants in *Papio anubis*. *Animal Behaviour*, 1980, *28*, 512–520.

Packer, C., and Pusey, A. E. Female aggression and male membership in troops of Japanese monkeys and olive baboons. *Folia Primatologica*, 1979, *31*, 212–218.

Pfeiffer, T. *Paternity and status in a rhesus monkey group reconsidered*. Paper presented at the annual meeting of the American Society of Primatologists, Atlanta, Ga., 1978.

Popp, J. *Male baboons and evolutionary principles*. Unpublished doctoral dissertation, Harvard University, 1978.

Quiatt, D. D. *Social dynamics of rhesus monkey groups*. Unpublished doctoral dissertation, University of Colorado, 1966.

Rowell, T. E., Hinde, R. A., and Spencer-Booth, Y. Aunt–infant interaction in captive rhesus monkeys. *Animal Behaviour*, 1964, *12*, 219–226.

Sackett, G. P., Holm, R. A., Davis, A. E., and Fahrenbruch, C. E. Prematurity and low birth weight in pigtail macaques: Incidence, prediction, and effect on infant development. In S. Kondo, M. Kawai, A. Ehara, and J. Kawamura (Eds.), *Contemporary primatology, Proceedings from the Fifth Congress of the International Primatological Society*. Tokyo: Japan Science Press, 1975.

Sade, D. S. Determinants of dominance in a group of free ranging rhesus monkeys. In S. A. Altmann (Ed.), *Social communication in primates*. Chicago: Univ. of Chicago Press, 1967.

Sade, D. S. A longitudinal study of social behavior of rhesus monkeys. In R. Tuttle (Ed.), *The functional and evolutionary biology of primates*. Chicago: Aldine, 1972a.

Sade, D. S. Sociometrics of *Macaca mulatta* I. Linkages and cliques in grooming matrices. *Folia Primatologica*, 1972b, *18*, 196–223.

Sade, D. S., Cushing, K., Cushing, C., Dunaif, J., Figueroa, A., Kaplan, J. R., Lauer, C., Rhodes, D., and Schneider, J. Population dynamics in relation to social structure on Cayo Santiago. *Yearbook of Physical Anthropology*, 1976, *20*, 253–262.

Samuels, A., Silk, J., and Rodman, P. S. Rank changes in male bonnet macaques. *American Journal of Physical Anthropology*, 1980, *52*, 275.

Schulman, S. R., and Chapais, B. Reproductive value and rank relations among macaque sisters. *American Naturalist*, 1980, *115*, 580–593.

Seger, J. A numerical method for estimating coefficients of relationship in a langur troop. In S. Blaffer Hrdy, *The langurs of Abu*. Cambridge, Mass.: Harvard Univ. Press, 1977.

Selander, R. K. Sexual selection and dimorphism in birds. In B. Campbell (Ed.), *Sexual selection and the descent of man, 1871–1971*. Chicago: Aldine, 1972.

Seyfarth, R. Social relationships in female baboons. *Animal Behaviour*, 1976, *24*, 917–938.

Silk, J. B. Kidnapping and female competition in captive bonnet macaques. *Primates*, 1980, *21*, 100–110.

Silk, J. B. *Social behavior of female* Macaca radiata*: The influence of kinship and rank on cooperation and competition*. Unpublished doctoral dissertation, University of California, 1981.

Silk, J. B. Altruism among female *Macaca radiata*: Explanations and analysis of patterns of grooming and coalition formation. *Behaviour*, 1982, *79*, 162–187.

Silk, J. B., Samuels, A., and Rodman, P. S. Rank, reproductive success, and skewed sex ratios in *Macaca radiata*. *American Journal of Physical Anthropology*, 1980, *52*, 279.

Silk, J. B., Samuels, A., and Rodman, P. S. Hierarchical organization of female *Macaca radiata*. *Primates*, 1981a, *22*, 84–95.

Silk, J. B., Clark-Wheatley, C., Rodman, P. S., and Samuels, A. Differential reproductive success and facultative adjustment of sex ratios among captive female bonnet macaques (*Macaca radiata*), *Animal Behaviour*, 1981b, *29*, 1106–1120.

Silk, J. B., Samuels, A., and Rodman, P. S. The influence of kinship, rank, and sex on affiliation aggression between adult female and immature bonnet macaques (*Macaca radiata*). *Behaviour,* 1981c, *78,* 112–137.

Simpson, M. J. A., Simpson, A., Hooley, J., and Zunz, J. Interbirth intervals in rhesus monkeys vary with sex of infant and early rejecting behaviour of mothers. *Nature,* in press.

Smith, D. G. The association between rank and reproductive success of male rhesus monkeys. *American Journal of Primatology,* 1981, *1,* 83–90.

Stephenson, G. R. Social structure of mating activity in Japanese macaques. In S. Kondo, M. Kawai, A. Ehara, and J. Kawamura (Eds.), *Contemporary primatology, Proceedings from the Fifth Congress of the International Primatological Society.* Tokyo: Japan Science Press, 1975.

Sugiyama, Y. Characteristics of the social life of bonnet macaques (*Macaca radiata*). *Primates,* 1971, *12,* 247–266.

Sugiyama, Y. Life histories of male Japanese monkeys. In J. S. Rosenblatt, R. A. Hinde, E. Shaw, and C. Beer (Eds.), *Advances in the study of behavior* (Vol. 7). New York: Academic Press, 1976.

Sugiyama, Y., and Ohsawa, H. Life histories of male Japanese macaques at Ryozenyama. In S. Kondo, M. Kawai, and A. Ehara (Eds.), *Contemporary primatology, Proceedings of the Fifth International Congress of Primatology, Nagoya.* Basel: Karger, 1975.

Symons, D. *Play and aggression: A study of rhesus monkeys.* New York: Columbia Univ. Press, 1978.

Takahata, Y. The reproductive biology of a free ranging troop of Japanese monkeys. *Primates,* 1980, *21,* 303–329.

Taub, D. M. *Aspects of the biology of the wild Barbary macaque (Primates, Circopithecinae,* Macaca sylvanus *L. 1758): Biogeography, the mating system, and male–infant associations.* Unpublished doctoral dissertation, University of California, 1978.

Taub, D. M. Female choice and mating strategies among wild Barbary macaques (*Macaca sylvanus* L.). In D. G. Lindburg (Ed.), *The macaques: Studies in ecology, behavior, and evolution.* New York: Von Nostrand Reinhold, 1980.

Teas, J., Richie, T., Taylor, H., and Southwick, C. Population patterns and behavioral ecology rhesus monkeys (*Macaca mulatta*) in Nepal. In D. G. Lindburg (Ed.), *The macaques: Studies in ecology, behavior, and evolution.* New York: Von Nostrand Reinhold, 1980.

Trivers, R. L. The evolution of reciprocal altruism. *Quarterly Review of Biology,* 1971, *46,* 35–57.

Trivers, R. L. Parental investment and sexual selection. In B. Campbell (Ed.), *Sexual selection and the descent of man, 1871–1971.* New York: Academic Press, 1972.

Trivers, R. L. Sexual selection and resource accruing abilities in *Anolis germani. Evolution,* 1976, *30,* 253–269.

Uyenoyama, M., and Feldman, M. W. Theories of kin and group selection: A population genetics perspective. *Journal of Theoretical Biology,* 1980, *17,* 380–414.

Vandenbergh, J. G. Environmental influences on breeding in rhesus monkeys. In C. H. Phoenix (Ed.), *Primate reproductive behavior, Symposium of the Fourth International Congress of Primatology.* (Vol. 2). Basel: Karger, 1973.

Vessey, S. H. Free ranging rhesus monkeys: Behavioral effects of removal, separation, and reintroduction of group members. *Behaviour,* 1971, *40,* 216–227.

Watanabe, K. Alliance formation in a free ranging troop of Japanese macaques. *Primates,* 1979, *20,* 459–474.

Williams, G. C. Sex and evolution. *Monographs in Population Biology,* 1975, *8.*

Wilson, M. E., Gordon, T. P., and Chikazawa, D. Female mating relationships in rhesus monkeys *American Journal of Primatology,* 1982, *2,* 21–27.

Wittenberger, J. F. Group size and polygamy in social mammals. *American Naturalist,* 1980, *115,* 197–222.

Wolfheim, J., Jensen, G. D., and Bobbit, R. A. Effects of group environment on the mother–infant relationship in pigtailed monkeys (*Macaca nemestrina*). *Primates,* 1970, *11,* 119–124.

Wrangham, R. W. An ecological model of female-bonded primate groups. *Behaviour,*1980, *75,* 262–300.

Yamada, M. Five natural troops of Japanese monkeys on Shodoshima Island. II. A comparison of social structure. *Primates,* 1971, *12,* 125–150.

13

Reproductive Competition and Cooperation among Female Yellow Baboons*

SAMUEL K. WASSER

*The Mikumi site and operations were supported by grants to Ramon J. Rhine from NSF (FMS74-17531 and BMS75-05732), NIMH (1 R01 MH30428), the Leakey Foundation, the Center for Social and Behavioral Sciences Research of the University of California, Riverside, and by Biomedical Sciences Support grants to the University of California, Riverside. The Department of Psychology, University of Washington, provided financial support for this project as well.

ISBN 0-12-735950-8

I. Introduction

The application of evolutionary biology to the study of social behavior has expanded considerably during the past 15 years, largely because of a few significant ideas. One area of particular importance has been sexual selection theory (Darwin, 1871; Fisher, 1930; Huxley, 1938), which relates sex differences in parental investment patterns and in gamete production to mate choice and intrasexual competition for mates (Bateman, 1948; Williams, 1966; Trivers, 1972). Mate choice has been particularly emphasized for females, whereas intrasexual competition has been particularly emphasized for males. In fact, Parker (1979, p. 123) states, "sexual selection can be divided into two major aspects: (1) direct intramale competition for females, and (2) female choice."

Another area of special importance has been inclusive fitness theory (Hamilton, 1964), which focuses on the benefits and costs of association among individuals who share genes by common descent. Association among closest kin has been particularly emphasized for female as opposed to male social mammals (Kurland, 1980) because sex differences in natal dispersal patterns tend to result in a higher average degree of relatedness among females in a group compared to males; female, as opposed to male, dispersal from the natal group is relatively uncommon among social mammals (see Greenwood, 1980; Wrangham, 1980). The reverse has been argued for birds; female dispersal is more prevalent in their case (Greenwood, 1980, but see Greenwood *et al.*, 1979).

Although these theories have been quite powerful, their strong appeal appears to have resulted in the neglect of other important evolutionary considerations. Two such considerations treated here are (*a*) reproductive competition among females and (*b*) the influence of interindividual differences in associate quality on intrasexual, social partner preferences. The quality of one associate to another, say of individual i to individual j, is defined here as the effect of a unit of investment by individual i on the survivorship and reproduction of individual j (Wasser, 1982). Defined as such, associate quality is synonymous to k in Hamilton's (1964) formulation, $k > 1/r$, for the evolution of altruism through kin selection and is independent of associate relatedness.

Following a description of research methods (Section II), Section III focuses on reproductive competition among females. This topic has been previously neglected because biologists have tended to focus on only one of two important forms of reproductive competition, that which increases the quantity, as opposed to quality (i.e., the relative survivorship and subsequent reproductive capabilities) of offspring conceived. The nearly continuous supply of gametes by males, coupled with the relatively small cyclic supply of gametes by females, has made the former type of competition particularly important for males relative to females (Bateman, 1948; Williams, 1966). This sex difference is further exaggerated by parental investment patterns (Trivers, 1972), and especially by gestation and lactation among mammals. On the other hand, competition to improve the quality of offspring is particularly important for females. The sex that contributes the greatest amount of parental investment per conception (usually females) has the most difficulty replacing wasted investment (Williams, 1966). This makes mechanisms that protect against wasting investment particularly adaptive for that sex (Goodman, 1979). Mate choice provides one such mechanism (Williams, 1966; Trivers, 1972). Competitive behavior that increases the survivorship and reproductive capabilities of offspring provides another. Differences in the times these forms of intrasexual competition are expressed may have facilitated neglect of female–female competition as well; competition to increase the quantity of conceptions tends to be concentrated around the time of mating, whereas competition to increase the quality of conceptions is often spread over the entire duration of a reproduction (i.e., from conception of the offspring until its independence; Section V,A).

The fourth section of this chapter focuses on how the associate qualities of prospective associates influence a female's social partner preferences. This topic has been previously neglected because biologists have tended to focus on the benefits derived from aiding closest kin (but see S. Altmann, 1979; Wrangham, 1980). Closest kin are said to form affiliative associations most frequently, since they may indirectly benefit from the net gains their associates receive and may also be less likely to cheat one another (Alexander, 1974; West-Eberhard, 1975). More recently, however, Axelrod and Hamilton (1981) described conditions under which cooperation between nonrelatives can be evolutionarily stable against cheaters. Taking a quantitative approach similar to theirs, I showed that this stability depends primarily on the qualities of associates to one another (Wasser, 1982). Thus, the qualities of associates should also be an important determinant of associate preferences among female baboons. Moreover, under some circumstances in which the qualities of associates vary independently of, or inversely with, associate kinship, preferential association between distant kin should be expected. To the extent that these conditions hold among nondispersing female social mammals, their association patterns may be much more flexible than that implied by current sociobiological literature.

I address the preceding ideas with empirical data by examining the effects of reproductive state on competitive and affiliative patterns of association among female yellow baboons (*Papio cynocephalus*). The reproductive states of adult female baboons can be easily subdivided into 10 or more visually distinct stages (Table I) (Gilman and Gilbert, 1946; Rowell, 1969; Hendrickx, 1967; Saayman, 1971), and are assumed to reflect, in part, their associate qualities to one another (see Section IV,B). Changes in these states are shown to affect the degree to which, and with whom, females compete and cooperate, as measured by short-term changes in both competitive and affiliative aspects of their association patterns. These short-term changes reflect a rich array of female–female competition directed at improving the relative quality of the female's own conceptions by suppressing the reproduction of others. They also identify a substantial degree of temporal flexibility in female social partner preferences that results from temporal variation in the reproductive states of associates. Moreover, since associate relatedness and, generally, female rank (Hausfater, 1975) do not change over time, they cannot, in themselves, be used to predict these short-term temporal changes in female association patterns.

II. Methods

A. Natural History of Yellow Baboons

Yellow baboons live in large, semiclosed troops of 10–200 animals, and the adult sex ratio (males to females) varies from .8 to .2. Baboons are polygynous, lack definite pair bonds (Hausfater, 1975), and show relatively little paternal care (Altmann, 1980; Packer, 1980). Females breed all year round, but may show a slight birth peak between May and September, at least in Kenya and Tanzania (Altmann, 1980; Rasmussen, 1979). They give birth to singletons, with approximately 1.5–2.0-year intervals between births, but an average of 3 years transpires between the birth of an infant and the birth of its next surviving sibling (Altmann *et al.*, 1978). Two or more females often give birth around the same time.

Males disperse from their natal troops, generally into other troops, by the time they reach sexual maturity at about 6 years. Females rarely disperse from their natal troops. They reach full sexual maturity at about 4.5 years (Altmann, 1980). Sex-specific natal dispersal patterns result in adult female baboons being much more closely related to one another, on the average, than are adult males in a troop. Females have a dependent-rank system, whereby newborn daughters take on ranks in the dominance hierarchy just below that of their mothers, and female ranks are generally quite stable over time. This tends to result in the most closely related individuals also being closest in rank (Hausfater, 1975; Walters, 1980).

B. The Study Troop

The study was conducted on a troop of 75 baboons (13 adult males; 19 adult females; 43 infants, juveniles, and subadults) at Mikumi National Park, Tanzania, from June through September 1979. Descriptions of the habitat can be found in Rhine and Westlund (1978) and in Rasmussen (1979). Over 500 hours were spent observing the troop, approximately 400 hours of which constituted systematic sampling time. Although data on relatedness of adult females were limited, D. R. Rasmussen (personal communication, 1979) documented 10 cases of female immigration into the three current study troops since 1975, five of which are in the troop I studied. Thus, five effective "nonrelatives" of known identity exist in the study troop. Rank differences between female yellow baboons can also serve as a rough indicator of female relatedness (Section II,A).

C. Reproductive State Divisions

The reproductive states of all adult females in the troop were recorded each morning as they descended from their sleeping trees. Females could either be cycling, pregnant, or lactating. Each of these states was further subdivided, giving 10 visually distinct reproductive stages. Table I gives the identifying characteristics and average duration of each of these stages. Ovulation typically occurs on the last or second to last day of estrus 2 (E2) (Hendrickx, 1967, Hausfater, 1975).

The only two stages that could not be visually distinguished from one another were the flat stage (F) characteristic of cycling females, and the first trimester of pregnancy (P1). These two stages were distinguished by backtracking data records once a female began to show signs of pregnancy in the pregnant 2 (P2) stage. The proportion of available troop females in each of the 10 reproductive states over the study period is shown in Table I.

TABLE I
Subdivisions of Female Reproductive States

	Cycling					Pregnant				Lactating		
Reproductive state (RS)	Flat (F) Postovulatory	Estrus 1 (E1) Preovulatory	Estrus 2 (E2) Ovulatory	Estrus 3 (E3) Postovulatory	IMPLANTATION	Pregnant 1 (P1)	Pregnant 2 (P2)	Pregnant 3 (P3)	PARTURITION	Lactation 1 (L1)	Lactation 2 (L2)	Lactation 3 (L3)
Identifying characters	Sex-skin flat and gray	Menstruation, followed by sex-skin red and swelling	Sex-skin swollen and red	Sex-skin deflating and red		Pericollosal gray	Pericollosal light pink	Pericollosal bright pink		Black infant	Black-brown infant	Brown infant
Approximate duration	9 days	7 days	9 days	7 days		2 months	2 months	2 months		3 months	3 months	4 months
Proportion in study troop[a]	.125	.113	.110	.042		.110	.093	.104		.122	.061	.122

[a]Proportion of available troop females in each of the reproductive states over the total study period.

D. Sampling Methods

Four to five focal females were followed on foot during any given day, 5 days per week. The first follow began each morning at the onset of troop foraging movements. The order of focal subjects was determined randomly without replacement at the beginning of each week. No female was followed a second time until all other females had been followed. Each focal animal was continuously sampled (Altmann, 1974) for 1.5 hr to record its involvement in grooming, infant handling, supplants, threats, attacks, chases, mounts, and presents (for a complete ethogram of yellow baboon behavior see Altmann [1980] and Hausfater [1975]). Identities of actors and recipients were recorded to measure directionality. The identities of individuals at various distances (0–1, 1–4, 4–6, 6–25, and 25+ m) from the focal subject were also recorded during the 1.5-hr sampling period, using focal animal scan sampling (Altmann, 1974) at 5-min intervals.

Grooming data were also collected using scan sampling each morning before the troop began its first foraging movements. I slowly walked from one end of the aggregated troop to the other, recording the identities of all groomers and groomees, never stopping until I reached the end of the troop. The procedure was then repeated.

Whenever females were observed either attacking in or being attacked by "attack coalitions" (two or more individuals simultaneously attacking another individual), data on their identities were collected. This was carried out on a behavior-dependent basis, as attack coalitions occurred too sporadically to employ focal animal sampling techniques. Fortunately, attack coalitions were extremely loud and fast-moving. This made them easily identifiable over several hundred meters and insured that a coalition was rarely missed. When coalitions occurred during a focal sample, observations were temporarily terminated so that coalition data could be collected. The focal sample was then resumed for the remainder of the 1.5 hr sampling period, plus the time lost, to make equal focal sample lengths.

E. Analyses

Data were summarized in 10 × 10 matrices whose rows showed the reproductive states of initiators and columns showed the reproductive states of recipients for the continuously scored behaviors; for the instantaneously scored proximity variables, rows correspond to the reproductive states of focal animals and columns to those of females in proximity to them. Observed values were then compared to those expected if the behavior or proximity scores occurred independently of the interactants' reproductive states; that is, if the proportion of total behavior or proximity scores of females in a given reproductive state corresponded to the proportion of females in the population that was available in that state. Each matrix was analyzed as a whole, as well as row by row, using Bowker's modified chi-square statistic for nonparametric distributions (Marascuilo and McSweeney, 1977). The binomial statistic was used to analyze differences between individual observed and expected frequencies of behavior received or donated, as well as of proximity scores.

III. Reproductive Competition

A. Female–Female competition via Reproductive Suppression

In their review of the literature, Wasser and Barash n.d. found that socioenvironmental conditions (e.g., resource availability in relation to the number, relative age, and sex of young present in the social group when a female gives birth) often cause substantial variance in female reproductive success, as well as in the age at which the infant reaches sexual maturity (e.g., Wynne-Edwards, 1962; Rudnai, 1973; Drickamer, 1974; Fox, 1975; Bronson, 1979; Dittus, 1979; Frame *et al.*, 1979; Hrdy, 1979; Kleiman, 1979; Packard and Mech, 1980; Rood, 1980; Jarvis, 1981; Reiter *et al.*, 1981; Sherman, 1981).

Whenever the successful reproduction of an offspring of one female is a liability to that of others, females can potentially improve their reproductive success by timing their reproduction of offspring so that their infants are not born at the same time as many others or when a large number of older infants are present (see also Fox, 1975). This can be accomplished by suppressing their own reproduction or that of others. Although self-inhibition does occur (Low, 1978; Shepard and Fantel, 1979; Hendrickx and Nelson, 1971; Lee, 1980), and is an adaptive strategy if future reproductive opportunities might be better (Hamilton, 1966), suppressing the reproduction of others is the competitively favored mechanism whenever possible (e.g., Fox, 1975; Frame *et al.*, 1979; Jarvis, 1981; Kleiman, 1979; Rood, 1980). This leads to the prediction that socially mediated reproductive suppression might be an important means through which females compete to improve the relative quality of their offspring.

B. Female Yellow Baboons: Hypotheses, Predictions, and Results

Data on yellow baboons suggest that socially dependent factors do affect female reproduction, and hence that it may be beneficial for them to suppress the reproduction of others. At Amboseli, Altmann (1980) found approximately 30% mortality for animals under the age of two years. Infant survival was very much dependent on the availability of easily obtained, eaten, and digested "weaning foods" relative to the number and ages of infants in the population. Data also suggest that paternal care through surveillance and protection of young infants (Packer, 1980; Altmann, 1980; Busse and Hamilton, 1981; see also the following) may also limit a female baboon's reproductive success. Males vary considerably in their tendency to show such paternal care and are also limited in the number of infants they can survey at any one time.

1. *Attack Coalition Behavior*

In my own observations of yellow baboons, I found that females interacted aggressively, particularly in attack coalitions. Two observations (Wasser, 1981) led me to suspect that females were utilizing attack coalition behavior to mediate reproductive inhibition in consexuals. First, the tendency of females to enter into attack coalitions was positively correlated with the number of females simultaneously in estrus. The frequency of attack coalition behavior remained low when 1–4 females were simultaneously in estrus, but abruptly increased when this number reached 5 or more (see also Hrdy, 1977). One might expect such a trend if (*a*) the number of females simultaneously in estrus reflects the extent of clumped births likely to occur 6 months later; and (*b*) it is advantageous to give birth when only a small number of infants are present, thereby reducing the probability of food and peer-group competition, yet enabling the social needs of the infant to be met (Section IV,C). Second, I observed an apparent inhibition in a low-ranking, preovulatory estrous female subjected to a series of coalition attacks. This occurred at a time when 8 of 19 females were simultaneously in estrus 1 (E1) or 2 (E2). Severe attacks began on this female the day before she started her estrous swelling. Three of the attackers were in E1, two in pregnant 3 (P3), one in lactation 1 (L1) and one a near-pubertal female. All attackers were of ranks 1–9, while the attack recipient was ranked 14. Attacks on this E1 female persisted for 2.5 days, at the end of which her estrous swelling was completely gone. Swelling usually lasts for approximately 23 days in these females, with ovulation occuring around Day 16 (Hendrickx, 1967). Thus, by persistent harrassment, the other females apparently managed to terminate this female's estrus long before her ovulation.

Although this was the only time in the 4-month study that attack by a coalition could definitely be said to have affected the recipient's reproduction, observations on old-world monkeys by other investigators have revealed such stress effects. DeVore (1965) reports a rapid decrease in the sex-skin swelling size of wild female baboons involved in fights. Rowell (1970) also reports such effects in baboons, but notes that estrous swellings were generally prolonged as a result of repeated attacks by females (see also Gilman and Gilbert, 1946). The end result, however, is the same: the probability of fertilization is reduced. Sackett *et al.* (1975) found an excessively high rate of spontaneous abortions in pregnant female pigtail macaques treated for bite wounds who were living in a colony under harem conditions. Such stress also increased the percentage of live-born, low-birth-weight infants. Dunbar and Dunbar (1977) and Dunbar (1980) found that low-ranking female gelada baboons have significantly more estrous cycles per conception than do high-ranking females, even though they obtained the same number of copulations with ejaculations as did high-ranking females. This lowered conception rate was correlated with the tendency of low-ranking females to be harrassed more often than were high-ranking females. Finally, Keverne (1979) has shown that dominant female talapoin monkeys aggressively place stress on subordinate females, elevating prolactin and cortisol levels in the latter and preventing the LH (luteinizing hormone) surge that induces their ovulation. Lipsett and Ross (1979) note that elevated prolactin levels in humans may also inhibit progesterone synthesis by the corpus luteum, shortening the luteal phase of cycling females and causing early abortion if conception occurs.

Predictions If attack coalitions function to inhibit reproduction in recipients, this should be revealed in the patterns of attack. Specifically, females should attack others, and receive attacks, when the probability of their bringing offspring to term in the near future is high. However, attackers should also be in those stages of reproduction that are difficult to inhibit, whereas attack recipients should be in those stages of reproduction that are easiest to inhibit. Based on these ideas, the following predictions were made regarding the reproductive states and ranks of attackers and attack recipients in yellow baboon coalitions.

1. Cycling females in the follicular stages of ovulation (E1 and E2) should attack others more than those in the postovulatory or luteal stages (E3 and F). However, ovulating (E2) females are protected from attacks by their consorts (Rowell, 1970; Hausfater, 1975) and are physiologically less susceptible to stress-induced ovulatory failure than are E1 females (Rowell, 1972; Matsumoto *et al.*, 1968). Thus, the tendency of follicular females to attack others should be greater when they are nearest to ovulation, whereas their tendency to receive attacks should be greatest at earlier preovulatory stages. Thus, using the reproductive state divisions given in Table I, E2 (ovulating) females were predicted to attack others more than E1 (preovulatory) females, both of whom were predicted to attack others more than E3 or F (postovulatory or luteal) females (E2 > E1 >> E3, F). On the other hand, E1 females were predicted to receive attacks more than E2, E3, and F females (E1 >> E2, E3, F). Estrus 1 females were also expected to receive more attacks than were pregnant females under the assumption that it is easier to suppress a female's reproduction of offspring prior to implantation (Shepard and Fantel, 1979).

2. Females in the last trimester of pregnancy should attack others more than those in the second trimester, and those in the second trimester should attack more than those in the first trimester (P3 > P2 ≫ P1). The tendency of pregnant females to receive attacks should show the reverse pattern (P1 ≫ P2 > P3). This assumes that (*a*) spontaneous abortion and other teratogenic effects are easiest to induce early in development when the embryo is differentiating most rapidly (Hendrickx and Nelson, 1971; Wilson, 1973; Shepard and Fantel, 1979); and (*b*) the reduced vulnerability of females in the later trimesters of pregnancy also makes them more likely to bring their offspring to term and hence to gain benefits of improved rearing conditions for their offspring by attacking others.

3. The high energetic and time demands of lactation (Hanwell and Peaker, 1977; Altmann, 1980) combined with the vulnerability of their young infants should cause lactating females to enter rarely into attack coalitions. In addition, lactating females should rarely receive attacks. This assumes that paternal vigilance buffers lactating females from attacks (Packer, 1980; Altmann, 1980; Busse and Hamilton, 1981) and that lactating females will respond in kind when their infants are endangered.

4. Females should attack others who are lower in rank than themselves because a victory would be more assured. (Rank was measured by

one-on-one supplant behavior, in which one female moves toward another, and the approached female moves away.) However, high-ranking females should attack others infrequently because their offspring already have a high probability of success, and attack coalition behavior is always costly (see also Drickamer, 1974; Seyfarth, 1976; Cheney, 1977; Altmann, 1980; Dunbar, 1980). This should also result in mid-ranking females initiating attacks most frequently. On the other hand, lower-ranking females should tend to receive more attacks than do high-ranking females. This not only follows because physical defeat of the former by the latter is most assured, but also because low-ranking females should be least resistant to having their reproduction inhibited. Low-ranking females can least well gain resources for their young when they are born at the same time as many other infants, making the relative probability of survival of their young particularly low at such times. This being the case, self-inhibition may actually be optimal behavior for low-ranking females until less competitive times for reproduction present themselves.

Results Nearly all of the predicted results were confirmed. Figure 1 illustrates the differences between observed frequencies of females of each reproductive state attacking others in the form of coalitions and those expected by chance alone (based on their availabilities). Overall differences between observed and expected values in Fig. 1 are significant ($p < .001$, Bowker's modified chi square). Both ovulating estrous (E2) and preovulating estrous females (E1) attacked others more than expected by chance alone ($p < .00003$, and $p < .01$, respectively), whereas post-ovulating cycling females (E3 and F) attacked others less than expected by chance alone—this difference being highly significant for E3 females ($p < .002$). Although E2 females did attack others more frequently than did E1 females, this difference was not statistically significant.

Females in the last trimester of pregnancy (P3) attacked others much more than expected by chance alone ($p < .00003$, binomial), as did P2 females ($p < .01$). However, the difference in attack frequencies between P3 and P2 females is again in the predicted direction, but not significant. Pregnant 1 females, on the other hand, attacked others much less than expected by chance alone ($p < .00003$). Females in the first two stages of lactation rarely attacked others ($p < .01$, $p < .002$, respectively). Contrary to predictions, however, females in the third stage of lactation (L3) attacked others slightly more than expected by chance alone, but this difference was not significant. Furthermore, the L3 females that did attack others also began cycling again soon afterwards. Finally, mid-ranking females entered into attack coalitions much more frequently than did low- or high-ranking females (Fig. 2).

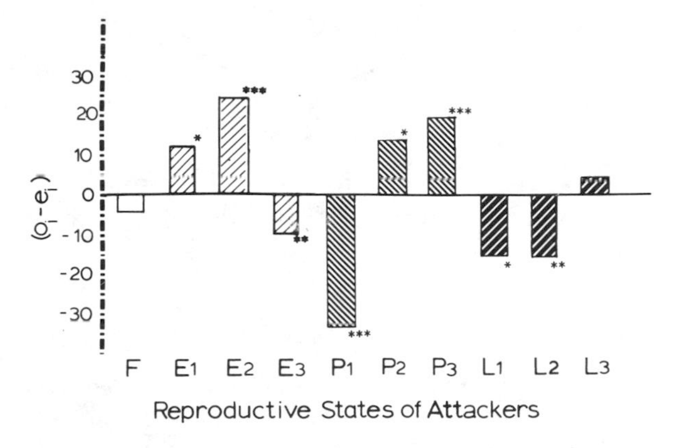

Fig. 1 Differences between observed (o_i) and expected (e_i) frequencies of coalition attacks given by adult females of each reproductive state. $e_i = Np_i$, where $N = 360$, the total number of coalition attacks given by all adult females; p_i is the proportion of females in the population in the ith reproductive state; $^*p \leq .01$; $^{**}p \leq .002$; $^{***}p \leq .00003$; binomial test.

Figure 3 illustrates the differences between observed frequencies of females of each reproductive state receiving coalition attacks and those expected by chance alone. Overall differences between observed and expected values in Fig. 3 are significant ($p < .001$, Bowker's modified chi square). Estrus 1 females were attacked significantly more than would be expected by chance alone ($p < .0001$, binomial). Estrus 2 and 3 and F females were attacked less than expected by chance alone, and significantly less than were El females. Estrus 1 females also received more attacks than did P1 females, although this difference was not significant. Females in the first trimester of pregnancy (P1) were attacked significantly

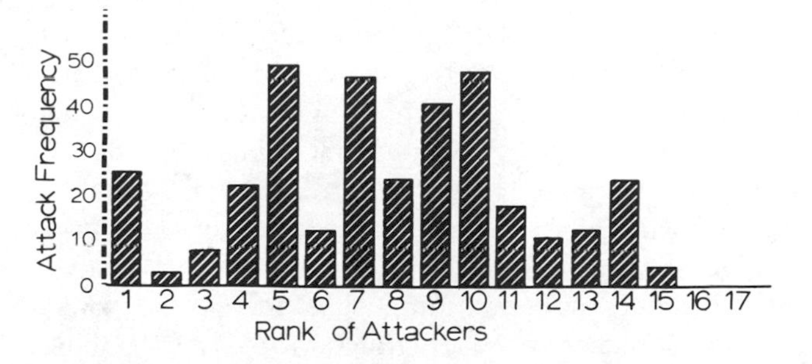

Fig. 2 Frequencies of attacks inflicted by females of each dominance rank upon adult females. ($N = 360$; 1 is the highest rank.)

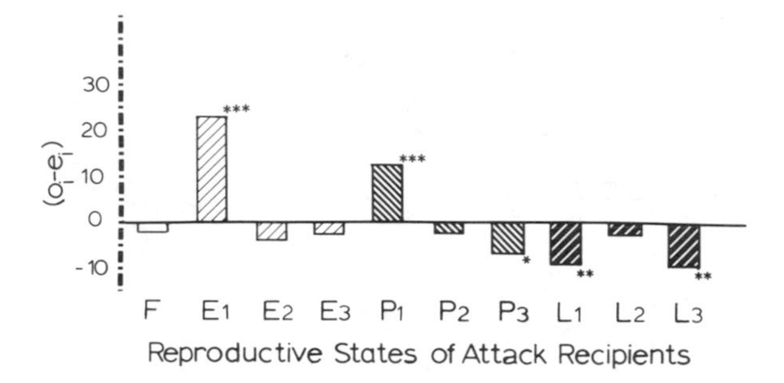

Fig. 3 Difference between observed (o_i) and expected (e_i) frequencies of coalition attacks received by adult females of each reproductive state. $e_i = Np_i$, where $N = 83$, the total number of attacks received by all adult females from adult females; $^{*}p \leq .05$; $^{**}p \leq .01$; $^{***}p \leq .0001$; binomial test.

more than expected by chance alone ($p < .0001$, binomial). The tendency of pregnant females to receive attacks then declined with successive trimesters of pregnancy. Pregnant 2 and 3 females received significantly fewer attacks than did P1 females, and P2 females received fewer attacks than did P3 females. However, the latter difference is not significant. Lactating females were rarely attacked (Fig. 3). It is also noteworthy that all attackers seemed to follow the same "rule"—attack those whose reproduction is easiest to inhibit (Table II). Figure 4 shows that high-ranking females were never attacked, whereas mid- to low-ranking females received attacks at fairly equal frequencies. The suggestion that E2 and newly lactating females (L1) are rarely attacked because of protection by adult males is supported by the selective attention these females received from them (Fig. 5).

Two additional analyses are also noteworthy. First, both small and large juvenile females received attacks and attacked others significantly more often than did males of the same age (in both cases, $p < .001$, binomial), suggesting that reproductive competition among females begins earlier in life than does such competition among males. Second, examination of one-on-one systematic attack data ($N = 54$) broken down by the reproductive states of attacker and recipient were not consistent with results from the attack coalition data analyses. This inconsistency is not surprising, as coalition attacks typically involve 2–4 attackers and are of much longer duration and greater intensities than one-on-one attacks. As a result, attack coalitions are probably much more stressful to the recipients, and hence more likely to mediate reproductive inhibition, than are one-on-one attacks.

TABLE II
Reproductive States of Females Most Frequently Attacked during Coalitions, Tabulated by the Reproductive States of Their Attackers

Attacker's reproductive state[a,b]	Recipient's reproductive state[a]	o^c	e^d	$o - e$
P3	E1	21	6.9	14.1
($N = 59$)	P1	7	6.49	.51
	F	5	8.02	−3.02
E2	E1	20	7.8	12.2
($N = 65$)	P1	15	7.54	7.46
	F	6	8.52	−2.52
	E2	4	4.1	−.1
P2	P1	15	4.75	10.25
($N = 48$)	E1	9	5.95	3.05
	F	4	6.62	−2.62
E1	E1	12	3.87	8.13
($N = 48$)	P1	12	6.2	5.8
	F	6	6.47	−.47
L3	E1	10	5.76	4.24
($N = 48$)	P1	8	5.57	2.43
F	E1	10	4.59	5.41
($N = 41$)	F	4	3.77	.23
L1	E1	9	3.36	5.64
($N = 28$)	P1	5	3.25	1.75
E3	E1	3	.6	2.4
($N = 6$)	L3	1	.8	.2

[a] Abbreviations for each reproductive state are explained in Table I.
[b] N is the total number of attacks involving females in the reproductive state listed immediately above.
[c] o is the observed attack frequency.
[d] e is the expected attack frequency based on the proportion of troop females in each of the reproductive states when the attacker was in the reproductive state shown.

Alternative Hypotheses for Attack Coalition Behavior (ACB). Several alternative hypotheses have been considered to explain ACB among adult female baboons.

1. *The rank attainment hypothesis.* ACB functions to establish, maintain, or improve the attacker's social rank as well as that of kin or other supporters (e.g., Sade, 1967; Cheney, 1977; Massey, 1977; Kaplan, 1978;

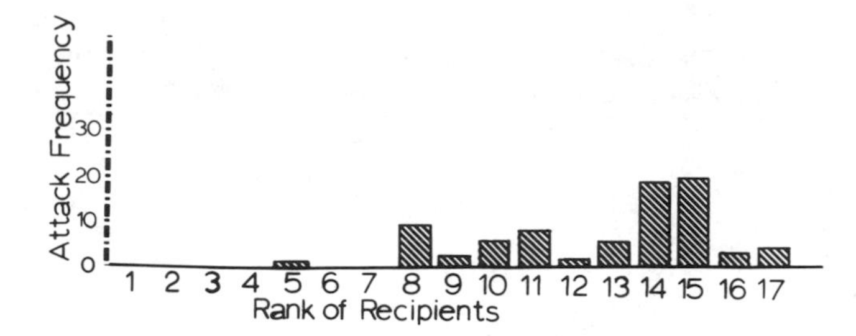

Fig. 4 Frequencies of attacks received by adult females of each dominance rank from adult females. $N = 83$; 1 is the highest rank.

Chapais and Schulman, 1980; Walters, 1980).

2. *The resource access hypothesis.* ACB functions to ensure access by the attacker, her kin, or other supporters to resources such as food (Wrangham, 1980).

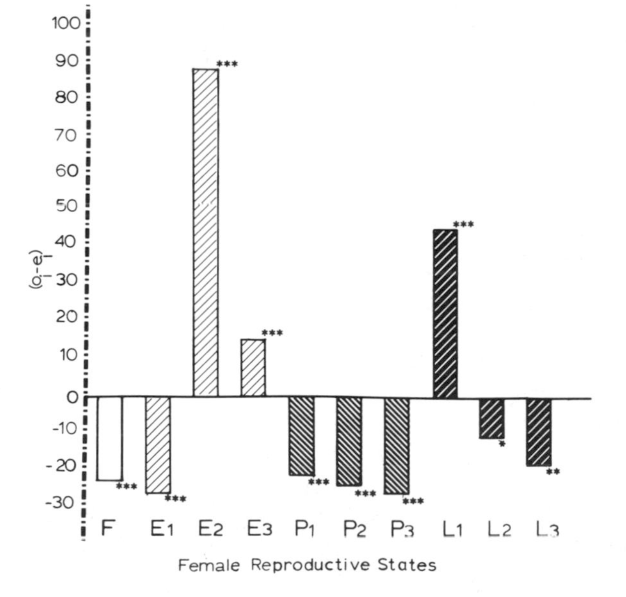

Fig. 5 Reproductive states of adult females found 0–1 m from adult males. Presented as differences between observed and expected frequencies, as in Fig. 1. $N = 324$, total number of five-minute samples in which all adult females were within 0–1 m of adult males; $^{*}p \leq .01$; $^{**}p \leq .001$; $^{***}p \leq .00001$.

3. *The excess aggression hypothesis.* ACB functions to vent aggression resulting from endocrinological changes during estrous cycles and pregnancy (Svare, 1981).

Although these alternatives are not necessarily mutually exclusive to the reproductive inhibition hypothesis, none of them successfully predict the observed reproductive states of the attackers and victims. The rank attainment hypothesis predicts that ACB will be largely independent of both the attackers' and their victims' reproductive states. Energetic demands of lactation and pregnancy might keep such females from joining attack coalitions, with late trimester females the least likely of the pregnant females to enter into attack coalitions. However, late trimester females were among the most frequent coalition participants observed. The rank attainment hypothesis also predicts that recipients will be close in rank to their attackers. However, recipient ranks were usually well below those of their attackers ($\bar{X} = 5.76 \pm 8.11$), almost always being the lowest of the females in a given reproductive state.

The resource access hypothesis predicts that: (*a*) ACB will be independent of interactants' reproductive states, with effectively permanent formation of rank alliances between particular individuals (Wrangham, 1980); or (*b*) that females attacked or attacking others most often will be those having the greatest nutritional requirements. Rank alliances were not permanent. Although females did tend to coalesce with females somewhat close in rank to themselves (the average rank difference between coalition partners, across all 19 adult females, was 3.65 ± 2.46), who coalesced with whom was also very much dependent on reproductive states. Females most likely to enter attack coalitions (i.e., those in reproductive states E2, P3, P2, E1, in that order, Fig. 1) did not tend to coalesce with others of the same reproductive state (Table III). In fact, females in the P3 and P2 states coalesced with females of the same reproductive state significantly less than expected by chance alone ($p < .02$, binomial, Table III). On the other hand, only females of the flat state coalesced with females of the same reproductive state significantly more than expected by chance alone ($p < .03$, binomial), although E2 females coalesced to some extent with one another as well ($p < .08$). The most frequently observed coalition partnership of all, however, was between E2 and P3 females ($p < .00001$; Wasser, 1981), even when considered independently of the relatively high tendencies of each to enter any coalition at all.

On the other hand, if females attacked and attacking others were those with the greatest nutritional requirements (Prediction (*b*) of the resource access hypothesis), attacks on pregnant females should have shown an increase, rather than the observed decrease, with successive stages of

TABLE III
z Scores of the Differences between Observed Frequencies with Which Females of the ith Reproductive State Coalesced with Females of the Same Reproductive State and Those Expected (e_i) by Chance Alone[a]

Reproductive state	z score and frequency	Reproductive state	z score and frequency
F	1.90 (54)	P2	−2.10 (74)
E1	−.61 (89)	P3	−2.12 (99)
E2	1.39 (104)	L1	−.04 (42)
E3	−.42 (7)	L2	−.73 (12)
P1	−.80 (9)	L3	−.74 (78)

[a] $e_i = N_i p_i$; N_i is the total frequency with which females of the ith reproductive state entered coalitions with all adult females, and is given in parentheses next to its respective z score; p_i = proportion of total study time that females of the ith reproductive state were available to females of the same reproductive state; $|z| = 1.65$, significant at $p < .05$.

pregnancy (see Post *et al*, 1980). Paternal vigilance could still come into play here, buffering attacks on lactation 1 (L1), and perhaps lactation 2 (L2), females. But this does not account for the observed lack of attacks on L3 females; their former mates were only weakly associated with them at this time (Fig. 5). Although these observations question Wrangham's (1980) implication that within-troop female–female coalitions function primarily to gain immediate resources over the short term, they in no way question his suggestion that resource access constitutes a long-term objective of female–female competition among primates, both within and between groups.

Regarding the excess aggression hypothesis, if aggression initiated by E2, E1, P3, and P2 females was merely a consequence of hormonal events, the following predictions would be made. Females would either be attacked (*a*) irrespective of their reproductive states; (*b*) when in the same reproductive states as their attackers, and hence when experiencing the same hormonal changes; or (*c*) when in reproductive states in which they were least likely to retaliate (i.e., F > E3 > E1 > E2; P1 > P2 > P3; L3 > L2 > L1).

Failure to confirm predictions from the alternative hypotheses regarding the reproductive states of attackers or victims does not mean that these alternatives are wrong, but rather that they do not fully explain all functions of ACB. Thus, attack coalitions are likely to mediate reproductive suppression. Yet, both ACB and reproductive suppression of consexuals may also provide means of ensuring access to resources over the long term. Attack coalition behavior is also likely to maintain rank order over the long term. Only the rank improvement portion of the first alternative hypothesis could be excluded by my data; coalescing females always attack others lower, rather than higher in rank, than themselves. This does not, however, exclude this explanation for ACB where immature females who have not yet fully established their adult rank are involved (e.g., Cheney, 1977; Walters, 1980).

2. *Indirect Reproductive Competition*

Females may also influence the reproduction of others through a variety of indirect means. For example, they may suppress the reproduction of others through pheromonal cues, by monopolizing access to mates at times when other females would be most likely to conceive, or by interfering with a newly lactating female's attempts to care for her young.

a. Synchronized or Prolonged Receptivity

Females may monopolize mates by synchronizing or prolonging their estrus or receptivity with respect to that of other females. Such a strategy would be particularly effective for females who have some mating advantage, such as high status, over others (e.g., Conaway and Koford, 1965; Hrdy, 1977; Rood, 1980), or who would otherwise not become pregnant at that time (e.g., pregnant or anovulatory females). Either case would enable the instigator to "buy time" for her offspring so as to improve its relative age advantage in subsequent competition.

The highest-ranking female in the study troop was in a constant state of peak estrus (E2) for the entire duration of my study period (which may explain why she was so highly ranked, but see Hausfater, 1975), and for at least two months afterward (G. Norton, personal communication, 1979). During this time, she was consorted by, and copulated with, high-ranking males on a regular basis, but never conceived. While it is unclear whether this particular female's case constitutes a pathological state, female–female competition, or both, it does show that such a female strategy can effectively monopolize high-ranking males. Rowell (1970) found that stressing adult female baboons generally caused them to prolong their estrous swellings, in one case for 100 or more days. Hendrickx and Nelson (1971) also report occasional sexual skin cyclicity during pregnancy in both chimpanzees and in yellow baboons, as does Loy (1981) in patas monkeys. Kleiman and Mack (1977) review cases of sexual receptivity by female primates during pregnancy. (See also Hrdy, 1977).

b. Infant Handling Behavior

Although newly lactating females are rarely recipients of coalition attacks (Fig. 3), their infants may be subjected to more subtle forms of harassment, such as other females handling them carelessly and preventing them from suckling (cases reviewed in Wasser and Barash, 1981; see also Mohnot, 1980; Silk and Boyd, this volume). Thus, infant handling behavior may constitute another indirect form of reproductive competition among females. Such handling can be particularly costly for baboon infants. Baboon milk is low in protein and perhaps in fat as well (Ben Shaul, 1962a and 1962b) relative to that of other mammals. This, combined with the threat of dehydration, requires newborns to have almost continuous access to the nipple (Blurton Jones, 1972; Hrdy, 1977). At Mikumi, accidental separation from their mothers of infants up to 6 months of age resulted in death within a matter of hours (Rhine *et al.*, 1980). The cause of death was presumably dehydration, compounded by severe depression.

Predictions. Confirmation of the following predictions would suggest that infant handling behavior is, in part, competitive among yellow baboons and may also serve as a means of gathering information about the young in the troop.

1. Newborns (0–3 months old) should be handled more than older infants because the former are more vulnerable to the nutritional consequences of maternal separation, as well as less able to defend themselves and/or run away.
2. Both nulliparous and multiparous females should handle infants, and treat them roughly, even though the latter have already gained experience at mothering.
3. Females in some reproductive states should handle infants more than do those in other states. Specifically, pregnant females about to give birth (P3) should handle infants most frequently. These infants are likely to be peer-group mates of the P3 females' prospective infants. Reducing the competitive abilities of these infants would therefore decrease any age advantage they might have over the pregnant females' prospective infants once the infants begin to compete with one another during weaning. Newly lactating females should also handle the infants of others to gather information about the immediate competitors and social partners of their own infants. However, having newborns of their own, they may not necessarily handle these infants roughly. Their own infants would be more vulnerable to retribution or other rough handling if the mothers were

preoccupied handling other infants (see also Section IV,A).

4. Mothers should resist the efforts of others attempting to handle their infants, especially those would-be handlers who would receive the most benefit from mistreating them—that is, P3 females (Prediction 3 above). It follows that:

5. High-ranking females should handle infants more than do low-ranking females; and

6. Infants of low-ranking mothers should be handled more frequently than infants of high-ranking mothers.

Results. All of the predictions were confirmed. Adult females attempted to handle newborns 0–3 months old significantly more than they handled older infants ($p < .00001$, Bowker's modified chi square). Both nulliparous and multiparous females handled infants of other females roughly. High-ranking females handled infants significantly more than did low-ranking females ($p < .0001$, Bowker's modified chi square), while infants of low-ranking mothers were handled significantly more than were those of high-ranking mothers ($p < .001$, Bowker's modified chi square). As their pregnancies progressed, there was also a significant increase in the tendency of females to handle infants; this tendency peaked at early lactation (Fig. 6). The tendency of cycling females to handle infants also increased as females approached ovulation, dropping off thereafter. These reproductive state differences continued to hold up when all females were divided into high-, medium-, and low-ranking groups as well. It is also interesting to note that with the exception of other L1 females, the reproductive states of females who handled infants most frequently (Fig. 6) were quite similar to those of females who most frequently initiated ACB (Fig. 1). Finally, efforts of late pregnant (P3) females to come into contact with newborns were resisted by the newborns' mothers (as measured by supplants, Fig. 7) significantly more than were those of females in any other reproductive state, even though other newly lactating females handled infants more frequently.

Data by Hrdy (1977) and Silk (1980) (see also Silk and Boyd, this volume) also confirm predictions 1–6 for wild langurs and captive bonnet macaques, respectively. Prediction 1 is additionally confirmed by data from a number of other primatologists (reviewed in Seyfarth, 1976 and Hrdy, 1977). Prediction 2 is additionally confirmed by data from Altmann (1980) and Kurland (1977). Altmann's data also illustrate the general competitive nature of infant handling behavior among yellow baboons. She describes two general types of mothers, the "laissez faire" and the "restrictive." Laissez faire mothers tend to be calm, high in rank, and

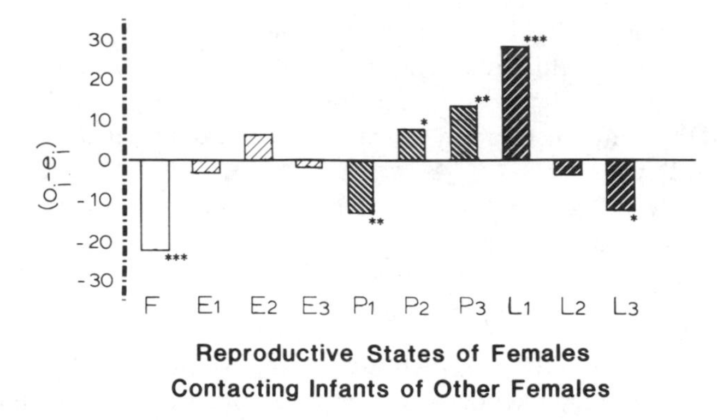

Fig. 6 Reproductive states of adult females contacting 0–3-month old infants other than their own. Presented as differences between observed and expected frequencies as in Fig. 1. $N = 364$, total number of times infants were contacted by adult females other than their mothers; significance levels as in Fig. 1.

tolerant of others playing with and grooming their infants. Restrictive mothers, on the other hand, tend to be "nervous," low in rank, and quite protective of their infants. Altmann (1980) clearly illustrates that avoiding infant handlers has costly effects on the newly lactating female's already

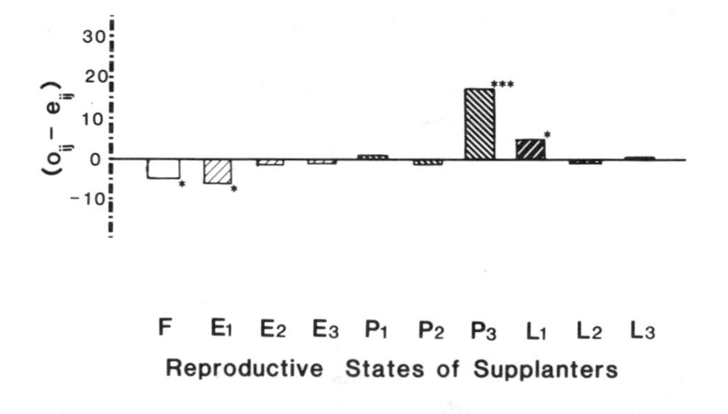

Fig. 7 Differences between observed (o_{ij}) and expected (e_{ij}) frequencies of supplants received by L1 females from females of reproductive state i. $e_{ij} = N_i p_{ij}$, where N_i is the total number of supplants delivered by females in reproductive state i (28, 24, 84, 73, 12, 43, 58, 71, 24, 49, respectively), and p_{ij} is the proportion of total study time in which L1 females were available to females in the ith reproductive state. Significance levels as in Fig. 1.

strained time budget. Laissez faire mothers also weaned their infants sooner and apparently had higher reproductive success than did the restrictive mothers. This should give laissez faire mothers much greater control over the timing of their subsequent reproductions as well.

Alternative Hypotheses for Infant Handling Behavior (IHB). The following hypotheses have also been advanced to explain infant handling, or "allomothering," behavior (for review, see Hrdy, 1976, 1977, 1979; see also Scollay and DeBold, 1980; Wasser and Barash, 1981).

1. *The maternal relief hypothesis.* IHB functions to free the mother for foraging or other maternal duties (Poirer, 1968; Hrdy, 1976; Altmann, 1980).
2. *The adoption hypothesis.* IHB facilitates adoption if the mother dies (Krege and Lucas, 1974; Hrdy, 1976).
3. *The infant socialization hypothesis.* IHB facilitates socialization of infants to other troop members (Krege and Lucas, 1974; Hrdy, 1976).
4. *The learning-to-mother hypothesis.* IHB provides nulliparous infant handlers with basic maternal skills before they give birth themselves (Gartlan, 1969; Lancaster, 1971; Krege and Lucas, 1974; Hrdy, 1976).
5. *The agonistic buffering hypothesis.* IHB serves to buffer the handler against agonism by other troop members (Deag and Crook, 1971).
6. *The "cute" hypothesis.* IHB is a "fortuitous outcome of selection for a behavioral orientation toward young conspecifics," promoting maternal attachment (Quiatt, 1979, p. 316; Scollay and DeBold, 1980).

None of these alternative hypotheses are strongly supported by the available data. All hypotheses predict that newborns will be handled most frequently. Contrary to the results, however, all six alternative hypotheses predict that multiparous females will handle infants carefully. The maternal relief, adoption, and infant socialization hypotheses predict that mothers will not resist the efforts of infant handlers because the behavior either benefits the mother or her infant. For the same reason, alternative hypotheses 1–3 predict that handlers will not necessarily be higher-ranking than the mothers of the handled infants; if anything, they would predict the reverse to be true. The adoption hypothesis predicts that infant handlers will most frequently be multiparous females who have the potential for lactation but may lack infants of their own. This somewhat conforms to the observation that P3 and newly lactating females handled infants most frequently, but does not conform to the rough treatment these infants received from them. Moreover, nearly all reported cases of adoption by baboons have been by adult males or prereproductive males and females (Hamilton *et al.*, 1982). The learning-to-mother hypothesis predicts that preadult and nulliparous adult females will handle infants more carelessly than do multiparous adults and that juvenile handling skills improve over time. However, qualitatively, many juveniles appeared to handle infants more carefully than did multiparous females (see also Hrdy, 1977; Altmann, 1980). This does not exclude the learning-to-mother hypothesis, as mistreatment by multiparous females may reflect their intentions (e.g., competitive) rather than their abilities. If anything, it shows the difficulties of ascribing a single function to this behavior. The agonistic buffering hypothesis predicts that allomothers will often be subordinate rather than dominant to the infants' mothers, or at least that infants are grabbed during aggressive encounters. This was never observed to be the case for female handlers. However, adult males did occasionally pick up infants when fighting with one another (but see Busse and Hamilton, 1981). Finally, the "cute" hypothesis predicts that all females will exhibit interest in infants. Such interest may be expected to vary with the reproductive state in the observed manner, but again, females with strongest interest are predicted to be the most careful handlers (see also Wasser and Barash, 1981). Predictions about the relatedness of infant handlers and the reciprocal nature of IHB can also be made (e.g., Kurland, 1977), but cannot be addressed with the present data base.

It is most likely that IHB evolved for several reasons, and therefore that all of the hypotheses presented above may apply in one form or another (Wasser and Barash, 1981). The competitive hypothesis may be most relevant for distantly related females or for all females whose offspring are to be born when resources are preditably limited and many older infants will be present. The alternative hypotheses, 1–3 may be particularly important when infant handlers are closely related to their subjects (e.g., Kurland, 1977), whereas alternative hypothesis 4 may apply to young juveniles. Furthermore, the relative important of any of these hypotheses is undoubtably species-specific (see Hrdy, 1976, 1977, 1979).

To summarize Section III, reproductive competition to improve the relative survivorship and reproductive capabilities of offspring does appear to be important among female yellow baboons. It is expressed in the form of attack coalition and rough infant handling behaviors and perhaps via prolonged or synchronized receptivity as well. Such competition depends to a large degree on the reproductive states and ranks of the interactants. This leads to the assumption that the reproductive states of associates reflect, in part, the associate qualities of interactants to one another (defined in Section I). Given this assumption, the reproductive states of associates should also affect the degree of affiliative association among them; this is investigated in Section IV.

IV. Affiliation

A. Reproductive-State-Dependent Shifts in Social Partner Preferences among Female Yellow Baboons

If association between females of the same reproductive state is favored for some states, but not for others, a shift in social partner preferences should occur as females pass from one state to the next. Females in the same state at Time 1 are also likely to be in the same state at Time 2. Thus, shifts in social partner preferences should also be least ambiguous when occurring between females of the same reproductive state. For this reason, changes in association between females of the same reproductive state are detailed below.

Up until lactation, females of the same reproductive state are expected to have little reason to associate with one another. In fact, the potential for competition among them may foster avoidance among females of the same reproductive state. Avoidance of females in the same reproductive state should be particularly common when females are in reproductive states likely to draw attacks from coalitions (e.g., E1 and P1 females, Section III,B,1); as associating with females of the same reproductive state could increase the probability of attack under such circumstances. Similar nutritional needs may also favor avoidance of each other by pregnant females in the same reproductive state. At early lactation (L1) however, this trend is expected to reverse, with females of the same reproductive state associating most affiliatively, largely for the benefit of their interacting young.

Although the presence of other infants may eventually result in competition between infants, association between them is also critical for optimal social development (Harlow, 1969, 1971). A large body of literature on humans, other primates, and carnivores suggests that the young tend to do best in social learning situations when they interact with peers most similar to themselves (e.g., Latane, 1966; Rosenkrans, 1967; Pratt and Sackett, 1967; van Lawick-Goodall, 1967; Baldwin, 1969; Masters, 1969; Fagen, 1974; Rhine and Hendy-Neely, 1978; Wasser, 1978).

Similarly aged playmates seem particularly beneficial for infants. Small infants playing with large infants may be consistently taken advantage of, while the large infants receive little challenge. Small infants are also likely to be competitively inferior to larger ones (Altmann, 1980; Dittus, 1979). Since newborns travel almost exclusively on their mothers' bodies, newly lactating females who associate with one another can ensure that their young are socialized to their optimal peers—those most similar in age. In this sense, L1 females each have something that the others need—an infant similar in age to their own. This should in turn elevate the relative associate qualities of L1 females to one another, favoring association between them. Association between lactating females should then break down during L2 or L3 when the young are being weaned. The young can then continue such peer associations on their own when they are not foraging with their mothers. This enables them to continue to fulfill their social needs, while minimizing peer-group competition for limited weaning foods. In theory then, the associate qualities of lactating females toward other females of the same reproductive state would decline at L2 or L3, as should be reflected by the affiliative association between them.

B. The Trade-off between Associate Quality and Relatedness in Social Partner Preferences

Assuming that the reproductive states of female associates do reflect their associate qualities toward one another in the prescribed manner, short-term changes in female association patterns are also predicted to contrast occasionally with predictions from unalloyed kin selection theory—that is, distant kin may associate preferentially. This should be particularly the case insofar as both of the following two criteria are met: (*a*) the qualities of associates vary independently of their relatedness, and (*b*) the group contains more distant than close kin. Under such conditions, individuals are more likely, on the average, to encounter distant rather than close kin of high quality, simply because more of the former are available from which to choose. If the associate qualities of distant kin are sufficiently high relative to those of close kin, association with the distant kin will enhance the individual's inclusive fitness more than would association with the closer kin (Wasser, 1982).

The reproductive states of associating females, and, by assumption, their associate qualities to one another, do appear to be independent of their relatedness (the first criterion). Using R. Rhine's data on the past four years of cycling records for the study troop, I showed that females who are closely ranked, and, by assumption, related (Section II,A) are no more likely than others to be either synchronous or asynchronous in their reproductive states. This independence of female reproductive state and associate relatedness probably occurs because of the many factors that can interfere with female reproductive synchrony over an extended period of time (e.g., general failure to conceive, inhibition of ovulation, spontaneous abortion, or loss of an infant by one of two associating individuals).

The large size of the study troop is also such that more distantly related than closely related adults are available in the troop as a whole (the second criterion). The average degree of relatedness among adults within a troop generally decreases as troop size increases (Chepko-Sade and Oliver, 1979). Furthermore, given a nearly 3-year interval between the birth of surviving sibs and an almost 50:50 sex ratio at birth (Altmann *et al.*, 1978), second-born females will have an average of one sexually mature sister 4–6 years older than themselves in their social group at any one time. Thus far, the Altmanns have found only one case of maternal sisters whose sexual maturity overlapped, although this may be somewhat more common among paternal sisters (J. Altmann, 1979). Moreover, most mothers died before their daughters reached full maturity (J. Altmann, personal communication, 1980).

If associate relatedness (and/or rank) is all-important, shifts in social partner preferences should either be restricted to closely related or ranked individuals in appropriate reproductive states, or they should not occur with shifts in associate reproductive states at all. Long-term association preferences between closest kin have been reported for a number of female old-world primates (reviewed in Walters, 1981). Seyfarth (1976, 1980) found a similar tendency in chacma baboons and in vervet monkeys, using associate rank as his independent variable. However, these long-term correlations do not preclude short-term changes in social partner preferences that favor more distant kin. Given (*a*) the benefit of association between newly lactating females, (*b*) the apparent independence of relatedness of the probability that any two females will give birth synchronously, and (*c*) the assumption that more distant than close kin are available in the social group from which to choose; then preferential association between distant kin should in fact be common. This does not imply, all else being equal, that a distantly related (or ranked) female will be preferred when simultaneously in L1 with two closely related (or ranked) females; indeed, data by Seyfarth (1976) suggest the reverse to be true. It does imply, however, that distantly related females are more likely to be simultaneously in L1 than are closely related females, and that given the choice between a distantly related L1 female and a related female in some other state, an L1 female will prefer to associate with the distant relative of the same state.

C. Results

Figures 8 and 9 illustrate the differences between observed and expected (by chance alone) frequencies with which females of each reproductive state were found 0–6 m and 0–1 m, respectively, from others of the same reproductive state (overall differences between observed and expected values are significant at both 0–6 and 0–1 m—$p < .001$, Bowker's modified chi square). Females of all reproductive states, except for postovulatory cycling females (F, E3) and newly lactating females (L1) associate with others of the same reproductive state less than expected by chance alone at both 0–6 m and 0–1 m. These differences are significant for the E1, P1–P3, and L2–L3 states at 0–6 m (Fig. 8), and for the E1–E2 and L3 states at 0–1 m (Fig. 9). Flat and newly lactating females (L1) associate with others of the same reproductive state more than expected by chance alone, based on their availabilities in the population. These differences are significant for L1 females at 0–6 m (Fig. 8), and for both F and L1 females at 0–1 m (Fig. 9). In fact, I observed that L1 females are found both 0–6 m and 0–1 m from other L1 females more than they are found these distances from females of any other reproductive state. The predicted shift in association preferences between females of the same reproductive state (Section IV,A) is quite apparent here.

Grooming data for these females show a similar trend (Table IV). F, L1, and L2 females groom others of the same reproductive state significantly more than expected by chance alone ($p < .001$, $p < .00001$, and $p < .005$, respectively, binomial). In fact, F and L1 females exchanged grooms with females of the same state more frequently than with females of any other state. The prediction that distant kin will associate preferentially at times (Section IV,B) was also confirmed. Thus, one of the most preferred grooming and proximity partners of L1 females was an L1 female among the effective "nonrelated" immigrants described earlier (Section II,B). Two other L1 females in association were of ranks 3 and 17, also likely to be distantly related.

Data on attack coalition partners correspond somewhat with proximity and grooming data as well. Recall that with the exception of females of the flat state, and less so the E2 state, females did not coalesce with females of the same reproductive state more than expected by chance

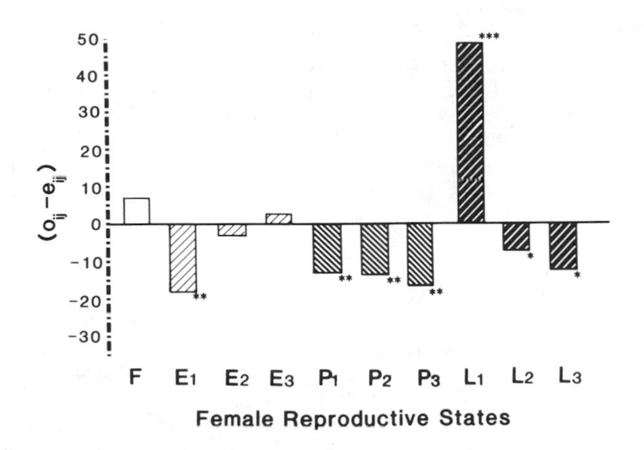

Fig. 8 Differences between observed frequencies with which females of the ith reproductive state were found 0–6 m from females of the same reproductive state and those expected by chance alone. N_i is the total frequency with which females in the ith reproductive state were 0–6 m from all adult females (375, 581, 612, 250, 424, 393, 502, 843, 304, 602, respectively). Significance levels are as in Fig. 1.

alone; furthermore, those pregnant females most likely to enter into attack coalitions (P3 and P2) coalesced with others of the same reproductive state significantly less than expected by chance (Section III,B,1). Also recall that E2 and P3 females coalesced with one another significantly more often than females of any other reproductive states, even after their expected frequencies of coalescing were adjusted for the greater tendencies of each to enter any coalition at all. In fact, one of two such females commonly

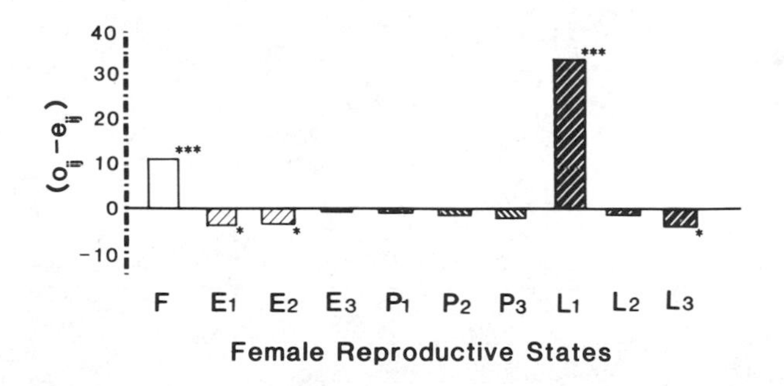

Fig. 9 Differences between observed frequencies with which females of the ith reproductive state were found 0–1 m from females of the same reproductive state and those expected by chance alone, as in Fig. 8. (N_i = 49, 70, 73, 19, 51, 56, 57, 213, 39, 80, respectively.) Significance levels as in Fig. 1.

TABLE IV
z scores of the Differences between Observed Frequencies with Which Females of the ith Reproductive State Groomed Females of the Same Reproductive State and Those Expected by Chance Alone[a]

Reproductive state	z score	(N_i, N'_i)	Reproductive state	z score	(N_i, N'_i)
F	3.94	(58, 48)	P2	−2.46	(102, 74)
E1	−.48	(84, 51)	P3	−1.93	(87, 74)
E2	−.20	(70, 36)	L1	19.29	(78, 231)
E3	1.31	(33, 11)	L2	2.58	(29, 46)
P1	−.01	(60, 46)	L3	−1.80	(127, 111)

[a] e_i calculated as in Table III. N_i is the total frequency with which females of the ith reproductive state groomed all adult females (the left-hand value in parentheses); N'_i is the total frequency with which females of the ith reproductive state received grooming from all adult females (the right-hand value in parentheses). Scan ($N = 428$) and focal animal continuous ($N = 300$) data are combined.

found in coalitions was an immigrant who was cycling after a recent abortion; her partner was frequently a P3 resident female, two rank units below her.

In summary then, females of the same reproductive state do not tend to associate with one another, with the exception of L1 and F females (who associate with others of the same reproductive state significantly more than expected by chance alone). This pattern is reflected by proximity data, both at 0–6 m and at 0–1 m, by grooming data, and somewhat by coalition partnership data as well. Some of the L1 females observed to associate preferentially were "unrelated." An "unrelated" pair consisting of an E2 and a P3 female was frequently observed to coalesce with one another as well.

V. Discussion

A. Reproductive Competition

Intrasexual reproductive competition to increase the quantity of conceptions appears to be most important for the sex that contributes the least amount of parental investment per reproduction (Williams, 1966; Trivers, 1972). However, the reverse seems to hold for competition to increase the quality of conceptions. Thus, male yellow baboons are very much involved in the former type of competition, attempting to obtain nearly exclusive access to females when they are maximally likely to be impregnated (Hausfater, 1975). Females, on the other hand, are particularly involved in the latter form of competition, suppressing the reproduction of others in order to reduce the number and ages of infants with whom their own infants must compete for resources, and perhaps for paternal care. Females also harass infants of others to improve the relative competitive abilities of their own infants. The above interpretation of the evolutionary basis for reproductive suppression among females differs from that of Dunbar (1980, p. 264), who labels such behavior "evolutionary spite."

Attack coalition behavior (ACB) appears to be one way in which reproductive competition is mediated between female baboons. Females most often attacking others appear to be those most likely to receive the highest net benefit from such behavior—mid-ranking to high-ranking females about to ovulate (i.e., E2, and to a lesser extent E1 females) or those about to give birth in the near future (primarily P3 and secondarily P2 females). Their recipients are females whose reproductions are most easily inhibited—low-ranking to mid-ranking females approaching ovulation (E1 females), or those maximally susceptible to spontaneous abortion (P1 females).

Whether a female is attacked or perhaps even terminates her own pregnancy may also depend on the sex of her offspring. Sackett *et al.* (1975) found that harem-living captive female pigtail macaques (in the last trimester of their pregnancies) carrying female fetuses were attacked significantly more frequently than were females carrying male fetuses. Unfortunately, these data were based on treatment records for bite wounds, so Sackett (personal communication, 1980) did not have data on the age or sex of their attackers. Consistent with this, however, are data of both Dittus (1979) and Altmann (1980), which indicate a trend by high-ranking females to produce more female offspring, and by low-ranking females to produce more male offspring.

Data also suggest that female–female reproductive competition continues after parturition as well, initially in the form of infant handling behavior (IHB). Infants handled most (0–3 months old) are those most vulnerable to the effects of maternal separation. They are treated roughly by both multiparous and nulliparous females, and their mothers tend to be subordinate to their handlers. With the exception of other newly lactating females, the reproductive states of handlers tend to be the same as those initiating attack coalitions (Figs. 1 and 6). Furthermore, mothers try to avoid attempts by such females to touch their infants (Fig. 7), often at considerable expense to their already strained time budgets (Altmann, 1980). Although my sample size was too small for analysis by sex of the infant, data for bonnet macaques by Silk and Boyd (this volume) show that female infants are handled significantly more than are male infants.

As the young grow older, they too become involved in ACB. Juvenile females both attack in and receive attacks from attack coalitions significantly more frequently than do similarly aged males, and this, along with the observations just mentioned of Sackett *et al.* (1975), Drickamer (1974), Sade *et al.* (1976),Dittus (1979), Altmann (1980), and Silk and Boyd (this volume), suggests that reproductive competition among females begins much earlier in life than does reproductive competition among males. Unlike male–male competition, female–female competition also tends to be spread out over the whole duration of the reproductive event, from before conception until well after parturition. Male–male competition tends to be concentrated around the time of mating (Hausfater, 1975), perhaps making it more conspicuous than that among females. This raises the important question of whether variance in male reproductive success is actually greater than that of females. The sex ratio pattern observed by Dittus (1979) and Altmann (1980) elicits this question even more so.

Trivers and Willard (1973) state that in species in which the male variance in reproductive success is greater than that among females, females in good condition will produce more males, whereas females in poor condition will produce more females. But baboons and toque macaques seem to show the reverse pattern.

Deviation of data from the predictions of Trivers and Willard (1973) could imply that the discrepancy between male and female variance in lifetime reproductive success is in actuality quite small. This quantitative comparison has yet to be made for any primate. However, data by some workers have shown substantial variance in female reproductive success for rhesus macaques (Drickamer, 1974; Sade *et al.*, 1976) and Japanese monkeys (Mori, 1975; see also Chapais and Schulman, 1980). This is probably the case for yellow baboons as well (Altmann, 1980). By contrast, Hausfater (1975, pp. 140–141) concludes that (*a*) depending on the number of ranks yellow baboon males occupy in their lifetimes and (*b*) their durations of occupancy in each rank, "the total lifetime reproductive success of all males may . . . be equal."

On the other hand, the Trivers and Willard (1973) model also requires that the following assumptions be met: (*a*) the condition of young at the end of parental investment (PI) tend to correlate with the condition of the mother during PI; (*b*) differences in the condition of young at the end of PI tend to endure into adulthood; and (*c*) adult males are differentially helped in reproductive success by slight advantages in condition, relative to that of females.

Assumptions (*a*) and (*b*) of Trivers and Willard are probably more true for females than for males in these old-world primates. Since females do not disperse, dominant females producing females can take advantage of their status to ensure that their daughters obtain adequate resources, comparable status, and presumably high reproductive value. Data by

Drickamer (1974) and Sade *et al.* (1976) show that higher-ranking genealogies do have a higher intrinsic rate of natural increase than do lower-ranking genealogies because daughters of high-ranking females are more likely to survive and give birth to their first offspring at an earlier age than do daughters of low-ranking females. On the other hand, subordinate females have difficulty ensuring that their offspring receive adequate resources. Giving birth to males alleviates this problem somewhat because the larger size of males at birth gives them a sex advantage in competition against similarly aged females, and this advantage lasts throughout their lives (Dittus, 1979). In addition, subsequent emigration by male offspring may insulate them from the long-term consequences of their mothers' low rank (e.g., Drickamer and Vessey, 1973). Thus, the condition (rank) of young females versus males at the end of PI is more likely to correlate with the condition (rank) of their mother during PI [see Assumption (*a*) preceding]. Furthermore, since female rank is stable throughout life, whereas male rank is not (Hausfater, 1975), differences in the condition (rank) of young at the end of PI is more lilely to endure into adulthood for females than it is for males [see Assumption (*b*) preceding]. Sex differences in rank stability should, however, comply with Trivers and Willard's (1973) third assumption.

In theory, local resource competition between mothers and daughters could also bias sex ratios in favor of males, as in Clark's (1978) bushbabies. However, such local mate competition between kin is unlikely here since, all else being equal, female baboons may benefit from the support of close kin in this group of multiple matrilines having a relatedness-dependent rank system (Cheney, 1977; Walters, 1980).

Clearly, the next data needed for such long-lived species are those that compare male and female variance in lifetime reproductive success (e.g., Payne, this volume). It would also be of value to know the effects of age on reproductive effort, and of ACB and IHB on female reproductive success and offspring maturation rate. Data on the effects of birth season on the frequencies of attack coalitions and of rough infant handling should be of value as well.

B. Females' Association Patterns as Determined by Their Reproductive States

Females suffer costs, but also benefit from the reproduction of others. All else being equal, as the number of infants present in a group increases, so should the degree of competition between them. On the other hand, association between infants of similar age, and probably sex, is optimal for their social development (see Section IV,A). In this way, competitive and cooperative needs of associating females are often in conflict, a conflict that is best characterized by the shifts in association patterns that occur among females of the same reproductive state. Newly lactating (L1) females, for example, associate in order to socialize their travel-dependent infants to others close to them in age—their infants' optimal available peers. Thus, a strong shift occurs just following parturition from avoidance to affiliative association between females of the same reproductive state. This pattern reverses to one of avoidance during L2 or L3 (Figs. 8 and 9), when the young are being weaned and freely moving about on their own. The significant avoidance between females of the same reproductive state, for most states other than L1 and F (postovulatory cycling females), is probably a reflection of the underlying reproductive competition among them (see Section IV,A).

The shifts in association patterns among females of the same reproductive state support the assumptions that the reproductive states of associating females reflect, in part, their associate qualities toward one another, and hence that changes in the qualities of associates can result in changes in their association patterns. Since relatedness, and presumably rank (Hausfater, 1975), do not change over time, these reproductive-state-dependent shifts in female association patterns could not be explained by associate relatedness or rank alone. Moreover, shifts in social partner preferences at L1 apparently favor distant kin.

Data on choice of attack coalition partners somewhat support this conclusion as well in that (*a*) females tended not to coalesce with females of the same reproductive state; (*b*) E2 and P3 females coalesced with one another significantly more frequently than expected by chance alone, even given the relatively high tendencies of each to enter coalitions at all; and (*c*) two such females who were apparently "unrelated" coalesced repeatedly. Attack coalitions in support of "nonkin" have been reported for a variety of old-world primates (Massey, 1977; Cheney, 1977; DeWaal, 1978; Kaplan, 1978; Walters, 1980). In fact, Walters notes that kin accounted for only 9% of interventions in coalitions among female yellow baboons at Amboseli, although this proportion was significantly larger than expected by chance.

The apparent reproductive-state-dependent preferences for distant kin can also be taken as tentative confirmation of the prediction that preferential association among distantly related female baboons will occasionally occur (see also Russell, this volume) as long as: (*a*) the qualities of associates vis-à-vis one another vary and are by and large independent of associate relatedness or ranks (see Section IV,B); and (*b*) there are more distant than close kin available in the group from which to choose (Wasser, 1982). In fact, future research that also takes into account the sex of the infants of newly lactating females may eventually find preferential association between more distantly related females to be even more common than suggested here. Newly lactating females may prefer their offspring to associate with similarly sexed peers. For female offspring, this would minimize the amount of resource competition they experience from larger males of the same age. For male offspring, this may provide them with potential dispersal partners. These being the case, it may be difficult for closely related females to be of optimal associate quality, even when they do give birth synchronously.

VI. Conclusion

The behavioral data presented here suggest that both female–female reproductive competition and the influence of associate quality on social partner preferences are important, at least among yellow baboons. As such, the data question some of the theoretical assumptions commonly employed in the behavioral biology literature. For example, a variety of models assumes little or no variance in female reproductive success or that sexual selection, mediated through reproductive competition, necessarily results in a large discrepancy between the variance in male versus female reproductive success among polygynous, social animals. Another common assumption is that preferential association between distant kin is unlikely. But these assumptions may not always hold.

Assumptions such as the preceding also influence the nature of predictions of a variety of theories, including those for the evolution of sex differences, sex ratio adjustments of offspring during reproduction, the influence of female choice on the evolution of male characters, and theories predicting the competitive and cooperative nature of association among particular individuals. These assumptions are often made with appropriate qualifiers (e.g., Wade and Arnold, 1980). However, insofar as these assumptions restrict the nature and interpretations of the data scientists collect in the future, it is essential that we now give serious attention to their validity. Thus, the results of the research presented here suggest that female association patterns are much more flexible than those patterns currently portrayed for social polygynous mammals by the behavioral biology literature. These results further suggest that our validation of the preceding assumptions may not only depend on where we look, but also on how we look for the data that address them. This, of course, assumes that we begin looking for such data in the first place.

VII. Summary

This chapter addresses both theoretical and empirical aspects of competitive and affiliative associations among females. The empirical study was conducted on female yellow baboons at Mikumi National Park, Tanzania.

Female–female reproductive competition (at least among social mammals) appears to be directed at improving the relative quality of conceptions—that is, the offsprings' survivorship and subsequent reproductive capabilities. When offspring quality is reduced by the number and relative ages of other young present in the group, a female may improve her offspring's lifetime reproductive success by suppressing the reproduction of others. This can enable the suppressor to minimize detrimental competition that her offspring would otherwise experience. Moreover, it may well be adaptive for females having low competitive abilities to be reproductively suppressed if breeding options are likely to be better at some later time. This form of reproductive competition contrasts with that among males, which is generally directed at increasing access to mates—that is, increasing the number of conceptions.

I found that female yellow baboons jointly attack other females whose reproduction is most easily inhibited, which is a function of the attacked animal's reproductive state and relative rank. Females attacking others are those who would benefit most from suppressing others, but who are themselves in those reproductive states that are difficult to suppress. Females also mistreat the 0–3-month-old infants of others, handling them roughly during their most vulnerable life stage. Older offspring become involved in ACB as well. However, juvenile females are attacked and attack others significantly more often than similarly aged males. In fact, data by other investigators suggest that female offspring consistently receive the brunt of female–female aggression from the time they can first be distinguished as females. Taken together, these data suggest that female–female behavioral competition among baboons begins earlier in life and is more spread out over a given reproductive event than that among males.

Affiliative association among particular females also appears to depend on their reproductive states. Most females avoid others in the same reproductive state, with the exception of postovulatory (luteal) cycling females and newly lactating females; these associate most affiliatively with others in the same reproductive state. Newly lactating females appear to associate with one another to facilitate socialization of their newborns to others of similar age. This should be optimal for both developmental and competitive reasons. Hence, even "unrelated" females who entered lactation together were observed to associate preferentially until their offspring became sufficiently mobile to be independent of their mothers. This pattern of avoidance, followed by affiliation, and then avoidance again, among females of the same reproductive state, could not be predicted by kin selection theory alone, but was predicted by the quantitative model for association preferences discussed in this chapter. It is concluded that association patterns among female yellow baboons appear to be much more flexible than those portrayed by current behavioral biology literature for nondispersing female social mammals.

Acknowledgments

I wish to thank Ramon Rhine, the Tanzanian National Scientific Research Council, the Serengeti Research Institute, and the Tanzanian National Parks for permission to conduct this research. In addition, I thank Guy Norton, Dinah Wilson, and our rangers, Charles Kidungho and Felix Siwezi, for assistance during the field portion of this research. This chapter has benefited considerably from comments by Robert Boyd, Irven DeVore, Glenn Hausfater, Sarah Hrdy, G. James Kenagy, Gordon Orians, Gene Sackett, Joan Silk, and Montgomery Slatkin. Special thanks go to my friend and former advisor, David P. Barash.

References

Alexander, R. D. The evolution of social behavior. *Annual Review of Ecology and Systematics*, 1974, *5*, 325–383.

Altmann, J. Observational study of behavior: sampling methods. *Behaviour*, 1974, *49*, 227–267.

Altmann, J. Age cohorts and paternal sibships. *Behavioral Ecology and Sociobiology*, 1979, *6*, 161–164.

Altmann, J. *Baboon mothers and infants*. Cambridge, Mass.: Harvard Univ. Press, 1980.

Altmann, J. Altmann, S., and Hausfater, G. Primate infant's effects on mother's future reproduction. *Science*, 1978, *201*, 1028–1030.

Altmann, S. Altruistic behavior: the fallacy of kin deployment. *Animal Behaviour*, 1979, *27*, 958–959.

Axelrod, R., and Hamilton, W. D. The evolution of cooperation. *Science*, 1981, *211*, 1390–1396.

Baldwin, J. D. The ontogeny of social behaviour of squirrel monkeys (*Saimiri sciureus*) in a seminatural environment. *Folia Primatologica*, 1969, *11*, 35–79.

Barlow, G. W., and Silverberg, J. (Eds.), *Sociobiology: Beyond nature/nurture? Reports, definitions and debates.* Boulder, Colo.: Westview Press, 1980.

Bateman, A. J. Intra-sexual selection in drosophila. *Heredity,* 1948, *2,* 349–368.

Ben Shaul, D. M. The composition of the milk of wild animals. *International Zoo Yearbook,* 1962a, *4,* 333–342.

Ben Shaul, D. M. Notes on hand-rearing various species of mammals. *International Zoo Yearbook,* 1962b, *4,* 300–332.

Blurton Jones, N. Comparative aspects of mother–child contact. In N. Blurton Jones (Ed.), *Ethological studies of child behaviour.* London/New York: Cambridge Univ. Press, 1972.

Bronson, F. H. The reproductive ecology of the house mouse. *Quarterly Review of Biology,* 1979, *54,* 265–299.

Busse, C., and Hamilton III, W. J. Infant carrying by male chacma baboons. *Science,* 1981, *212,* 1281–1283.

Chapais, B., and Schulman, S. R. An evolutionary model of female dominance relations in primates. *Journal of Theoretical Biology,* 1980, *82,* 47–89.

Cheney, D. L. The acquisition of rank and the development of reciprocal alliances among free-ranging immature baboons. *Behavioral Ecology and Sociobiology,* 1977, *2,* 303–318.

Chepko-Sade, B. D., and Oliver, T. J. Coefficient of genetic relationship and the probability of intrageneological fission in *Macaca mulatta. Behavioral Ecology and Sociobiology,* 1979, *5,* 263–278.

Clark, A. B. Sex ratio and local resource competition in a prosimian primate. *Science,* 1978, *201,* 163–165.

Conaway, C. H., and Koford, C. B. Estrous cycles and mating behavior in a free-ranging band of rhesus monkeys. *Journal of Mammology,* 1965, *45,* 577–588.

Darwin, C. *The descent of man, and selection in relation to sex* (Vols. 1 and 2). New York: Appleton, 1871.

Deag, J. M., and Crook, J. H. Social behavior and "agonistic buffering" in the wild Babary macaque, *Macaca sylvania* L. *Folia Primatologica,* 1971, *15,* 183–200.

DeVore, I. Male dominance and mating behavior in baboons. N. F. Beach (Ed.), *Sex and behavior.* New York: Wiley, 1965.

DeWaal, F. B. M. Exploitative and familiarity-dependent support strategies in a colony of semi-free living chimpanzees. *Behaviour,* 1978, *66,* 268–312.

Dittus, W. P. J. The evolution of behaviors regulating density and age-specific sex ratios in a primate population. *Behaviour,* 1979, *49,* 265–302.

Drickamer, L. C. A ten-year summary of reproductive data for free-ranging *Macaca mulatta. Folia Primatologica,* 1974, *21,* 61–80.

Drickamer, L. C., and Vessey, S. H. Group changing in free-ranging male rhesus monkeys. *Primates,* 1973, *14,* 359–368.

Dunbar, R. I. M. Determinants and evolutionary consequences of dominance among female gelada baboons. *Behavioral Ecology and Sociobiology,* 1980, *7,* 253–265.

Dunbar, R. I. M., and Dunbar, E. P. Dominance and reproductive success among female gelada baboons. *Nature,* 1977, *266,* 351–352.

Fagen, R. Selective and evolutionary aspects of animal play. *American Naturalist,* 1974, *108,* 850–858.

Fisher, R. A. The genetical theory of natural selection. London/New York: Oxford Univ. Press (Clarendon), 1930.

Fox, L. R. Cannibalism in natural populations. *Annual Review of Ecology and Systematics,* 1975, *6,* 87–106.

Frame, L. H., Malcolm, J. R., Frame, G. W., and van Lawick, H. Social organization of African wild dogs *(Lycaon pictus)* on the Serengeti plains, Tanzania (1967–1978). *Zeitschrift für Tierpsychologie,* 1979, *50,* 225–249.

Gartlan, J. S. Sexual and maternal behavior of the vervet monkey, *Cercopithecus aethiops. Journal of Reproduction and Fertility,* 1969, Suppl. 6, 137–150.

Gilman, J., and Gilbert, C. The reproductive cycle of the chacma baboon (*Papio ursinus*) with special reference to the problems of menstrual irregularities as assessed by the behavior of the sex skin. *South African Journal of Medical Science,* 1946, *11,* 1–54.

Goodman, D. Regulating reproductive effort in a changing environment. *American Naturalist,* 1979, *113,* 735–748.

Greenwood, P. J. Mating systems, philopatry and dispersal in birds and mammals. *Animal Behaviour,* 1980, *28,* 1140–1162.

Greenwood, P. J., Harvey, P. H., and Perrins, C. M. Kin selection and territoriality in birds? A test. *Animal Behaviour,* 1979, *27,* 645–651.

Hamilton, W. D. The genetical evolution of social behavior, I, II. *Journal of Theoretical Biology,* 1964, *7,* 1–52.

Hamilton, W. D. The moulding of senescence by natural selection. *Journal of Theoretical Biology,* 1966, *12,* 12–45.

Hamilton, W. J., III, Busse, C., and Smith, K. S. Adoption of infant orphan chacma baboons. *Animal Behaviour,* 1982, *30,* 29–34.

Hanwell, A., and Peaker, M. Physiological effects of lactation on the mother. *Symposium of the Zoological Society of London,* 1977, *41,* 297–312.

Harlow, H. F. Age-mate or peer affectional system. In D. S. Lehrman, R. A. Hinde, and E. Shaw (Eds.), *Advances in the study of behavior* (Vol. 2). New York: Academic Press, 1969.

Harlow, H. F. *Learning to love.* New York: Albion, 1971.

Hausfater, G. Dominance and reproduction in baboons (*Papio cynocephalus*): A quantitative analysis. *Contributions to Primatology,* 1975, No. 7 (Monograph)

Hendrickx, A. G. The menstrual cycle of the baboon as determined by vaginal smear, vaginal biopsy, and perineal swelling. In J. Vagtborg (Ed.), *The baboon in medical research* (Vol. 2). Austin: Univ. of Texas Press, 1967.

Hendrickx, A. G., and Nelson, G. Reproductive failure. In E. S. E. Hafez (Ed.), *Comparative reproduction of nonhuman primates.* Springfield, Ill.: Charles Thomas, 1971.

Hrdy, S. B. Care and exploitation of nonhuman primate infants by conspecifics other than the mother. *Advances in the study of behavior,* 1976, *6,* 101–158.

Hrdy, S. B. *The langurs of Abu: Female and male strategies of reproduction.* Cambridge, Mass.: Harvard Univ. Press, 1977.

Hrdy, S. B. Infanticide among mammals: A review, classification, and examination of the implications for the reproductive strategies of females. *Ethology and Sociobiology,* 1979, *1,* 13–40.

Huxley, J. S. The present standing of the theory of sexual selection. In G. R. deBeer (Ed.), *Evolution: essays on aspects of evolutionary biology presented to Professor E. S. Goodrich on his seventieth birthday.* London/New York: Oxford Univ. Press (Clarendon), 1938.

Jarvis, J. V. M. Eusociality in a mammal: Cooperative breeding in naked mole-rat colonies. *Science,* 1981, *212,* 571–573.

Kaplan, J. R. Fight interference in rhesus monkeys. *American Journal of Physical Anthropology,* 1978, *49,* 241–250.

Keverne, E. B. Sexual and aggressive behavior in social groups of talapoin monkeys. In *Sex, hormones, and behavior. Ciba Foundation Symposium* (No. 62). Amsterdam/New York: Elsevier, 1979.

Kleiman, D. G. Parent–offspring conflict and sibling competition in a monogamous primate. *American Naturalist,* 1979, *114,* 753–760.

Kleiman, D. G., and Mack, D. S. A peak in sexual activity during mid-pregnancy in the golden lion tamarin, *Leontopithecus rosalia* (Primates: Callithricidae). *Journal of Mammology,* 1977, *58,* 657–660.

Krege, P. D., and Lucas, J. Aunting behavior in an urban troop of Cercopithecus aethiops. *Journal of Behavioral Science,* 1974, *2,* 55–61.

Kurland, J. A. Kin selection in the Japanese monkey. *Contributions to Primatology* (Vol. 12). Basel: Karger, 1977.

Kurland, J. A. Kin selection theory: a review and selective bibiography. *Ethology and Sociobiology,* 1980, *1,* 255–274.

Lancaster, J. Play-mothering: The relations between juvenile and young infants among free-ranging vervet monkeys (*Cercopithecus aethiops*). *Folia Primatologica,* 1971, *15,* 161–182.

Latane, B. (Ed.). Studies in social comparison. *Journal of Experimental Social Psychology* (Suppl. 1). 1966.

Lee, R. B. Lactation, ovulation, infanticide, and women's work: a study of hunter–gatherer population regulation. In M. N. Cohen, R. S. Malpass, H. G. Klein (Eds.), *Biosocial mechanisms of population regulation.* New Haven, Conn.: Yale Univ. Press, 1980.

Lipsett, M. B., and Ross, G. T. The ovary. *Contemporary Endocrinology,* 1979, *1,* 119–134.

Low, B. S. Environmental uncertainty and the parental strategies of marsupials and placentals. *American Naturalist,* 1978, *112,* 197–213.

Loy, J. The reproductive and heterosexual behaviours of adult patas monkeys in captivity. *Animal Behaviour,* 1981, *29,* 714–726.

Marascuilo, L. A., and McSweeney, M. *Nonparametric and distribution free methods for the social sciences.* Monterey, Calif.: Brooks/Cole Publ., 1977.

Massey, A. Agonistic aids and kinship in a group of pigtail macaques. *Behavioral Ecology and Sociobiology,* 1977, *2,* 31–40.

Masters, J. C. Social comparison by young children. In W. W. Hartup (Ed.), *The young child* (Vol. 2). Washington, D.C.: National Association for the Education of Children, 1969.

Matsumoto, S., Igarishi, M., and Nagaoka, Y. Environmental anovulatory cycles. *International Journal of Fertility,* 1968, *13,* 15–23.

Mohnot, S. M. Intergroup infant kidnapping in Hanuman langurs. *Folia Primatologica,* 1980, *34,* 259–277.

Mori, A. Signals found in the grooming interactions of wild Japanese macaques of the Koshima troop. *Primates,* 1975, *16,* 107–140.

Packard, J. M., and Mech, L. D. Population regulation in wolves. In M. N. Cohen, R. S. Malpas, and H. G. Klein (Eds.), *Biosocial mechanisms of population regulation.* New Haven, Conn.: Yale Univ. Press, 1980.

Packer, C. Male care and exploitation of infants *Papio anubis. Animal Behaviour,* 1980, *28,* 512–520.

Parker, G. A. Sexual selection and sexual conflict. In M. S. Blum and N. A. Blum (Eds.), *Sexual selection and reproductive competition in insects.* New York: Academic Press, 1979.

Poirer, F. E. Nilgiri langur (*Presbytis johnii*) mother infant dyad. *Primates,* 1968, *9,* 45–68.

Post, D. G., Hausfater, G., and McCuskey, S. A. Feeding behavior of yellow baboons (*Papio cynocephalus*). Relationship to age, gender and dominance rank. *Folia Primatologica,* 1980, *34,* 170–195.

Pratt, C. L., and Sackett, G. P. Selection of social partners as a function of peer contact during rearing. *Science,* 1967, *155,* 1133–1135.

Quiatt, D. Aunts and mothers: Adaptive implications of allomaternal behavior of nonhuman primates. *American Anthropologist,* 1979, *81,* 310–319.

Rasmussen, D. R. Correlates of patterns of range use of a troop of yellow baboons (*Papio cynocephalus*). I. Sleeping sites, impregnable females, births, and male emigrations. *Animal Behaviour*, 1979, *27*, 1098–1112.

Reiter, K. J., Panken, J., and LeBoeuf, B. J. Female competition and reproductive success in northern elephant seals. *Animal Behaviour*, 1981, *29*, 670–687.

Rhine, R. J., and Hendy-Neely, H. Social development of stumptail macaques (*Macaca arctoidees*): Momentary touching, play, and other interactions with aunts and immatures during the infants first 60 days of life. *Primates*, 1978, *19*, 115–123.

Rhine, R. J., Klein, H. D., Norton, D. W., and Roertgen, W. J. The brief survival of free-ranging baboon infants (*Papio cynocephalus*) after separation from their mothers. *International Journal of Primatology*, 1980, *1*(4), 401–409.

Rhine, R. J., and Westlund, B. J. The nature of the primary feeding habit in different age–sex classes of yellow baboons (*Papio cynocephalus*). *Folia Primatologica*, 1978, *30*, 64–79.

Rood, J. P. Dwarf mongoose helpers at the den. *Zeitschrift für Tierpsychologie*, 1978, *48*, 277–287.

Rood, J. P. Mating relations and breeding suppression in the dwarf mongoose. *Animal Behaviour*, 1980, *28*, 143–150.

Rosenkrans, M. Imitation in children as a function of perceived similarity to a school model and vicarious reinforcement. *Journal of Personality and Social Psychology*, 1967, *7*, 307–315.

Rowell, T. E. Intra-sexual behavior and female reproductive cycles of baboons. *Animal Behaviour*, 1969, *17*, 159–167.

Rowell, T. E. Baboon menstrual cycles affected by social environment. *Journal of Reproduction and Fertility*, 1970, *21*, 131–141.

Rowell, T. E. Female reproductive cycles and social behaviour in primates. *Advances in the Study of Behavior*, 1972, *4*, 69–105.

Rudnai, J. *The social life of the lion*. Wallingford, Penna.: Washington Square East Publishers, 1973.

Saayman, G. S. Grooming behavior in a troop of free-ranging chacma baboons (*Papio ursinus*). *Folia Primatologica*, 1971, *16*, 161–178.

Sackett, G. P., Holm, R. A., Landesman-Dwyer, S. Vulnerability for abnormal development: Pregnancy outcomes and sex-differences in macaque monkeys. In N. R. Ellis (Ed.), *Aberrant development in infancy, human and animal studies*. New York: Wiley, 1975.

Sade, D. S. Determinants of dominance in a group of free-ranging rhesus monkeys. In S. A. Altmann (Ed.), *Social communication among primates*. Chicago: Univ. of Chicago Press, 1967.

Sade, D. S., Cushing, K., Cushing, P., Dunaif, J., Figueroa, A., Kaplan, J. R., Lauer, C., Rhodes, D., and Schneider, J. Population dynamics in relation to social structure on Cayo Santiago. *Yearbook of Physical Anthropology*, 1976, *20*, 253–262.

Scollay, P. A., and DeBold, P. Allomothering in a captive colony of Hanuman langurs (*Presbytis entellus*). *Ethology and Sociobiology*, 1980, *1*, 291–299.

Seyfarth, R. M. Social relationships among adult female baboons. *Animal Behaviour*, 1976, *24*, 917–938.

Seyfarth, R. M. A model of social grooming among adult female monkeys. *Journal of Theoretical Biology*, 1977, *65*, 671–698.

Seyfarth, R. M. The distribution of grooming and related behaviours among adult female vervet monkeys. *Animal Behaviour*, 1980, *28*, 798–813.

Shepard, T. H., and Fantel, A. G. Embryonic and early fetal loss. *Clinics in Perinatology*, 1979, *6*(2), 219–243.

Sherman, P. W. Reproductive competition and infanticide in Belding's ground squirrels and other animals. In R. D. Alexander and W. D. Tinkle (Eds.), *Natural selection and social behavior. Recent research and new theory*. Concord, Mass.: Chiron Press, 1981.

Silk, J. B. Kidnapping and female competition in captive bonnet macaques. *Primates*, 1980, *21*, 100–110.

Svare, B. B. Maternal aggression in mammals. In D. J. Gubernick and P. H. Klopfer (Eds.), *Parental care in mammals*. New York: Plenum, 1981.

Trivers, R. L. Parental investment and sexual selection. In B. Campbell (Ed.), *Sexual selection and the descent of man*. Chicago: Aldine, 1972.

Trivers, R. L., and Willard, D. E. Natural selection of parental ability to vary the sex ratio of offspring. *Science*, 1973, *179*, 90–92.

van Lawick-Goodall, J. *My friends the wild chimpanzees*. Washington D.C.: Nat. Geog. Soc., 1967.

Wade, M., and Arnold, S. The intensity of sexual selection in relation to male sexual behaviour, female choice and sperm precedence. *Animal Behaviour*, 1980, *28*, 446–461.

Walters, J. Interventions and the development of dominance relationships in female baboons. *Folia Primatologica*, 1980, *34*, 61–89.

Walters, J. Inferring kinship from behaviour: maternity determinations in yellow baboons. *Animal Behaviour*, 1981, *29*, 126–136.

Wasser, S. K. Structure and function of play in the tiger. *Carnivore*, 1978, *1*(3), 27–40.

Wasser, S. K. Reproductive competition and cooperation: General theory and a field study of female yellow baboons (*Papio cynocephalus*). Unpublished doctoral dissertation, University of Washington, Seattle, 1981.

Wasser, S. K. Reciprocity and the trade off between associate quality and relatedness. *American Naturalist*, 1982, *119*, 720–731.

Wasser, S. K., and Barash, D. P. The "selfish" allomother. *Ethology and Sociobiology*, 1981, *2*, 91–93.

Wasser, S. K., and Barash, D. P. *Reproductive suppression among female mammals: Implications for biomedicine and sexual selection theory*. Manuscript in preparation, n.d.

West-Eberhard, M. J. The evolution of social behavior by kin selection. *Quarterly Review of Biology*, 1975, *50*, 1–33.

Williams, G. C. *Adaptation and natural selection: A critique of current evolutionary thought*. Princeton, N.J.: Princeton Univ. Press, 1966.

Wilson, J. G. *Environment and birth defects*. New York: Academic Press, 1973.

Wrangham, R. C. An ecological model of female-bonded primate groups. *Behaviour*, 1980, *75*, 262–300.

Wynne-Edwards, V. C. *Animal dispersion in relation to social behavior*. Edinburgh: Oliver & Boyd, 1962.

Index

B

C

Q

R